To - Our DSTL Friends
To ongoing sharing
of common interests !
Alan Porter

TECH MINING

TECH MINING
EXPLOITING NEW TECHNOLOGIES FOR COMPETITIVE ADVANTAGE

ALAN L. PORTER
SCOTT W. CUNNINGHAM

WILEY-INTERSCIENCE

A JOHN WILEY & SONS, INC., PUBLICATION

Library of Congress Cataloging-in-Publication Data:
Porter, Alan L.
    Tech mining : exploiting new technologies for competitive advantage / by Alan L. Porter and Scott W. Cunningham.
        p. cm.
    Includes bibliographical references and index.
    ISBN 0-471-47567-X (cloth)
    1. Data mining. 2. Technological innovations—Economic aspects. 3. Research, Industrial. I. Cunningham, Scott W. II. Title.

QA76.9.D343P67 2005
005.74—dc22

                                                                                            2004042241

Printed in the United States of America

10 9 8 7 6 5 4 3 2 1

Table of Contents

List of Figures

Preface

In the "information economy," we recognize the increasing availability of information. On the one hand, we can be intimidated by the overwhelming amount of information bearing down on us. On the other hand, we now have tools to enable us to garner great value from that information quite readily. New information products can better inform decision processes. As businesses are making decisions under tremendous competitive pressures, they increasingly seek better information.

This book addresses how to inform technology management by mining a particularly rich information resource—the publicly accessible databases on science and technology. These include amazing compilations of the world's open R&D literature, patents, and attendant business and public aspects. This information, when integrated with other data sources (the Internet) and expert review, can improve decisions concerning development, licensing, and adoption of new technology.

"Tech mining" presents particular challenges. Most fundamentally, it uses information resources in unfamiliar ways. In the past, we searched abstract databases to find a few articles worth reading. However, when there are literally thousands of relevant articles or patents, we also need to present the "big picture." This book helps understand the value in "profiling research domains," mapping topic relationships, and discerning overall trends. This is a qualitatively different way to use technology information.

We wrote *Tech Mining* for those whose jobs engage emerging technologies. This includes two groups. Part I addresses those who use such studies, rather than perform them. We seek to help such professionals and managers become better informed consumers of tech mining. We inform engineers, researchers, product developers, business analysts, marketing professionals, and various technology managers on effective ways they can exploit these information resources. Part II adds "how to" details for those who analyze, or directly manage the analysis of, changing technologies. This includes information professionals, patent analysts, competitive intelligence specialists, R&D managers, and strategic planners.

This book is a primer. It sets forth the basic objectives and tools of tech mining. Chapters 1–5 aim to provide conceptual bases for practical tech mining actions. The conceptual foundations reside in understanding of how science and technology leads to successful technology commercialization (the innovation process) more than in information science. Chapters 6–16 provide practical advice on performing tech mining. These treat basic and advanced analyses but also process management considerations vital to effective implementation. We carry through to point to products of tech mining analyses and indicate how they can serve particular technology management functions. Chapter 13 arrays technology management issues and questions along with candidate "innovation indicators" to answer them.

Each chapter focuses on a particular aspect of tech mining. It explains the relevant aims, presents the basic steps in accomplishing those aims, and provides pointers to those who want further details. We illustrate the content with experiential cases slanted toward practical implementation issues and how results can be used. Some chapters work through a "chapter challenge" to think through application of the concepts presented.

Chapters 4 and 16 together step through a concrete analytical example. This applies *VantagePoint* software to actual abstract records obtained from three databases (Derwent World Patent Index, INSPEC, and Web of Science) on the topic of "fuel cells." Chapter 4 spotlights sample tech mining results to get you thinking of ways you could gain value from tech mining. Chapter 16 illustrates the analytical progression and notes pitfalls. The Wiley website ftp://ftp.wiley.com/public/sci_tech_med/technology_management offers a sample data set in *VantagePoint Reader* to experience the tech mining analyses directly.

The book does not require any statistics or artificial intelligence background. It is not specific to a particular technology domain (e.g., information technology). In addition to practitioners and managers, we believe it can benefit technology analysis workshops and graduate courses.

Acknowledgments

We expressly thank IEEE and Thomson Scientific for their permission to utilize their data in our sample analyses reported in the book and illustrated via sample data on the web at ftp://ftp.wiley.com/public/sci_tech_med/technology_management. We thank Wiley for taking this venture to publication. Our series editor, Andy Sage, has exerted major influence in forming the content toward a meaningful composition.

We thank our kind colleagues who took the time and energy to review one or more of the draft chapters. Some were kindly and said nice things; we liked that. Others were mean and found lots of problems; we didn't enjoy those nearly as much, but we and you owe them a hearty "Thanks!" for making the book better.

Erik Ayers
Kevin Boyack
Tony Breitzman
Merrill Brenner
Linda Carton
Joe Coates
Patrick Duin
Paul Frey
Arnaud Gasnier
Luke Georghiou
Russ Heikes
Leon Hermans
Diana Hicks
Katherine Jakielski
Sylvan Katz
Alisa Kongthon

Ron Kostoff
Loet Leydesdorff
Hal Linstone
Vincent Marchau
Brian Minsk
Nils Newman
Doug Porter
Scott Radeker
David Roessner
Fred Rossini
Phil Shapira
Robert Tijssen
Tony Trippe
Robert Watts
Julie Yang

Acronyms & Shorthands— Glossary

(We also try to define these on first use in a chapter; this is just for ready reference.)

AI	Artificial intelligence
ALP	Alan Porter, author
CTI	Competitive technological intelligence
DWPI	Derwent World Patent Index, or "Derwent" for short
EPO	European Patent Office
GTEL	Georgia Tech Electronic Library
INSPEC	A major database (covering physical and information sciences and engineering, especially electrical engineering; provided by IEE)
IP	Intellectual property
IT	Information technology
JPO	Japanese Patent Office
KDD	Knowledge discovery in databases
NLP	Natural language processing
PCA	Principal components analysis
PCD	Principal components decomposition
PCT	Patent Cooperation Treaty
R&D	Research and development
S&T	Science and technology
SCI	Science Citation Index (a database covering fundamental research, found at Thomson Scientific's Web of Knowledge)
SWC	Scott Cunningham, author

TIPs	Technology information (or intelligence) products (outputs of tech mining analyses)
Tech mining	Our shorthand for text mining applied to records, particularly science and technology abstracts retrieved from databases
USPTO	United States Patent and Trademark Office
VP	*VantagePoint* tech mining software; also available under the names TechOasis or Derwent Analytics.
WIPO	World Intellectual Property Organization

Part **I**

Understanding Tech Mining

Chapter **1**

Technological Innovation and the Need for Tech Mining

"Tech mining" is our shorthand for exploiting information about emerging technologies to inform technology management (see the Preface). This chapter anchors tech mining to technological innovation processes and payoffs. It keys on the two contextual forces that drive the book: "emerging technologies" and the "information economy." Chapter 2 builds on this to explain what tech mining entails and to the describe the book's organization.

1.1. WHY INNOVATION IS SIGNIFICANT

We use "innovation" to mean technological change. We are concerned with technological change resulting in practical implementation or commercialization, not just idea generation. This section addresses the importance of technological innovation to today's competitive economy and polity.

Today's worldwide economy depends on technology and technological innovation to an extraordinary degree:

- We perform a lot of research—for one thing, American companies spend over $100 billion annually on R&D; for another data point, the Organisation for Economic Cooperation and Development (OECD) countries spent over $550 billion in 1999 (about 70% by companies, 30%

Tech Mining: Exploiting New Technologies for Competitive Advantage, Edited by Alan L. Porter and Scott W. Cunningham.
ISBN 0-471-47567-X Copyright © 2005 by John Wiley & Sons, Inc.

by government).* That research pays off—participating companies in the U.S. Industrial Research Institute estimate average new sales ratio—the percentage of sales attributable to products newly designed in the past five years—at roughly 35%. In other words, $1 of every $3 in their revenue comes from recent innovations.

• National economies depend critically on technology. "High-Tech Indicators" (http://tpac.gatech.edu) show that the U.S. once dominated technology-based export competition. Then Japan raced up to become a staunch competitor. Now, other countries are advancing dramatically. Tiny Singapore now exports technology-based products at the level of the European powers. China is advancing dramatically in technology-based exports, but also in R&D that will drive future generations of products and services. And they are not the only ones looking ahead. The 371 expert panelists anticipate that, in another 15 years, essentially all of the 33 countries tracked will be significant high-tech competitors (Fig. 1-1).

Technological innovation impacts our lives in many ways, some direct and some not so direct. High-technology companies are a significant and growing component of the economy, contributing over 20 million jobs in the U.S. (Hecker, 1999). The competitiveness of those companies depends on innovation, credited with being the main economic growth factor in the western world.

Innovation delivers substantial public and private returns. Mansfield's classic survey (1982) on 37 innovations concluded that the firm's median return on investment was close to 25 cents on every dollar. And the public benefits of innovation far outweigh the firm's benefits—70 cents on every dollar spent on R&D is returned to society. Despite these rosy average returns, Mansfield and others find that innovation is highly risky, and failure can be immensely costly. In some cases, companies bet their existence on the success of an innovation.

Innovation is improving our standard of living. Developments in medical and pharmaceutical technologies have delivered extensive returns in health and life span. The toddler of today can expect 25 more years of life than the newborn of the year 1900. Death rates from infectious disease have been reduced 10-fold over the course of the previous century. However, we remain engaged in an evolutionary war with infectious disease and continue to struggle with cancer and vascular diseases (Lederberg, 1997). Our health and welfare is intimately linked to innovation.

Without belaboring it, innovation is vitally important to scientists and engineers, private and public organizations, and society. Our key underlying premise is that tech mining facilitates innovation. To accomplish this, tech

*OECD Science, Technology and Industry Scoreboard 2001: Towards a knowledge-based economy [http://www1.oecd.org/publications/e-book/92-2001-04-1-2987/]

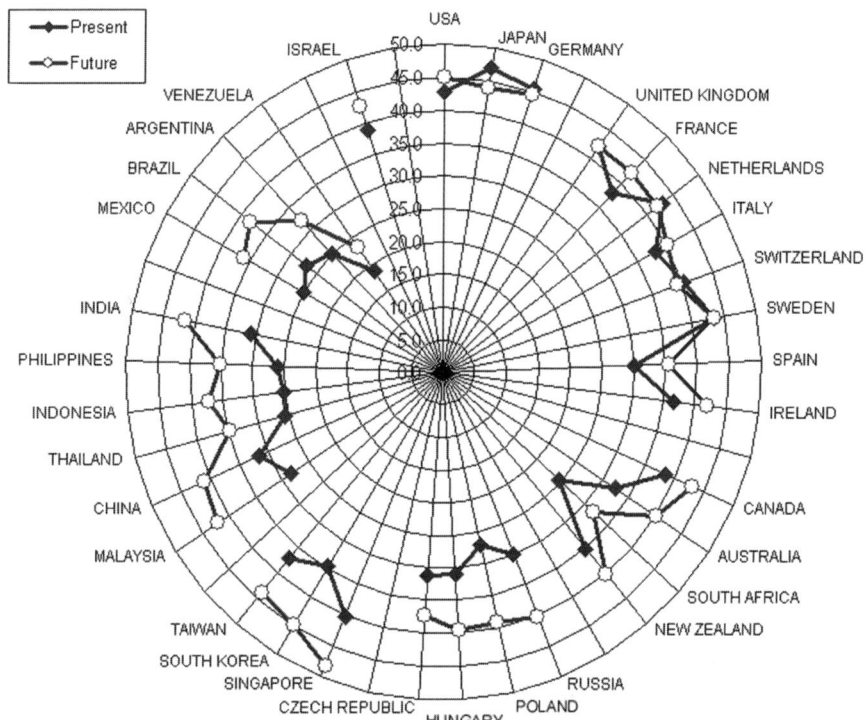

Figure 1-1. Increasing national technological competitiveness. This radar chart summarizes scaled opinions of knowledgeable observers on the relative ability of each country to compete in high-tech-based exports. Scores farther out from the center reflect relatively stronger competitiveness. Note the almost universal expectation of increasing competitiveness. For details see "High-Tech Indicators" at //tpac.gatech.edu.

mining relies on understanding technological innovation processes to track them effectively and to inform decisions about R&D and subsequent implementation and adoption choices.

1.2. INNOVATION PROCESSES

Our colleague Mary Mogee defines innovation (1993) as "the process by which technological ideas are generated, developed and transformed into new business products, processes and services that are used to make a profit and establish marketplace advantage." Let's explore this *process* to figure out empirical measures deriving from innovation activities to generate actionable technological intelligence (tech mining).

We briefly scan the rich history of *models* of technological innovation processes (Fig. 1-2). Dating from the 1950s, the *technology push* model focused

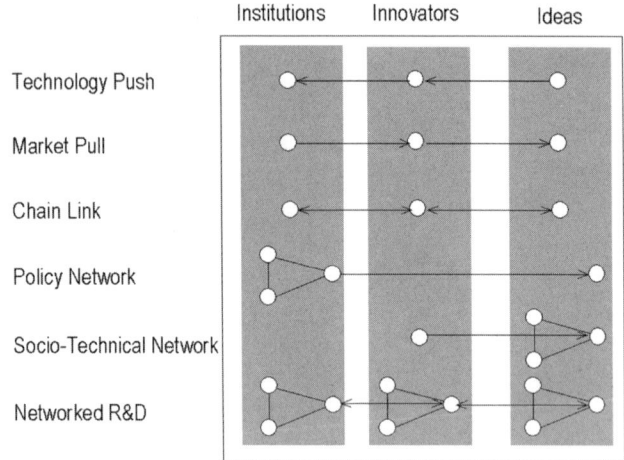

Figure 1-2. Comparison of R&D models

on R&D as generating the essential push that prompts new product development, which the marketplace then accepts. The realization that many innovators and institutions deliberately frame R&D to meet perceived market opportunities suggested the *market pull* model. This reversed the main influence pathway to begin from the customer end. The *chain link* model offered a compromise between the two, acknowledging that flows between technology and the marketplace are iterative and multidirectional. This first class of models is basically singular in nature—one organization generates new technology and takes it to market.

A second class of models recognizes interplay among institutions in generating and acting on science and technology. The *policy network* approach acknowledged that institutions exist in a framework of competitive and collaborative relationships. As the great central R&D laboratories (e.g., Bell Labs, IBM) shrank and distributed activities among operating divisions, companies turned outward for science and technology inputs. Governmental and academic R&D eased from the isolated, single-investigator model of science. Organized research units fostered interdisciplinary and interinstitutional collaborations. Institutions share and compete for the R&D findings of innovators, with significant knowledge spillovers. Notions such as regional innovation centers emerged to bolster purposeful interchange of science and technology approaches and results. A complimentary perspective, the *socio-technical systems* approach, examined how different innovators link and unify ideas. This evolution points toward *networks* of concepts and material objects ("artifacts") forming a stratum for the creation and dissemination of new science and technology knowledge, resulting in technological innovation (Fig. 1-2). (Chapter Resources adds pointers to continuing refinement of such networking models.)

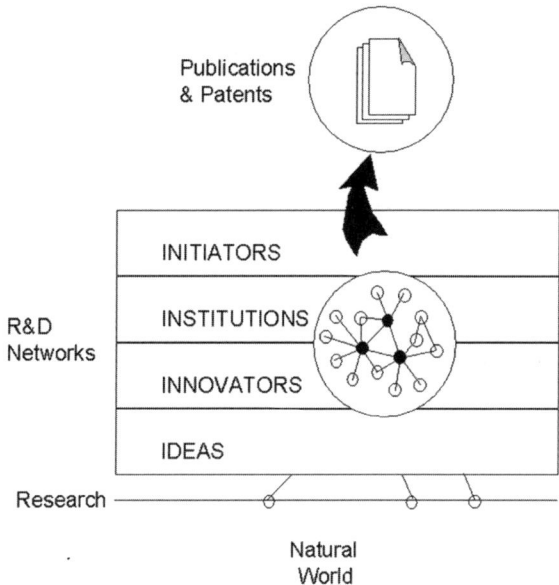

Figure 1-3. A networked, instrumented model of innovation

We see networks again and again—networks of researchers and networks of ideas. These "knowledge networks" are woven by many individuals—certainly by scientists and engineers—and also by the many institutions that support and fund new R&D activity—"initiators." Like the webs of knowledge they create, individuals and institutions find themselves in complex and interwoven relationships with other innovators. We distinguish four layers of networking activity (Fig. 1-3). *Ideas* compete and become interlinked. *Innovators* select, vary, and propagate the successful ideas. *Institutions* construct teams of innovators and cooperate and compete with other institutions. *Initiators* fund the research and development activities of institutions. At the foundation of the system lies the *natural world*. Ideas are tested constantly against the facts and needs of the real world.

This networking interchange provides the essential opportunity for tech mining. The various exchanges of science and technology information effectively *instrument* (document) knowledge at all four levels. How so? Innovators (scientists and technologists) produce findings. Institutions provide incentives for innovators to publish or patent those findings. The ideas used by the innovators are reflected in their publications and patents. Relationships among innovators can also be discerned from papers (journals and conferences) and patents. Also, the institutional arrangements, in funding, conducting, and disseminating R&D, often are reflected in the details of those publications and patents. So, publications and patents—as by-products of the exploitation and exploration of science and technology—provide a lot of insight into actual practices leading to technological innovation.

Innovation is significant! But how can tech mining assist in the innovation process? In the next two sections we examine innovators and innovative institutions in society. Our brief survey suggests challenges and needs faced by these groups and individuals.

1.3. INNOVATION INSTITUTIONS AND THEIR INTERESTS

Let's examine the institutions that fund and perform research, with an eye toward how tech mining can further their interests. At least five sources fund research—industry, government, education, nonprofit, and cross-national funding. Recent sources place industrial funding of R&D at more than 63 percent of the total (OECD, 2003). In the United States, the Federal government is the largest single source of R&D funding. "High-technology" manufacturers fund the highest portion of industrial R&D activity. Service-related R&D spending is much smaller, but a rapidly rising proportion of the total. Most industrial R&D focuses on "development"-related efforts. Notably, industry is the largest performer of R&D. Most of that is done by the largest companies (NSF, 2000). Data for 1997 show five leading U.S. companies contracting for $3 billion or more: GM, Ford, IBM, Lucent, and H-P.

Companies face multiple challenges in making those huge R&D investments (Tassey, 1999). Technology investment is inherently risky, and one's R&D often results in spillovers whereby others accrue benefits from it. So, before diving into an R&D program, the company needs to ascertain what existing knowledge might be capitalized upon. Tech mining can uncover external research results to save rediscovering that wheel. It can identify intellectual property ("IP") land mines before a substantial technology development program finds itself blocked.

If new development activities need to be initiated, one method of reducing risk is strategic partnership. This allows individual partners to leverage their resources, reduce costs, and enable activities that might not otherwise have been possible. Additional benefits for corporations may include speeding up development and reduced competition when the developed product reaches the marketplace (NSF, 2003). Tech mining can find out what R&D others are pursuing and pertinent IP so that you can determine the best route to your goals, possibly via partnering in some form.

Academia is a significant source of public science and the dominant generator of basic research findings. Government also plays a very substantial role—particularly through defense funding—in the support of new technology development. Industry is often involved in carrying technological developments to fruition, so it pays to keep tabs on university and governmental lab research activities.

Significant issues for innovators and their institutions include:

• How can we recognize and reward new and innovative ideas in our organization?
• How do we capitalize on the strengths of our knowledge to attract new funding?
• Can we attain new knowledge before our competitors?
• Can existing, publicly available knowledge provide us with needed solutions?

Note the extent to which these issues demand knowledge of others' science and technology activities—and tech mining can provide this.

The recognition that new products and processes are central to corporate renewal underlies these issues and why we care about them (Danneel, 2002). A careful balance must be sought between exploiting existing competencies and discovering and developing new competencies. New products close to existing core capabilities have a greater chance of success. Unfortunately, however, existing competencies can crowd out opportunities for growth—resulting in inflexibility and missed opportunities. Positive feedback causes innovative "path dependencies," that is, technological choices that lock a firm, agency, or academic unit in or out of specific development trajectories. With respect to tech mining, we need to track both internal and external technological capabilities. Procter & Gamble tells a story on itself. After submission of a patent application, they got back good news and bad. The bad—a patent had already been issued. The good—they held it. It's hard to keep track of your own technology, much less everyone else's.

The customer is also critical to new product development. New products build on a match of new ideas to existing competencies. Successful products stem from the intersection of customer need and technological competencies. Institutions that are successful in new product development must understand their customer—building upon the existing customer base and learning about new customers, or formerly unrecognized needs of old customers. This is another form of intelligence essential to successful innovation.

March (1991) characterizes the options of building upon old knowledge or reaching out to find new knowledge as *"exploitation"* versus *"exploration."* Exploitation is a process of linking—integrating existing knowledge, combining and recombining core competencies to meet market need. Exploration leverages what is known to gain new knowledge. Tech mining seeks to contribute to both.

1.4. INNOVATORS AND THEIR INTERESTS

Where do you find the innovators (largely scientists and engineers)? The United States has the greatest concentration. Interestingly, nearly one in eight U.S. scientists or engineers was born abroad, coming particularly from Asia.

The European Union and Japan also have large scientific and technical work-forces. An estimated two out of three scientists and engineers today are men; however, this situation is rapidly changing. By the year 2020, there may be as many women as men engaged in innovative activities. Most of today's technology innovators work in industry, taking on R&D and various other roles. Although hard to categorize (much is not considered R&D), innovative activities of the advanced service sectors are ascending rapidly.

In tech mining we often want to know "who's doing what." Tracking ideas and individuals provides vital intelligence that serves various innovators (and technology managers).

The output of successful industrial innovation is reflected in new products, processes, and services. The IP involved may be protected through patenting. Some sectors patent more than others as discussed in Chapter 12. On average, an industrial scientist innovator patents once every six years. As we shall see, however, averages can be misleading—the majority don't patent at all, whereas a very few patent a very lot. Industrial researchers are also avid consumers of science and technology information.

Academic innovators contribute to public knowledge mainly through publication (journal and conference papers). An academic scientist or engineer is 45 times more likely to publish his/her research than an industrial counterpart. Not surprisingly, academic researchers contribute about three-quarters of all publicly available R&D. And, although comparatively infrequent, academic patenting is rapidly growing (Hicks et al., 2001). Academic innovators have traditionally also been the greatest consumers of science and technology information.

Too often, innovators reproduce solutions that are known elsewhere. One Russian researcher, Genrich Altshuller, established that roughly 25% of all patents solved problems well known in other disciplines or industries. Another 35% are merely minor extensions to established technologies. Less than 1% of all patents involve creation of foundational knowledge. Altshuller developed tools to stimulate invention—"TRIZ"—that we introduce in Chapter 12.

Finding the science and technology information needed to inform R&D presents a vexing challenge to innovators. Innovations are increasingly dependent on new science, and new sciences such as nanotechnology are increasingly multidisciplinary and therefore distributed across multiple fields. Observers have called for a device called the "memex"—a machine that could link and correlate ideas, finding connections and recommending relevant resources. Today, we have the Internet, but the vision of a seamless network of knowledge seems as distant as it was in the 1930s when the memex notion was conceived (discussed further in Chapter 2).

The nature of the emerging technologies of interest is itself changing. Many of our technology analysis tools were generated for an era of industrial (manufacturing) technologies dominated by defense interests (during the Cold War era) (Technology Futures Analysis Methods Working Group, 2004). In addition to information technologies, we see emergence of the "molecular technologies"—biotech, advanced materials, and nanotechnology. Managing these emerging technologies requires adaptation to "science-based industry"—these are industries whose competitiveness is rooted in their appropriation of basic research knowledge. Besides challenges in identifying and fulfilling opportunities in these emerging technologies, there are corresponding challenges in identifying threats and abating the attendant risks. Cross-disciplinary and cross-sector knowledge awareness and integration is mandatory (Rasmussen, 1997). Tech mining can play a major role in forming these integrative links.

New scientific knowledge is generated primarily in specialist communities. In some ways the quality of this new knowledge can only be assessed by others in the same community. But the knowledge itself must be utilized and integrated by others who often have vastly different skills. Recognizing expert specialists can prove challenging. Studies of the social network of innovators demonstrate that a few scientists or engineers serve as hubs for much of the collaboration. Others are connectors, individuals who synthesize and integrate knowledge from diverse groups. Identification of, and communication with, these hubs and connectors can be extremely useful in exploiting the knowledge of the community.

In this section we have presented a broad-brush portrait of the innovators— particularly researchers and inventors. We recognize that others play essential roles in achieving innovation, for example, new product developers, software engineers, and technology managers. Keeping abreast of science and technology information is vital in creating new knowledge, as well as in rediscovering existing solutions to problems. Challenges arise from science and technology specialization and the increasing volume of R&D findings.

Innovative individuals and institutions face a number of challenges—challenges that tech mining can help address. Among these challenges, we have identified the following:

- Innovation is essential, but risk management practices are needed.
- Innovation necessitates identifying both individuals and institutions with complementary knowledge; R&D partnerships are commonplace.
- Knowledge creates spillovers; protecting knowledge as well as accessing all publicly available knowledge is crucial to success.
- Knowledge is often specialized in character, yet it must be synthesized and integrated by other nonspecialists.
- Too many organizations "reinvent" the wheel.
- Innovation draws on knowledge of customers and the marketplace.
- Science-based industries are making very direct connections between basic knowledge and the marketplace.

These first sections of the chapter have addressed the first of the two drivers of this book—"emerging technologies." The next section transitions to the second driver—the "information economy." Out of the confluence of the two comes the hope for exploiting electronic forms of science and technology information in new and powerful ways—tech mining.

1.5. TECHNOLOGICAL INNOVATION IN AN INFORMATION AGE

We truly live in an "information economy"—the developed nations have largely transitioned from manufacturing to service (information) economies. The service sector contributes no less than 64% of the U.S. GDP as of 1997. The U.S. National Science Foundation shows U.S. high-tech service valued at over $3 trillion for 1997. Amazingly, U.S. commercial service exports were valued at $240 billion in 1998 (do you even think of "service" as exportable?). Look around to see how computers and telecommunications permeate our lives at work, school, and play.

Pervasive waves of increasing information accessibility characterize recent decades. In the 1980s we did two studies for the U.S. Industrial Research Institute (IRI) (Rossini et al., 1988). These documented the remarkable transition in how engineers and other technical professionals worked. As of 1980, only about 10 percent of them routinely used a computer in day-to-day activity; by the end of that decade, most did so (about 80 percent in large U.S. companies—first curve of Fig. 1-4).

On the heels of this transition, another S-curve (not shown) would track how others came to routinely use personal computers—students, office workers, and managers. In the 1990s, the developed economies of the world experienced another transition—those computers became connected. The telecommunications industry saw data transmission overtaking voice trans-

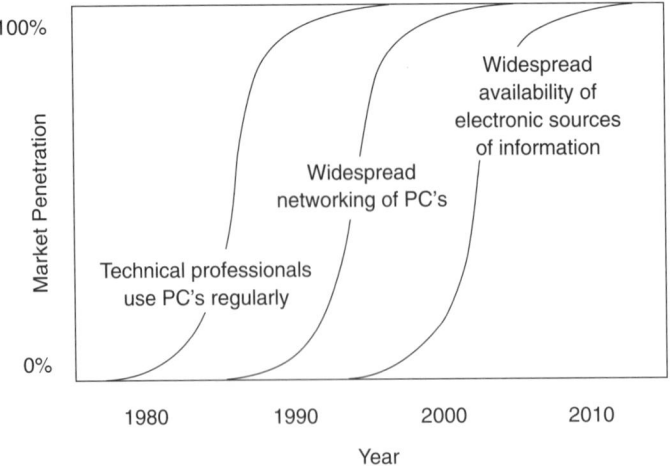

Figure 1-4. Information/technology growth curves

mission. We experienced the takeover of our lives by the Internet and E-mail (second curve of Fig. 1-4). Similar S-shaped curves could be drawn indicating the penetration of these phenomena into our work and home lives (not shown). Information technology ("IT") literally changed lives around the globe, of course, unevenly.

What is the corresponding transition of this decade? We believe it is not an "IT" phenomenon as such (see *Sidebar*), but, rather, it derives from the wide adoption of computers in the 1980s and networking in the 1990's—namely, the increased availability of information (third curve of Fig. 1-4). In the words of Bill Gates, we have "information at our fingertips."

Sidebar: Technology Information does not Equal Technology

This book does not restrict "technology" to "information technology." The changing technology to be managed certainly includes computing and telecommunications, but it also includes transportation, energy, chemicals, pharmaceuticals, aerospace, military systems, and so forth. The critical resource of the Information Age is not IT, it is information.

The message implicit in Figure 1-4 is that new capabilities get adopted—first by lead users, then by others. These new capabilities are used—first to do our old tasks faster, then to do them better, then to do entirely new tasks. We are now beginning to make fuller use of electronic information. Information worker productivity is finally increasing after decades of investment in IT (Greenspan, 1998). Moreover, we are going beyond the stage of doing our old tasks more efficiently, to contriving new tasks that take advantage of the easily available information. This book is about one such opportunity—exploiting electronic information resources that pertain to changing technologies in new and powerful ways.

Technology information is particularly vital to those who would manage better. Consider the information implications of the fact that we do a lot of R&D: Knowing about others' research (and development) is critical—few companies can do technology development all by themselves today. We anticipate that by about 2010, virtually all successful technology managers will avail themselves of empirical, not just intuitive, knowledge. Those who don't "get it" will get gone.

1.6. INFORMATION ABOUT EMERGING TECHNOLOGIES

Consider the following propositions (which will constitute this chapter's "take-home messages"):

- Innovation is a significant force in today's high-technology organization.
- Technological innovation can be measured and understood.
- Innovation can be tracked—we have the tools to monitor new developments in science and technology.
- Innovation involves ideas, generated by innovators, and institutions that initiate, perform, and apply R&D.
- Organizations that track science and technology gain significant market advantages over those who do not.
- There are advantages in tracking innovation for all participants in the innovation cycle.

The importance of networking to innovation implies payoffs from pursuing technological intelligence. Information technology networks are a part of these social networks, and we can make fuller use of electronic science and technology information.

The quantity of science and technology information is huge:

- Research generates a tremendous amount of information—for example, the two databases, EI Compendex and INSPEC, capture a sizable portion of the world's major engineering journal and conference papers— together they add about 500,000 papers each year.
- Other databases capture other research segments—for instance, Web of Science provides access to some 12,500,000 fundamental research papers and MEDLINE has compiled over 11,000,000 medical research abstracts (as of 2002).
- Americans applied for 125,000 U.S. patents in 1997; foreigners filed for another 110,000 U.S. patents; more amazing is that Americans applied for 1,500,000 patents in other countries in that one year.

Besides R&D, there are standards, press releases, business news, and signposts of popular acceptance. The information spews out through myriad pipes—print news, electronic media, human contacts, and the Internet. It can seem like an information onslaught. Nonetheless, you need to track this technology information and extract knowledge pertinent to your work. Whether or not you do, your competitor will.

We're saying that you need to know about changing technologies and that there are daunting amounts of pertinent technology information pouring forth. Intimidating? It need not be. Our message is that you can get at the needed information with powerful new tools that make this relatively fast, affordable, and easy.

The search for emerging science and technology information is not unbounded. There are over 10,000 publicly accessible databases. Of those, on the order of 400 relate to technology. However, 15, or so, key databases filter and compile major segments of the key publication and patent information.

Depending on core interests (e.g., pharmaceuticals, chemicals, aerospace, consumer goods), a company's competitive technological intelligence and tech foresight professionals might only access one to five databases to get excellent coverage of what they need to know. The world's leading center of "scientometrics"—the Center for Science and Technology Studies at the University of Leiden—accomplishes most of its work of profiling national and organizational R&D publication by using just four databases: SCI, INSPEC, EI Compendex, and MEDLINE. The information contained in such databases is rich and suitable for mining.

Assessment and monitoring of innovation is crucial to success. Innovative individuals and institutions have core challenges in managing R&D. What are needed are tools to help manage this wealth of science and technology information. In the next chapter we introduce a suite of tools to do so—tech mining. We'll show how the innovation management challenges can be met head on by exploiting publicly available sources of science and technology information.

CHAPTER 1 TAKE-HOME MESSAGES

- Innovation is a significant force in today's high-technology organization.
- Technological innovation can be measured and understood.
- Innovation can be tracked—we have the tools to monitor new developments in science and technology.
- Innovation involves ideas, generated by innovators, and institutions that initiate, perform, and apply R&D.
- Organizations that track science and technology gain significant market advantages over those who do not.
- There are advantages in tracking innovation for all participants in the innovation cycle.

CHAPTER 1 RESOURCES

This *networked model of innovation* is, in many ways, a synthesis of previous R&D models. Some interesting extensions follow: Van Bueren (Van Bueren et al., 2003) discusses the role of networked activity in societal and institutional decision-making; Hagedoorn (Hagedoorn et al., 2002) discusses the spillover in technological learning in the marketplace. Similarly, Stuart (Stuart et al., 2003) empirically confirms the role of social and geographic networks in the capitalization of intellectual property, whereas Deroian (2002) provides an analytically explicit model of how collective action occurs in the evaluation and diffusion of new technologies.

Chapter **2**

How Tech Mining Works

Chapter 1 overviewed the significance of technological innovation in modern society and made the case for tech mining tools to provide intelligence in support of technology management. This chapter takes a closer look at tech mining by providing a broad overview and summary of the issues associated with tech mining. This chapter introduces tech mining through six broad sections that consider:

- What is tech mining?
- Why do tech mining?
- When did tech mining first start?
- How is tech mining performed?
- Who performs tech mining?
- Where is tech mining most needed?

2.1. WHAT IS TECH MINING?

Tech miners are technology watchers, meaning we analyze changing technologies. That is, we address what is happening now and likely to happen in the future with regard to development of particular technologies. To do so, we compile and analyze information from multiple sources, particularly exploit-

Tech Mining: Exploiting New Technologies for Competitive Advantage, Edited by Alan L. Porter and Scott W. Cunningham.
ISBN 0-471-47567-X Copyright © 2005 by John Wiley & Sons, Inc.

TABLE 2-1. Sample Tech Mining Questions

- What R&D is being done on this technology?
- Who is doing this R&D? Toward what probable market objectives? (sometimes we focus on a particular competitor to profile what it is doing in multiple areas)
- How does this technology fit organizational aims?
- What are the prospects for successful commercialization?

TABLE 2-2. Various Types of Technology Analyses That Can be Aided by Tech Mining

(A) *Technology Monitoring* (also known as technology watch or environmental scanning)—cataloguing, characterizing, and interpreting technology development activities

(B) *Competitive Technological Intelligence* (CTI)—finding out "Who is doing what?"

(C) *Technology Forecasting*—anticipating possible future development paths for particular technologies

(D) *Technology Roadmapping*—tracking evolutionary steps in related technologies and, sometimes, product families

(E) *Technology Assessment*—anticipating the possible unintended, indirect, and delayed consequences of particular technological changes

(F) *Technology Foresight*—strategic planning (especially national) with emphasis on technology roles and priorities

(G) *Technology Process Management*—getting people involved to make decisions about technology

(H) *Science and Technology Indicators*—time series that track advances in national (or other) technological capabilities

ing the science and technology databases. We do so to answer a wide variety of questions—Table 2-1 gives a few. We want to answer variations on the questions of Table 2-1 (Chapter 13 poses a more exhaustive set of 39 management of technology—"MOT"—questions).

Such information can facilitate a variety of technology analyses (Table 2-2). Tech mining provides strong empirical bases for many of these; it does not take their place.

"Chapter Resources" offers pointers to pursue these analyses. "Knowledge management" overtones come into play as well. It is important that tech mining information compilation and analyses cumulate into useful bodies of knowledge over time. The payoffs far outweigh those from one-off studies that start from scratch every time.

Later chapters will show "how" to do tech mining. For now, we emphasize that you really can do more with all this information than warehouse it and retrieve bits here and there. Tech mining is about exploiting this information to see patterns, detect associations, and foresee opportunities. The derived knowledge can help make better plans, designs, and decisions, thereby gaining significant competitive advantage.

Tech mining is the application of text mining tools to science and technology information, informed by understanding of technological innovation processes. We distinguish tech mining from data mining and text mining by its reliance on science and technology domain knowledge to inform its practice (see also Section 2.3).

To illustrate, U.S. Army, Navy, and Air Force colleagues track foreign R&D to identify common interests. The aim is to foster collaboration with American researchers to advance technologies of military interest. Tech mining software can digest searches, say in Japanese, to overview a technological domain of interest that can be processed through automatic translation software to spotlight activities of interest for detailed investigation much more quickly and efficiently than by other means.*

With regard to tech mining, note several major threads running through these types of analyses (Table 2-2).

- *Intelligence*—external information gathering and interpretation to serve particular organizational interests—is central to A and B and contributes to all the others.
- *Futures research*—anticipating likely future developments, for us, emphasizing technological developments—is central to C, D, E, and F.
- Concern about *socioeconomic contextual factors* as these influence, and are affected by, changing technologies is central to E and F and important to B, G, and H.
- *Opportunities analysis*—interpreting technological change with regard to threats and opportunities for our organization—is vital to D, F, and G.
- *Process* considerations—involvement of stakeholders in determining actions, in our case, regarding technology—are central to G and especially salient to D and F.

These questions, analyses, and the threads through them suggest the content of tech mining. There are also important process dimensions to tech mining. Figure 2-1 associates four tech mining outcomes with the technology analysis types. The associations are not one-to-one, only suggestive. To varying degrees the technology analyses can be served by one or more of the "4 Ps" we seek from tech mining:

(P1) Product—information and analyses delivered as some form of report
(P2) Process—engaging tech mining doers and users (see Section 2.5) to collaborate in focusing, interpreting, and acting on tech mining results

*TechOASIS is the "twin" version of VantagePoint software that we use in this book's examples. TechOASIS (available free for U.S. Government use) works with SYSTRAN, to achieve such purposes.

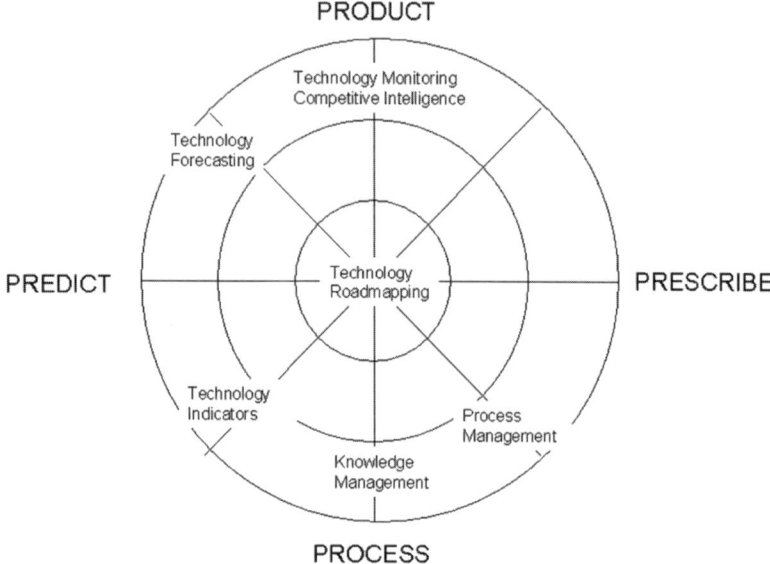

PRODUCT

Technology Monitoring
Competitive Intelligence

Technology
Forecasting

PREDICT

Technology
Roadmapping

PRESCRIBE

Technology
Indicators

Process
Management

Knowledge
Management

PROCESS

Figure 2-1. Tech mining outcomes (4 Ps) and technology analyses

(P3) Prediction—indicating more probable future development pathways, generally in an "extrapolative" mode

(P4) Prescription—advocating actions to affect future technology and business development pathways in a "normative" mode

This book emphasizes that tech mining is more than P1; in particular, the radical message to some analysts is that P2, interpersonal process, merits major attention.

Keep these variations in mind as you design your own tech mining program. Types "A–H," plus knowledge management, together with the 4 Ps, can point you to craft a tech mining effort to deliver the best value. For instance, if you determine that action recommendations are the target bull's-eye, you surely want to think in terms of P4 (prescription) and P2 (process management).

Tech mining activities vary considerably as you adapt them to your organization's needs. The focus is generally on technology, but in a contextually rich way. Many tech mining users want the contextual aspects fully integrated with the technological. For instance, they may demand that cost information, market projections, and transition plans be combined to inform decisions about technological choices. Scan Tables 2-1 and 2-3 and observe how much pertains to the future (see Chapter Resources for more). A key principle: Good tech mining exploits multiple information sources with multiple tools. In particular, empirical sources (database mining) need to be complemented by expert opinion.

TABLE 2-3. Some Reasons to Do Tech Mining

- Forecast likely development paths for emerging technologies
- Identify competitors, or collaborators, at the "fuzzy front end" of new product development
- Identify potential customers for your intellectual property ("IP")
- Discover additional application arenas for the outputs of your R&D
- Gauge market potential for prospective technology-based products and services
- Be a wiser consumer of others' science and technology
- Manage the risks of technology development and implementation based on better information.

TABLE 2-4. Six Principles of Networked R&D

- Multiple selection
- Indirection
- Multiple scales
- Networked organization
- Success accrues to the successful
- Networked externality

Tech mining works by gauging the direct outputs of R&D—articles and patents. We can measure R&D activity many ways, including both input (e.g., funding, personnel) and output measures.

2.2. WHY DO TECH MINING?

Table 2-3 presents some reasons why you need to do tech mining.

Fulfilling these promises requires that we attune to the challenges and opportunities of networked R&D, as per Chapter 1. Think about six underlying principles (Table 2-4) that orient us in answering tech mining questions like those posed in Table 2-1.

Multiple selection observes that innovative new knowledge and technologies are forged from multiple selective principles. Selective pressures are exerted by investors, the initiators of much R&D, but also by the research institutions, individual innovators, and communities of relevant knowledge. Insightful tech mining must attend to multiple dimensions, not zoom in too tightly on a singular research emphasis.

Indirection alerts us to inquire as to where new innovations are likely to arise: typically not in a single place. Innovations reside in changes in the structure and composition of the network itself. For instance, in the sample tech mining of Chapter 16 concerning fuel cells, we are alerted to consider infrastructures and roles played by various interests, not just technical aspects.

A related principle is the idea of *multiple scales*. Networks deliver innovations of all sizes and shapes—from small, incremental changes that occur on a daily basis, to rarer, expansive "transformational" changes that reshape whole economies and fields of knowledge (Kash and Rycroft, 2000). Studies of dynamic systems indicate a rich interplay between micro- and macroeffects. Again, tech mining needs to look down from "40,000 feet" to see the big picture, but also to zoom in to particulars to understand advances and their potential payoffs.

A fourth principle is the concept of *networked organizations*. Networks are increasingly acknowledged as an economically significant facet of science and technology organization. Relationships are neither exclusively competitive nor fully cooperative in character. Risks are shared and diffused throughout the community. In today's technological environment, R&D contributing to a technological innovation often draws on multiple sources. Tech mining can help array collaborative options to help an organization build a winning team to generate high-payoff innovation.

A fifth principle of innovation systems is that *success accrues to the successful*. The highly structured social and intellectual networks that stimulate innovation tend to reward the ideas, innovators, and institutions that have been successful in the past. Wonder how to be nominated into the U.S. or other national academies of science or engineering? Hobnob with the members—Caltech or Harvard faculty are far more likely to be discovered than those at colleges lacking current national academy members. Identifying capstone nodes—the individual and institutional leaders in the field—is one way tech mining provides valuable intelligence.

A sixth principle is that of *networked externality*. Advances achieved by any one innovator or institution diffuse throughout the community, to varying degrees. The most effective R&D networks may also be those that actively transmit and receive information. A paradox of this situation is that it is very difficult for the leaders to exclusively capitalize on new science and technology developments. Again, this places a premium on tech mining, to take advantage of others' R&D.

Chapter 1's model identifies levels—ideas, innovators (individuals), and institutions. We can further distinguish institutions that initiate (fund) R&D, those that perform it, and those that use it to generate technological innovations (see Fig. 1-3). This "cast of characters" constitutes both the subjects of tech mining investigations and the important users of the results of those investigations.

Tech mining interests arise both at a given level and cross-level. For instance, researchers can benefit from identifying other researchers. But, also, organizations may study individuals (innovators) to understand competitor strengths or to recruit talent they want. Initiator institutions may study R&D performers to determine where their scarce funds can exert the biggest effect. In many cases, interests follow the "success to the successful" motif in supporting or collaborating with the best in the field.

We repeat mention of two essential network operations–*exploration* and *exploitation.* Some authors, drawing on an "evolutionary" analogy, call these two operations *variation* and *selection.* Exploration means discovering the context of ideas, innovators, and institutions. Sometimes exploration involves varying ideas that have been successful in the past (cf. TRIZ in Chapter 12). Exploitation involves fully utilizing existing knowledge and relationships; this includes building connections where they previously may have not existed. Exploitation often involves selecting, and supporting, some components of the innovative process at the expense of others. In tech mining, we see variations on this theme. In assessing a body of R&D, we can study the core—seeking ways to exploit the major stream of activities (e.g., we direct management attention to the "hot" aspect of fuel cell development). On the other hand, we also pursue exploration when we spot an unusual juxtaposition of research findings or novel tool application. Effective tech mining addresses both exploration and exploitation.

2.3. WHAT IS TECH MINING'S ANCESTRY?

In coining the term tech mining we owe a nod toward its forebears. Content analysis (finding patterns in term usage to infer intents) has roots in the pre-electronic information era. Talmudic scholars have tabulated content patterns in the Bible for ages. In Western society, the dawn of the modern era foresaw an interest in rationalism, empiricism, and the application of the scientific method. Applying the growing knowledge resources was a major concern of the great Encyclopedists of the Enlightenment. Philosophers such as Diderot, Voltaire, and Rousseau all anticipated the need to muster newly found knowledge toward the betterment of society.

Vannevar Bush, the instigator of major government research funding, discussed a device he called the "memex," in the 1930s. The memex was depicted in the *Atlantic Monthly* as "a device in which an individual stores all his books, records, and communications, and which is mechanized so that it may be consulted with exceeding speed and flexibility." The modern notions of hypertext and a hyperlinked Internet bear similarity! More importantly for tech mining, the memex conceived of a system by which ideas themselves could be linked together in a complex and evolving system. With considerable foresight, Bush anticipated the coming glut of information and the need for technological aids in its management. His ideas are important predecessors of data mining.

Social scientists have applied methods of content analysis for decades. Counting scientific publication activity dates back at least to the pioneering work of Derek DeSolla Price (1963). In another domain, intelligence agencies might analyze a stream of press releases concerning a particular foreign politician to anticipate where his or her interests are heading. With the advent of electronic text sources and analytical software, content analysis has matured into text mining.

Bibliometrics means counting (metrics) of bibliographic content. It has been energetically applied to science ("scientometrics"—cf. *Current Science*, 2000) and technology ("technometrics"). Modern text mining for technology foresight counts among its pioneers Henry Small, Anthony van Raan, and Michel Callon.

Data mining seeks to extract useful information from any form of data, but common usage emphasizes numeric data analyses (e.g., linking your credit card purchases to your demographic profile). Text data mining or text mining exploits text sources of various sorts (cf. Kostoff and Giesler, 1999). Specialties in text understanding are rapidly evolving with fine distinctions in emphasis. For present purposes, we just note emerging areas that concern tech mining: computational linguistics, natural language processing, and "KDD" (knowledge discovery in databases).

Tantalizing glimpses are emerging of the next wave of technologies beyond natural language processing and KDD. These technologies are modular and decentralized; they exploit probabilities and patterns across diverse data sets. They are good at taking general tasks and solving them, even as they improve on previous generations of technology in exploiting the current knowledge of human users. Machine learning is one such technological expression of ideas about collecting, summarizing, and managing exhaustive sources of new data.

2.4. HOW TO CONDUCT THE TECH MINING PROCESS?

Tech mining is a problem-solving process. Like other problem-solving processes, it relies on a structured set of phases (steps, stages) that guide the user through the process. Generic problem-solving—like tech mining—is well treated as iterative in nature. It builds upon, and expands the results of, previous efforts. Indeed, the process might be more appropriately called a problem-solving *cycle*. When learning and problem-solving is a continuous activity, it makes little difference at which step in the process you start.

In 1960 the Nobel laureate Herbert Simon presented a three-phase process of decision-making. Broadly adopted and even extended (Mintzberg et al., 1976; Huber, 1986), Simon's decision-making process has been applied across diverse domains including information system design. Tech mining also follows this three-step process; the process has the advantage of describing both how tech mining ought to be done in practice as well as how tech mining is actually performed. The steps are (1) *intelligence*, (2) *design* (we include analysis), and (3) *choice*.

Let's survey tech mining in terms of these three phases, associating chapters to each.

The intelligence phase involves planning for and collecting the data to be mined. In addition, reviewing, surveying and integrating the collected data are all important processes in this phase of activity. Chapter 6 discusses how to find the right sources to mine. Chapter 7 discusses how to formulate the right queries to collect the data. Chapter 8 discusses the mechanics of collecting the data. Chapter 9 discusses surveying and cleaning the collected data. Tech mining often draws on multiple sources of information to create a robust picture of developments in science, technology, and innovation. There tends to be much noise and redundancy (e.g., manifold variations in topic description or affiliation naming). Data description can also be considered part of this phase in tech mining. Chapter 8 therefore also examines data cleaning and the use of lists and tables in creating short, useful summaries of the data you have collected.

The design and analysis phase of tech mining involves deriving knowledge from the data collected to solve specific problems of innovation or technology management. This often mixes deductive and inductive analyses. Analysts may test specific hypotheses about the data through analytical models. Model building can serve multiple purposes: It can describe science and technology activities but also predict coming developments. Modeling serves both confirmatory and exploratory purposes. Chapter 10 introduces the topic of designing tech mining models and producing advanced analyses. Chapters 11 and 12 examine certain types of analyses in greater detail. Trends over time are a particularly interesting source of tech mining analysis and are discussed in Chapter 11. Patents share much in common with other sources of technology information but have some idiosyncratic characteristics of their own. Patent analysis is therefore discussed in a chapter of its own, Chapter 12.

The choice phase involves nominating options via tech mining and then selecting the right innovative opportunities for your organization. Key to this effort is creating specific metrics or scorecards to gauge available options using the criteria that are most salient for your target users. We discuss the generation and presentation of "innovation indicators" in Chapter 13. When successful, tech mining projects create actionable results for organizations. Chapter 14 therefore discusses how to manage and distribute tech mining results across organizations. Chapter 15 discusses techniques for measuring the impact and value of tech mining. Chapter 16 is a case study that puts all the pieces together—intelligence, design and analysis, as well as choice phases of mining.

Table 2-5 provides an overview of the book. Scanning suggests the book's aims of both telling about tech mining and describing how to do it. We won't belabor the three-phase process, but this maps the chapters to that process to help you gain perspective. The chapters (1–5) you read now are about tech mining; they provide an overview of key ideas and the significance of tech mining in technology management. The details of how to do tech mining await Chapter 6.

In later chapters we present a more detailed tech mining process (Tables 4-2 and 16-1). A more detailed process is helpful in tracking a tech mining study. Goals and deliverables for each stage are clear; transparency of the process is improved.

Table 2-6 introduces this tech mining process and relates it to phases of decision-making. Each phase of decision-making is in itself is a decision that

TABLE 2-5. The Structure of the Book

Purpose	Tech Mining Phase	Tech Mining Book
About Tech Mining		Chapter 1— Technological Innovation and The Need for Tech Mining Chapter 2— How Tech Mining Works Chapter 3— What Tech Mining Can Do for You Chapter 4— Example Results: Fuel Cell Tech Mining Chapter 5— What to Watch For in Tech Mining
Doing Tech Mining	Intelligence	Chapter 6— Finding the Right Sources Chapter 7— Forming the Right Query Chapter 8— Getting the Data Chapter 9— Basic Analyses
	Analysis and Design	Chapter 10—Advanced Analyses Chapter 11—Trend Analyses Chapter 12—Patent Analyses
	Choice	Chapter 13—Generating and Presenting Innovation Indicators Chapter 14—Managing the Tech Mining Process Chapter 15—Measuring Tech Mining Results Chapter 16—Example Process: Tech Mining on Fuel Cells

TABLE 2-6. Decision Phases and the Tech Mining Process

Intelligence	1. Issue Identification
	2. Selection of Information Sources
	3. Search Refinement and Data Retrieval
Analysis and Design	4. Data Cleaning
	5. Basic Analyses
	6. Advanced Analyses
Choice	7. Representation
	8. Interpretation
	9. Utilization

can benefit from a clear and structured approach. The goals at this stage are simply to introduce the process—Chapters 4 and 16 give practical examples of the process in action.

2.5. WHO DOES TECH MINING?

To sharpen our thinking as to what knowledge is needed, think in terms of tech mining users and doers. We look at the role of tech mining in organizations, large and small, private and public. Finally, we consider how

TABLE 2-7. Tech Mining Users

- Strategic planners (to target emerging technologies of core importance)
- R&D managers and funders (e.g., to identify portfolio gaps, assess merits of new proposals)
- Researchers, inventors, and project managers (e.g., to keep informed and facilitate networking)
- New product developers and designers (to help select technological alternatives)
- Procurement (to help assess alternative products and suppliers)
- Process managers (to advise technology insertion)
- Product managers (to roadmap technologies contributing to the product line)
- Product service managers (to detect causes underlying maintenance problems)
- Marketing experts (e.g., to identify new opportunities to leverage products and services)
- Information professionals and librarians (to help justify acquisitions)
- IP managers and specialists (e.g., to help assess desirability of patenting a disclosure)

tech mining should be performed—in house by a single unit or distributed, or outsourced.

This book is for users as well as doers. Knowledgeable tech mining consumers will get more from tech mining, thereby strengthening the case for doing it. We largely assume that, for the present, users and doers are likely to be distinct, but we later explore the case where they are one and the same.

Users

Within larger organizations, many can gain value from tech mining findings (Table 2-7).

How about in smaller organizations? We are less sanguine about tech mining here. In terms of potential payoff, tech mining might well be greatest for small companies. Entrepreneurial companies may put everything at stake in technology decisions, and they often engage emerging technologies and new markets. They need as much sound information as possible to decide where and how to proceed. However, today they often suffer from a lack of time (too harried to do analyses), information resources (not familiar with the databases), and financial resources (too costly to have others analyze technologies for them). We see the need, but also see multiple, difficult hurdles impeding small companies and nonprofits from doing tech mining. This presents an opportunity for consultancies to assess technologies on their behalf.

One of us (SWC) once worked with start-up companies in Silicon Valley. Now that the bubble has burst on the "dot.com" industry, software and internet companies are a lot leaner, and a whole lot more frugal. Nonetheless, even as belts tighten and employment declines among many small, privately held software companies, the role of IP managers is ensured. Why? IP is key to the valuation and, therefore, the long-term health of emerging high-tech companies. Managing IP means attending closely to technology information; patent analysis is the leading application of *VantagePoint*, or *Derwent Analytics*, software that we exemplify in this book.

Emphases vary. We find the private sector to be energetic pursuers of CTI and increasingly involved with technology forecasting, and also with technology and product roadmapping. The public sector has led the way in technology assessment and technology foresight activities. Much public policy has a technology component (consider medicine, defense, economic development), and tech mining helps track activities to help allocate scarce R&D resources.

Academia presents an intriguing venue for tech mining. University offices of technology licensing are increasingly active in pursuing IP opportunities, demanding tech mining. Research planning and "nudging" (the academic counterpart of managing?) warrant tech mining to identify university strengths and weaknesses (gap analysis). Most universities have extensive access to the science and technology repositories, and youthful students (and faculty) comfortable with electronic information access. Researchers can gain richer perspective on how their work links to that within and at the boundaries of their R&D domains (Porter et al., 2002). We see great promise for university tech mining, but little action to date.

Doers

Who performs tech mining? We find an intriguing array of "pros"—including information professionals, technology analysts, business analysts, and planners. We also find diverse "amateurs" doing aspects of tech mining—students, researchers, project leaders, and managers.

Information specialists and librarians have long been the gurus of information search and retrieval (and data is essential for tech mining), along with warehousing information collections. With the rise of electronic information resources, their roles must change. Consider the university library. Students and faculty visit the library less than they once did, so they interact less with the librarians. They do much of their own searching on-line. So, the library's information professionals need to go to the faculty and students, not wait for them to come to the library. They need to specialize in particular scholarly domains (e.g., one works with aerospace engineers, another with chemists). They need to know about and train others in use of the tools for exploiting information resources (i.e., text mining). That implies that they need to become analysts as well as information searchers and managers. We see particular appeal in their joining research teams (Newman et al., 2001). As an analogy, consider how statisticians often participate in research teams, sharing their specialized methodological knowledge with subject matter experts.

We use the term "business analysts" to represent the diverse organizational functions beyond the traditional technical side (R&D, engineering, patenting, etc.). Without pretending to catalog all the roles, those who develop and assess marketing, sales, service, finance, and so forth, need access to tech mining. "Concurrent design" exemplifies the desirability of forming multifunctional teams to develop new products, processes, and services from the research stage on. Knowledge of changing technology is vital to such work.

Technology analysts, on the other hand, find it increasingly advantageous to be able to get their own data. They seek to develop their information search and retrieval skills, as well as to learn how to exploit various forms of retrieved information. In "lean" organizations, analysts need to work closely with managers. It is too slow and costly to pass reports up through multiple human filters. Sometimes those human filters make findings more relevant and credible to decision-makers, but at the expense of making the process less direct. So, analysts have to stretch their role from the comfort zone of solitary study to that of active communicator. They need to sell their studies, push managers to participate in the analyses (generally uncomfortable for both sides), and present findings powerfully and interactively.

Choosing the right organization with which to collaborate on a development is immensely important. Small companies cannot hope to perform all the tasks necessary to develop successful, technology-intensive product lines. Some organizations may be able to design and build a new technology-based product yet require other companies to market it. Larger corporations entertain patent arrangements to share IP. Litigation is expensive, but so is allowing IP to fall undefended into the hands of competitors. Tech mining can assist in finding corporations with related interests with whom profitable collaborations may be possible.

"Planners" come in many colors. We mention them here to stretch the tech mining doer community—to involve new product designers, strategic corporate planners, technology developers, and others. Their attention and skill sets need to incorporate tech mining because it dramatically expands their science and technology information base.

We also nominate "managers"—our prototypical users. They need to be well aware of, and conversant with, tech mining to become good users. Beyond that, as mentioned earlier, doing some tech mining themselves can pay off in quick, on-the-money intelligence.

In essence, we need to reduce the distinctions among the tech mining-related roles. "Everyone" needs to stretch his/her skill sets to better understand the others' specialties. This enhances the ability to collaborate on teams to implement technological changes. And, admitting to slight bias, we advocate that they all learn some tech mining.

Users as Doers

Distinctions often blur. Some of the most avid users of tech mining results have been analysts. The notion of do-it-yourself tech mining (the doer is the user) warrants some attention. Do-it-yourself tech mining holds a compelling advantage—it reduces the information chain's length to zero (i.e., how many heads have to exchange information along the way from "finder" to "keeper"? The fewer, the better.) Unfortunately, tech mining often suffers from an exceptionally long chain.

Why is that? Because those most open to, and able at, mining electronic information tend to be the youngest and most junior. Imagine that the senior manager has a problem to resolve. His assistant translates that into a request to the Engineering Department Head; she passes it to the Director of Tech Mining, who gives a roughly corresponding assignment to the junior associate (who also knows least about organizational culture). The chances for tech mining failure—meaning unused results—are high because of lags and communication noise. In contrast, suppose that senior manager can directly search a patent database and roughly profile "who's doing what" on the technology in question. Within 5 minutes, he has a ballpark answer that may resolve the problem on the spot. And, if not, he can pose a more precise assignment that is more likely to be fulfilled effectively.

Given the increasing accessibility of information, many users could certainly do a little, or even a lot, of tech mining themselves. We are seeing companies providing desktop database access and expecting researchers and technologists to know what's happening in their technology domains. Our vision for tech mining reaches out even further toward the omega point of "all" scientists and engineers, faculty and students, and technology managers doing their own tech mining as they initiate and justify projects. Someday soon!

2.6. WHERE IS TECH MINING MOST NEEDED?

The message is clear—managers need to consider how changing technology affects their business. Tech mining is certainly needed by anyone conducting innovative activities, defined broadly. Many companies depend on technological change to generate competitive products and services. They also need timely technology insertion to improve their processes. Additionally, academic researchers are significant contributors to the production of knowledge and face significant challenges in managing a runaway supply of science and technology information. National governments have strong needs for R&D—particularly with regard to health, defense, and economic competitiveness issues. Changing technology poses concerns for many players, including:

- Technology-producing and science-based companies
- Companies that consume technology produced by others
- Companies that must assess the market value of other companies
- Government agencies, with interests ranging across research (U.S. National Institutes of Health) to operations (defense, World Health Organization) to regulation (state Environmental Protection Agencies)

- Universities
- Nonprofits (e.g., research, environmental, and policy organizations)

Another "where" dimension concerns whether tech mining functions should be centralized, distributed, or outsourced. Chapter 14 examines how best to institutionalize tech mining. So, what's best? The answer naturally depends on your organizational culture, resources, and priorities. A combination of a small centralized tech mining unit providing support to a wider contingent of part-time technology miners is particularly attractive. Such arrangements could be further enriched by judicious use of external studies (both general and tailored). And all this should be embedded in well-thought-out processes, with incentives, to ensure that tech mining information is used.

At this juncture, let's review how the book's pieces fit together. Part I concerns "Understanding" tech mining. Chapter 1 examined the role of innovation in society. Chapter 2 addressed how tech mining can offer its users significant competitive advantages in tracking that innovation. Chapter 3 scans Part II content from a user (non-doer) perspective. Chapter 4 illustrates tech mining results through a case example (fuel cells). Chapter 5 explores likely future tech mining developments. Part II then details "Doing" tech mining. Those eleven chapters step through each phase of the tech mining process. The goal is to exploit the accumulated knowledge resident in the large science and technology databases.

CHAPTER 2 TAKE-HOME MESSAGES

- Tech mining is the application of text mining tools to science and technology information, informed by understanding of technological innovation processes.
- Tech mining has roots in content analysis, bibliometrics, and text mining.
- Tech mining has matured to the stage that it has something real to offer in support of technology management and various technology analyses.
- Identify prospective tech mining users in your organization; we nominate a good number of candidates, especially in larger organizations.
- Tech mining is not just done by information specialists and IT professionals, it is widely done by technology analysts and is also accessible to sometime doers, particularly researchers, technologists, and business analysts.
- Tech mining offers critical competitive advantages to a wide range of institutions.
- Empirically grounded technology management is better than solely intuitive technology decision-making.
- Tech mining draws on networked R&D sources and users; we offer six principles to keep in mind.

- Tech mining is interested in both "input" and "output" measures of science and technology activity as potential indicators of technological development. (We emphasize output measures, particularly R&D publications and patents.)
- Don't attend just to technical analyses; effective tech mining comes from holistic treatment of Product, Process, Prediction, and Prescription.
- Consider tech mining as a three-phase process: intelligence, design and analysis, and choice.

CHAPTER 2 RESOURCES

We suggest entrees into the various types of technology analyses introduced in the chapter. Major types of published analyses include:

- *Science and technology road maps*
- *Technology indicators*
- *Technology foresight studies*
- *Competitive intelligence*
- *Technology forecasting*
- *Technology assessment*
- *Process management studies*

Multiple views on these perspectives are offered by Glenn and Gordon (2002). The Technology Futures Analysis Working Group (2004) is working to integrate understanding of diverse techniques. Kostoff (Kostoff, 2004; Kostoff et al., 1994) provides an authoritative starting point for the practice and content of science and technology road maps. Institutions engaged in producing technology indicators include the National Science Board of the U.S. National Science Foundation (National Science Board, 2002), as well as the Georgia Institute of Technology. The European Community Directorate General for Research provides an in-roads into the literature on technology foresight. Martin (1989), Salo (Salo et al., 2003), and Cuhls (2003) are also good sources on the topic of technology foresight. Ashton and Kavens (1997) provide a handbook of materials on competitive technology assessment. Porter (Porter et al., 1980; Porter et al., 1991) provide guidebooks and references for technology assessment and technology forecasting. Other excellent technology forecasting resources include a book by Martino (1993) and a survey article (Coates et al., 2001). De Bruijn (De Bruijn et al., 2002) discusses why better methods are not enough; process management is also required.

Chapter **3**

What Tech Mining Can Do for You

This chapter largely excerpts important tech mining features from the "how to" section that we feel are vital to understanding the process and what it can provide to you.

3.1. TECH MINING BASICS

We want you to know what tech mining is and what it can do for you. If you are in a position to receive empirical technology intelligence, some feel for the processes involved in generating it will make you a better consumer. Table 3-1 suggests the sort of questions you might ask about the data.

Section 6.3 discusses the characteristics of information sources that make them suitable for tech mining. We are mainly talking about databases—compilations of certain kinds of information. These are essential because so much science and technology information is being generated. For instance, each year there are about 1.5 million new patents issued (of which some 650,000 are distinct; the others reflect filing of the same invention in multiple locations). You do not want to track these down yourself and try to digest them one by one.

You may find tracking the "Chapter Challenges" an interesting way to get a sense of various tech mining activities. Several of these follow the adventures of a hypothetical philanthropic foundation, the DeBrand Foundation, as it exploits science and technology information resources. In Chapter 7 we

Tech Mining: Exploiting New Technologies for Competitive Advantage, Edited by Alan L. Porter and Scott W. Cunningham.
ISBN 0-471-47567-X   Copyright © 2005 by John Wiley & Sons, Inc.

TABLE 3-1. Informed Consumer Questions

- Which information resources were used, and why? Do these cover all the facets that should be integrated, especially contextual elements such as business information? Which sources were considered but not used, and why? (Chapter 6 discusses databases.)
- Are the tech mining study assumptions and boundaries clear? What is included, what is left out, and why?
- Are the searches suitable? How inclusive are they? What is excluded? Can we estimate how much relevant material we have missed? And how much extraneous material we've had to deal with? (Chapter 7 explores searching.)
- How clean are the data?

consider the issues they confront in deciding how to monitor the literature to expand DeBrand's grants program.

3.2. TECH MINING ANALYSES

Don't let analysts or advocates buffalo you with tech mining imperatives. Confirm that the analysts have checked critical results with suitable experts. Make sure you are clear on what those results reflect. In simplest terms, most tech mining can be reduced to four levels:

1. Lists—i.e., simple activity counts that tell how much of something is taking place
2. Breakouts from those lists—i.e., take the "Top 10 Patent Assignees" (or however many interest you of whatever data field you want to know about) and tell about what each of those leaders is doing. Put another way, combine two lists to make a matrix or a depiction we find often very informative—a "profile."
3. Maps—showing relationships among a chosen type of data, such as keywords or authors
4. Trends

Chapters 9–12 address tech mining analyses. We suggest you browse these chapters to get a feel for what is involved. This section can guide such browsing.

In Chapter 9, DeBrand tracks literature citing a key figure in a research arena central to their interests. We see how the R&D abstract records are manipulated to determine how much various researchers have contributed. We see "two-dimensional" breakouts to characterize the emphases of leading researchers. Tech mining software helps elicit relationships among data fields such as authors and the topics about which they write. It does so based on co-occurrence. That is, if particular terms tend to occur in the same records more than expected by chance, we presume a possible relationship. Table 9-5 arrays many possible relationships that can be explored. Tech mining users should be generally familiar with these so as to request any of interest that the analysts might not have pursued.

An intriguing variation on co-occurrence is to cross a field with itself. Table 9-6 notes several interesting possibilities. For instance, we could look at inventors by inventors to see who teams with whom. This is one form of intelligence that can help uncover knowledge networks. Such awareness can inform decisions on collaborating, acquiring intellectual property, or countering competitor strengths (e.g., hiring away key personnel).

Chapters 9 and 10 ponder the implications of the fact that distributions of science and technology activity tend to be highly skewed. Whether we are considering prolific authors, leading patent assignees, or subject matter terms, these tend to be highly concentrated. That is, the leaders tend to be extremely prolific, while many others occur in "ones and twos." One implication for your analysts is they can build very effective thesauri to consolidate information for the leaders, neglecting the bit players for most purposes.

Chapter 10 introduces the notion of looking beyond the obvious. Underlying constructs can be postulated and explored in the data. For instance, Table 10-1 notes that the "hidden variable" of researchers' prestige can be gauged by the measurable amount of citations to their work. In particular, we assert rich interpretive possibilities based on consideration of technological innovation processes. After the introduction in Chapter 1, we go on to develop a detailed tech mining framework in Chapter 13 based on our understanding of how technological innovation comes about.

Patent or publication *mapping* is a popular way to present relationships among particular aspects of a data set. Chapter 10 explains principal components analysis (PCA) as an important way to model text data. Familiarity with PCA can aid in understanding the strengths and limitations of these and other mapping approaches. The discussion of ways to assess the merits of particular data models can enable you to probe into a given tech mining model to gauge how well it fits the data.

PCA actually represents one of three important families of mapping. You will want to recognize these because they emphasize different ways in which items can be grouped and graphically depicted:

1. Dimensional Analyses—emphasize reducing detail to aid comprehension. For instance, we might take 150 or so leading keywords from the thousands of records on a topic and use factor analysis (PCA is a basic variant) to consolidate subtopics based on keyword co-occurrence patterns down to a dozen factors. We can then map these to see relationships among the factors.
2. Clustering Techniques—also seek natural groupings in the data based on a chosen similarity measure. Some approaches allow items (terms or records) to belong to multiple clusters, and some do not.

3. Tree-Based Techniques—successively divide data into classes. Items are usually exclusively placed on a particular branch. Trees may be generated top-down or bottom-up. They lend themselves to presentation as text outlines or through statement as explicit rules. Whereas dimensional and clustering approaches key on similarities among the data, trees can offer a nice way to accentuate differences. Trees also can partition data through multiple levels.

These approaches can become hard to distinguish. Quite often, analysts could use more than one approach to achieve similar aims in grouping the data. For instance, both hierarchical clustering and tree-based techniques help identify parent-child relationships (dependencies) in the data. Also, be warned that terminology is imprecise. This book, which highlights the dimensional approach, PCA, often refers to the resulting principal components as "factors" or "clusters."

Mapping can take many forms. Chapter 10 illustrates clustering (as opposed to PCA) for DeBrand Foundation uses. A sidebar mentions regional innovation clusters that are a natural for geographic mapping. Note that most of our other maps show term or record proximity locations; often the axes have no inherent meaning. The bottom line is, if you don't find a particular formulation helpful, explore whether other approaches might be more illuminating.

Chapter 10 offers a useful sidebar on "threats to validity." These provide a set of "detective questions" with which to approach a given analysis. Four different threat types are distinguished:

- Unfair (biased) comparisons being drawn
- Effects likely due to chance
- Poor labeling of effects (e.g., calling more papers being published "technological progress")
- Overstated claims for generalizability

Chapter 11 discusses *trend analyses*. Challenge your tech mining findings to ensure that you are getting the information most salient to the decision at hand, not the information that makes the prettiest trend line. Lots of choices go into selection of which data to analyze, whether and how to transform raw data, and how to plot the trends most informatively. Chapter 11 illustrates a range of trend formulations for nanotechnology publication activity. S-shaped growth curves are the most prevalent models of technological advance, but they aren't the only ones.

Get comfortable with the idea that trend growth modeling fits the technological and contextual realities of your case. If trends are extrapolated into the future, understand how limits were set and internalize what are reasonable high and low projections. Test whether interpretations stand up to challenge. For instance, the nanotrends illustration infers three distinct research fronts successively advancing (Fig. 11-10). Can you pose alternatives that fit the data and reflection as well?

Patents are arcane. Whereas publications and their abstracts strive to convey what the research is all about, patents are highly circumspect. Make sure your tech mining analysts are highly familiar with interpreting patent information. Determine whether you can gain the knowledge you seek as effectively from relatively raw patent data (e.g., the information as provided by patent authorities) as from that compiled and standardized by databases (e.g., MicroPatent and Delphion) or if you are best served by having patents abstracted and indexed by technical specialists (as in the Derwent World Patent Index).

In tech mining, we emphasize patent macroanalyses, that is, analyzing large sets of patent abstract records to profile activity in a target domain. These differ from the microanalyses needed to get exacting and comprehensive coverage to resolve legal matters, such as prior art and patentability. Section 12.7 contrasts the different patent analyses for different uses.

You may find Figure 12-1 helpful in benchmarking other organizations on two dimensions—patent activity and rate of change therein. Chapter 12 continues the sample case analysis on fuel cells with patent data. It shows ways to get at technologies or competitors in terms of who, where, and when, what, and why.

Patent and publication *citation* analyses go a step beyond activity measures to generate indicators of utilization. Figure 12-6 quickly shows the bidirectional aspects of looking back from a given patent (or publication) to see what knowledge it builds upon (what it references) and looking forward to see who builds upon it (by citing it).

TRIZ builds upon patent analysis in totally different ways. Section 12.8 introduces this creativity tool that systematically explores alternative ways to solve a given problem. Particularly interesting is that TRIZ is spreading from the purely technical domain to help assess promising innovation directions to pursue broader technology management issues.

3.3. PUTTING TECH MINING INFORMATION TO GOOD USE

Chapter 13 is the cornerstone of our approach to tech mining. It arrays management of technology (MOT) issues that cascade into more specific MOT questions. Those can be answered through purposive mining of the empirical science and technology information resources in what we call "innovation indicators." Chapter 4 gives a taste of this approach and what it can deliver, but do browse Chapter 13 to familiarize yourself with innovation indicators as a base resource to support MOT decision processes. Table 13-2 arrays 39 MOT questions together with some 200 candidate indicators.

An important facet of tech mining is information representation. Sprinkled through the analytical chapters (Chapters 9–12), Chapter 13, and the case analyses (Chapters 4 and 16) are lots of alternative information visualizations. Our pride and joy is the "one-pager," illustrated in Figures 4-5 and 13-1. This digests volumes of tech mining information to an executive-level condensation. It aims to bring to bear key intelligence to aid a particular MOT decision.

As explored in Section 13.3, we see "paradigm-changing" prospects in the systematization of business decision processes. This has profound ramifications for tech mining. By mandating inclusion of specific tech mining findings, in specific forms, in those strategic processes, we set the stage for two momentous changes. First, we can sensibly expedite the generation of repeat analysts by scripting search, data cleaning, analysis, and representation steps. Second, the outputs become much more valuable because they become familiar. No manager wants to rely on unfamiliar and untested sources for critical decisions. But through systematization, technology intelligence products (TIPs) can be vetted. They then will come to be relied on because they provide better information on which to base MOT decisions than traditional, highly intuitive knowledge sources.

3.4. MANAGING AND MEASURING TECH MINING

Chapter 14 turns our attention to issues in managing tech mining. It explores options in locating and integrating tech mining activities within an organization. The challenges deserve empathy. Moreover, they raise significant infrastructure requirements and particular sensitivities. Successful tech mining can pay off tremendously, but it does not come about casually. The chapter ends with reflections on how the tech mining community can facilitate learning by sharing experiences within and across organizations.

Consider the five different tech mining "players" introduced in Figure 14-1. If you are interested in establishing tech mining capabilities, or taking fuller advantage of existing ones, reflect on who is doing what. Gaps in capabilities are a concern. But one also needs to adjudicate turf and balance perspectives. We don't want to make this process unduly complicated, but recognizing multiple interests is vital to dealing effectively with the players.

By this point in the book, we are hoping our readers are on board and enthusiastic about tech mining. But we recognize that buy-in is not automatic. The "Information Age" means ready access to more and more electronic information. Those resources portend increasingly information-based decision processes. That means dramatic changes in how technology managers function. We believe the successful ones will exploit tech mining and similar capabilities and thoroughly outperform more intuitive managers. We also need to recognize the implications for change in how information professionals and researchers, in particular, work. "Information at your fingertips" means new forms of interpersonal exchange and new job functions. We don't believe you can overestimate the ramifications.

Chapter 14 also promotes the idea of "process management." As technology analysts, we have had to learn the hard way that attaining effective utilization of tech mining takes additional steps beyond good analyses. Sections 14.3 and 14.4 offer situational analysis diagnostics that point to distinct actions to take to enhance tech mining uptake. Table 14-1 packages these into an eight-point checklist for managers and analysts to use together to work out suitable study strategy.

Section 14.5 compares three broad approaches to obtaining tech mining—centralization, diffused tech mining, or outsourcing. Of course, "it all depends," but we make the case for at least partial centralization of tech mining function.

Assessing the effectiveness of tech mining is a phenomenon deserving attention in its own right. Chapter 15 lays out factors to consider measuring and approaches to do this. For those of you not directly concerned with such matters, this chapter provides a different use. It offers a number of "meta" reflections about tech mining that can provide perspective on the overall endeavor. What can it offer to whom? Our overall orientation is to consider two key dimensions: Validity treats whether the tech mining findings are correct; utility concerns whether they prove useful.

For those who are concerned with assessing the worth of tech mining efforts, Chapter 15 suggests evaluation approaches and points to cover. Two vital bases for assessment deserve attention. One concerns the value of establishing comparisons rather than seeking absolute, stand-alone valuations. Oversimplifying, how are we better off with tech mining than without it? The second premise is that multiple measures are a whole lot better than single ones. Tech mining interjects itself into complex business processes concerning MOT, so efforts to judge it on singular grounds are inadequate. Effective evaluation requires attention in its own right.

To show we're not short on chutzpah, Chapter 15 wraps up with our rendition of the Ten Commandments. These compress what we hope is a lot of insight on what is entailed in doing tech mining well into a handy list. After this chapter, Part I continues with example tech mining findings from fuel cell analyses (Chapter 4, whereas Chapter 16 steps through their generation) and a perspective on possible advances in tech mining to watch for (Chapter 5).

So—wondering what it takes folks to initiate tech mining activities? A VP of a Singapore small enterprise related how they came to start active intellectual asset management. The day after successful introduction of their software to the U.S. market at a trade show, they were served a "cease and desist" order. After sinking half of their initial profits in attorneys' fees, they abandoned North America. They went to Australia and were slammed with a lawsuit. In both situations, the VP felt the cases against them were weak, but, no matter, the threats served to derail their marketing ventures. They now carefully assess the IP landscape with an eye toward offensive action themselves. Tech mining is just what it takes to gain perspective on technological threats and opportunities.

CHAPTER 3 TAKE-HOME MESSAGES

- For those of you not regularly engaged in doing or using tech mining, we suggest that gaining passing familiarity can alert you to situations in which you can call on it fruitfully.
- Although the book is arranged for Part I to provide general familiarity, this chapter can guide you on a useful browse of Part II to deepen your understanding by selective perusal.

For those of you more deeply involved with managing and/or performing tech mining, this chapter spotlights elements of special importance.

Chapter **4**

Example Results: Fuel Cells Tech Mining

What do tech mining results look like? This chapter illustrates a case technology analysis.

4.1. OVERVIEW OF FUEL CELLS

We have selected fuel cells as the sample topic of an emerging technology with broad promise and a considerable literature and patent record base. Fuel cells fulfill a fundamental need—the storage of power. We expect you have a general sense of the value of the technology, despite the often advanced chemistry involved. In general:

> A fuel cell converts hydrogen and oxygen into water, producing electricity and heat in the process. It functions as an electrochemical device, like a battery, where it is "recharged" with hydrogen and oxygen instead of electricity.

We will mention several types of fuel cells distinguished by the type of electrolyte they use (Table 4-1). The oxidant is nearly always oxygen, and the main effluent is water. The lower-temperature cells tend to use more highly active, expensive catalysts (e.g., precious metals). These cells can be smaller and start up more quickly, thereby making them more suitable as portable power sources.*

*The website http://auto.howstuffworks.com/fuel-cell2.html gives an overview of fuel cells; http://fuelcells.si.edu/ also provides background information.

Tech Mining: Exploiting New Technologies for Competitive Advantage, Edited by Alan L. Porter and Scott W. Cunningham.
ISBN 0-471-47567-X   Copyright © 2005 by John Wiley & Sons, Inc.

TABLE 4-1. Five Main Fuel Cell Types

Main Fuel Cell Types	Operating Temperatures (°C)	Sample Leading Applications	Primary Fuels
Proton exchange membrane (PEM)	90–100	Automotive	Hydrogen or methanol
Alkali	100–250	Space program	Hydrogen or methanol
Phosphoric acid	150–220	Power plants	Hydrogen or methanol
Molten carbonate	500–700	Power plants	Hydrocarbons
Solid oxide (SOFC)	700–1000	Power plants	Hydrocarbons

TABLE 4-2. The Nine-Step Tech Mining Process

1. Issue identification
2. Selection of information sources
3. Search refinement and data retrieval
4. Data cleaning
5. Basic analyses
6. Advanced analyses
7. Representation
8. Interpretation
9. Utilization

4.2. TECH MINING ANALYSES

Chapter 16 carries through a nine-step analytical process (Table 4-2). This process begins with specification of the questions to be answered. Because this case analysis is done to illustrate many facets of tech mining, it is not motivated by a specific technology management target. Therefore, the findings are not as sharply focused as they usually would be in practice.

Chapter 12 further expounds on patent analyses. If you would like to see the data in the tech mining software that we use in these analyses—*Vantage-Point* (also available as *Derwent Analytics*)—visit www.theVantagePoint.com.

As laid out in Chapter 13, we offer a tech mining framework of:

- 13 Technology management issues
- 39 Technology management questions
- ~200+ Innovation indicators

Tech mining addresses one or more particular issues, focusing on the related questions, to inform the decisions being posed. By analyzing science and technology as well as other data, it generates indicators of progress toward practical innovation for a given emerging technology. We illustrate mainly with science and technology data here, but mention the importance of complementary contextual information and expert opinion.

4.3. TECH MINING RESULTS

We begin with a few basic measures from searches on "fuel cells" performed in 2002 and 2003. We go on to show a few more intricate innovation indicators (derived measures). Those searches resulted in two files of abstract records. One contains 9724 patent family records from the *Derwent World Patents Index* ("Derwent" for short). The other file includes 11,764 abstracts of journal and conference research papers gathered from the *Science Citation Index* and *INSPEC*.

The first tech mining "result" indicates how much R&D activity is taking place—the overall numbers of patents and papers on fuel cells—a lot! Tables 16-3 and 16-4 break these out by data fields to show, for instance, that we have found articles in 1367 journals and could list 3311 patent assignees. You might well wonder if such tallies are so exact. Not really—they depend on the scope of the data set—which search phrases, applied to what databases, over what time frame. They also depend on how thoroughly the data are cleaned. For instance, further work would surely consolidate many of those 3311 assignees as being the same or closely related organizations. However, one trades off precision for analysis speed and cost. Good analysts will balance these wisely to provide "clean enough" results for the issue at hand. As a user, you are usually looking for the order of magnitude in tech mining—much or little R&D? many or few competitors?

We can break out any such tally. Consider two main types:

- *"What"* information that serves to characterize the technology
- *"Who"* information that serves to portray competitor (or collaborator) involvement

At the simplest level, these yield counts or *lists*. For instance, here are a couple of *"What"* breakouts:

- *How many articles address solid oxide fuel cells (SOFC)?*
  We scan the articles file using *VantagePoint* and find a striking 11% (1286) of the article *titles* address solid oxide fuel cells (SOFC).
- *How many articles mention particular materials?*
  Table 4-3 lists a few prominent materials from the *keywords* field; reviewing these with your organization's materials scientists might uncover gaps in your capabilities. You might want to know, "what we do have ongoing in zirconium?"

TABLE 4-3. Sample Fuel Cell Materials

Material	Main Keyword	All Variants
Zirconium variants	673	1050
Lanthanum variants	647	830
Yttrium variants	586	1328
Nickel variants	348	724

Chapter 16 gives more examples of handy lists—for example, leading publishing universities, leading publishing firms, and leading European patenting firms on fuel cells for cars.

Tech mining software can help take the next steps beyond just counting how much activity we find. One way is to combine two lists. Chapter 16 tallies priority vs. family patents, by country, to help distinguish R&D locations from market interests. It is also easy to count and segregate records on particular subtopics in the data set to pursue specific concerns. For instance, we could investigate what a particular competitor is doing with zirconium in fuel cells (Julich GmbH leads our publishers on this topic).

The Sidebar sets up a "Who" illustration to show that tech mining software can combine and manipulate data to get at one's particular interests.

Sidebar: Aussie Solid Oxide Fuel Cell (SOFC) Partner?

Imagine a scenario in which we are an American company initiating operations in Australia. These involve an innovation that needs a power supply for remote settings. We have investigated technologies and determined that SOFCs appear most promising, but will need customization. So, we would like to find an Aussie partner to work with us on this development.

We first check who is involved with SOFCs Down Under. We locate 41 Aussie SOFC articles, then spotlight all authors with 4 or more papers each. Figure 4-1 portrays the highly interconnected "knowledge network" based on co-authorship patterns.

To pursue the Aussies further, we might "profile" these researchers. Table 4-4 combines several fields of data to give some sense of where these folks are located and what they are up to. They appear to be located at three organizations—one company, one government lab, and one university. Just from their titles (leaving out common terms such as SOFC for this group), Foger and Zhang appear to be close collaborators, even though affiliated with different organizations. Love is the most recent player; all, except possibly

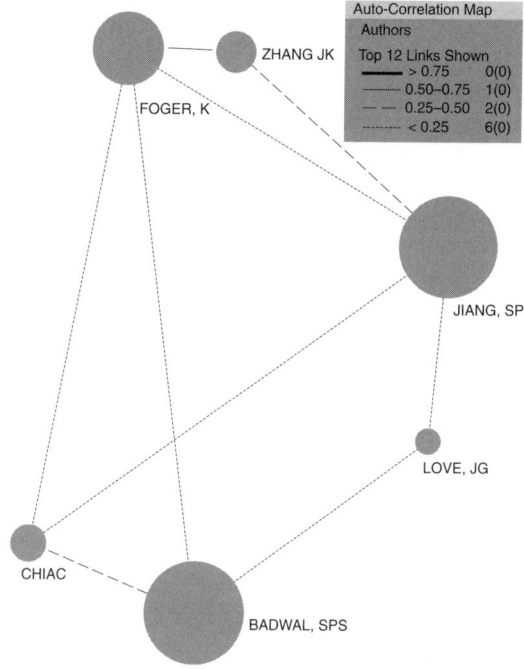

Figure 4-1. Australian knowledge network on SOFCs

TABLE 4-4. Quick Profile of Leading Australian SOFC Researchers

Authors [# of papers]	Affiliation	Interesting Title Phrases	Publication Trends
Badwal, SPS[22]	Ceramic Fuel Cells Ltd.	chemical diffusion [3]; fabrication [2]; perovskite cathodes [2]	
Jiang, SP[22]	CSIRO	chromium species [7]; deposition [7]	
Foger, K[16]	Ceramic Fuel Cells Ltd.	chromium species [7]; deposition [7]; Sr-doped LaMnO/sub 3/ electrodes [3]	
Zhang, JX[9]	CSIRO	chromium species [7]; deposition [7]; Sr-doped LaMnO/sub 3/ electrodes [3]	
CIACCHI, FT[8]	Ceramic Fuel Cells Ltd.	commercial zirconia powders [2]; DC magnetron [2]	

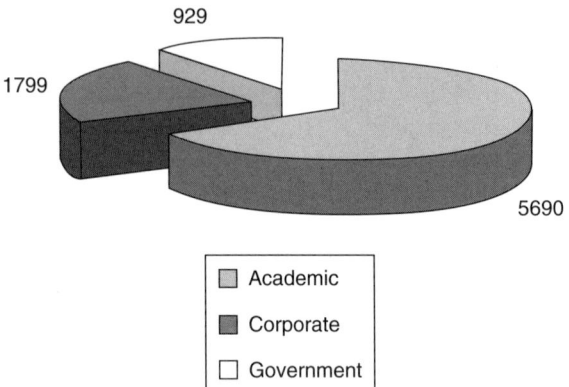

Figure 4-2. Organization types publishing on fuel cells

Ciacchi, seem to remain active in this research area. We will return to this case shortly.

Chapter 16 shows other fuel cell profiles—leading researchers and topics at the three top American universities; leading inventors and topics at the five top patenting firms.

Figure 4-2 illustrates one way to consolidate activity to add interpretive value. This is done by using a thesaurus that one can continually improve. In this case it identifies organization types by clues such as "univ" for academic, in addition to knowing particular names as academic ("MIT"), etc. The portion of R&D publishing by industry is an innovation indicator. Benchmarking the typical extent of publishing by certain sectors, companies, and/or technologies contributes toward determination of how likely rapid commercialization of the target technology is. In this case, 21 percent of the papers were generated by companies, suggesting very strong industrial interest in fuel cell development.

Trend analysis provides another basic piece of intelligence. Figure 4-3 presents patenting trends for organizations whose patent abstracts mention SOFC.

4.4. TECH MINING INFORMATION PROCESSES

As mentioned earlier, Chapter 13 arrays a tech mining framework of technology management issues, questions, and indicators. A major leap forward takes place when an organization systematizes the way it integrates these into its standard business decision processes. Systematization provides many benefits, particularly the semiautomation of much of the tech mining analyses—that is, specification of usual data sources, scripting of software analytical steps, and provision of well-recognized output forms. One adopts such systematization while needing to guard against overdoing it, thereby straitjacketing tech mining creativity, adaptiveness, and insight.

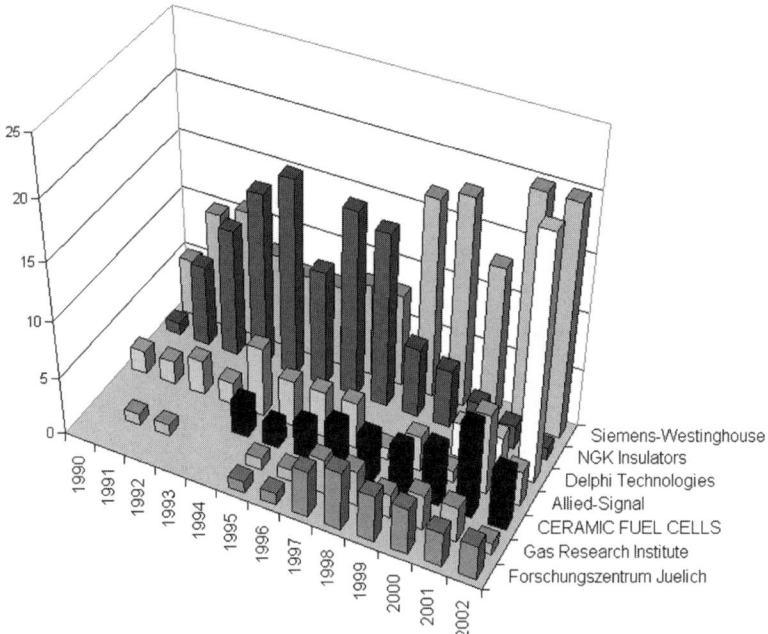

Figure 4-3. SOFC patenting trends

To give the tenor of a happy middle way, we propose that organizations explicitly consider how tech mining can best contribute to stage-gate and similar systematic decision processes. Experience will need to refine these decision-aiding processes, but we suggest an incremental approach. This might focus on a particular technology management question that is faced repeatedly.

For example, suppose our multinational company frequently assesses other companies as potential collaborators for joint technology development. We provide a tech mining template illustrated as Figure 4-4. Ideally, the target user and tech mining analysts would review this together and adapt to the issue at hand. Adaptation could involve selecting alternative indicator options [e.g., for the "Informative chart on what they do best," let's substitute Option (3) Patent map]. We might also simplify by choosing fewer items to address. Or we might enrich by including additional elements. Note that these need not be restricted to presentation on one page!

As experience builds, new templates would be generated for other frequently faced technology management questions. The favored indicators to help answer these questions would evolve through experience as to what information truly adds value to decision-making. Likewise, information product formats would evolve based on what works.

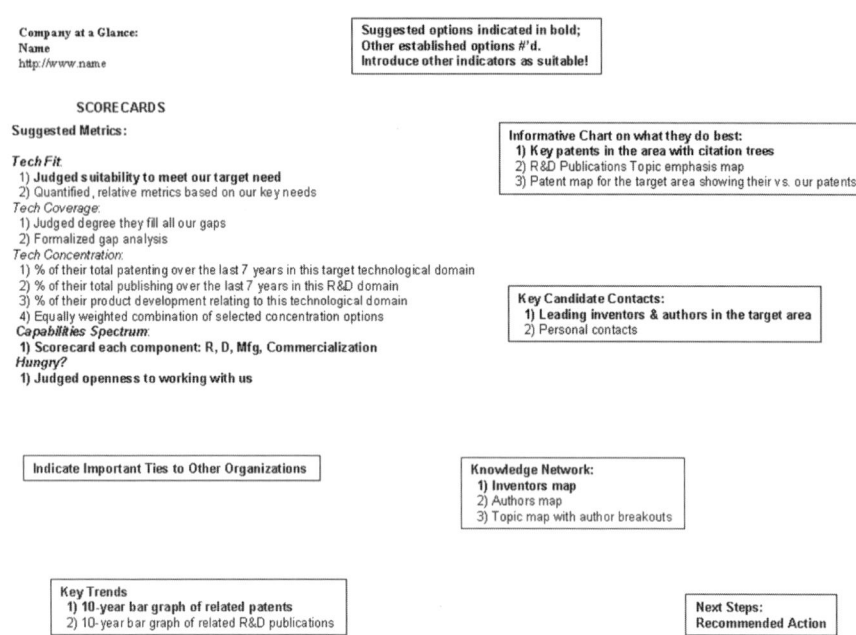

Figure 4-4. Tech mining template to assess a potential collaborating organization

4.5. TECH MINING INFORMATION PRODUCTS

Presentation of findings is vital. Here, we consolidate key information pertinent to the immediate decision as a "one-pager." Figure 4-5 illustrates the instantiation of Figure 4-4 (use of color considerably helps convey this information). We have taken "poetic license" here, building on the real data with hypothetical extensions. The bottom-line question is whether to pursue contact with Ceramic Fuel Cells, Ltd., as a potential collaborator. We follow the nominal "default" choices of the template (Fig. 4-4), except for including all five candidate scorecard elements.

To support this decision-making, we provide a variety of indicators. Starting from the upper left, the "Scorecard" helps assess whether we may have a good fit. It is deliberately not quantified. Some of the components are judgmental—for example, "Tech Fit" with our needs, whether the company is likely to be "hungry" for opportunities. Others are quantitative—for example, "Tech Concentration" reflects the portion of their patents pertinent to our interest in working with them. "Tech Coverage" concerns whether they appear apt to meet all our needs. "Capabilities Spectrum" indicates their apparent strengths in R&D, manufacturing, and commercialization (they have just started marketing products). Just below is a note from their website indicating possibly important relationships. Scorecards intend to facilitate cross-comparisons—for example, if we are considering three prime candidates for

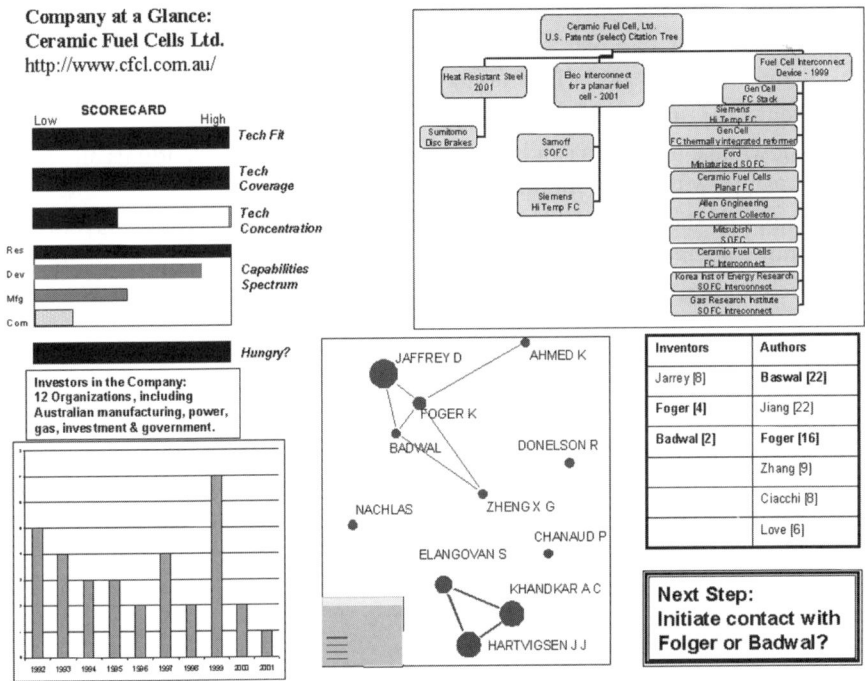

Figure 4-5. One-pager on a potential collaborating organization

joint development activity, those should help executives quickly gauge their relative strengths.

Much of this particular report keys on patenting. At the bottom left is the trend in their priority patents. Swinging to the upper right, we see a patent citation tree for three of their patents. These are drawing attention as cited prior art by other notable players, including Siemens, Ford, Sumitomo, etc. The "knowledge network" starts with a coinvention map (lower center) that identifies two teams. We combine that with the R&D publication authorship information presented just earlier. We have sifted out the inventor team (near the bottom of the map) associated with earlier patent activity at Cerametec that Derwent codes together with Ceramic Fuel Cells. (In a real analysis, we must investigate this further.) As per the lists of inventors and authors, we see a key SOFC R&D team. We suggest initiating contact with Badwal or Foger, who are both actively patenting and publishing.

CHAPTER 4 TAKE-HOME MESSAGES

- This chapter illustrates what tech mining can do for you. It is necessarily very selective, but should convey the flavor of what is possible. You can see more results in Chapters 12 and 16.

- Lists convey the amount of R&D activity.
- We can cross two lists to add a second information dimension to show many variations on "who's doing what."
- Trends capture various emerging technology patterns.
- Knowledge networks can be depicted by focusing on individuals and relationships within and among organizations.
- Mapping shows relations among *terms* (e.g., technologies, players) or among *documents* (cf. Fig. 16-8).
- Tech mining should be tailored to answer your prime questions.

CHAPTER 4 RESOURCES

We recommend study of this chapter together with use of the *VantagePoint Reader** and sample fuel cell data† to try out those methods that seem potentially useful to you.

---

*Available at http://www.thevantagepoint.com.
†Available at ftp://ftp.wiley.com/public/sci_tech_med/technology_management.

Chapter 5

What to Watch for in Tech Mining

What does the future hold for tech mining? This chapter suggests ways to raise our visions of what is possible and ways to lower the barriers to attain those visions.

The chapter consists of seven sections that address, in turn: better tech mining basics, alternative perspectives on the data, more informative products, knowledge discovery, knowledge management, new tech mining markets, and dangers.

5.1. BETTER BASICS

We are on the right path. Many of the basic requirements to deliver on the promise of tech mining are coming together. With work by various players, we could dramatically advance tech mining effectiveness and diffusion. Let's look first at tech mining inputs, then at generating outputs.

Better Ingredients and Infrastructure

This is the "push" side of the equation—the data mining perspective. We key on nurturing more available information resources and better tools to exploit them.

As illustrated in the fuel cell analyses in this book (Chapters 4, 12, 16), R&D databases offer a great resource to mine. But, oh, could things be more con-

Tech Mining: Exploiting New Technologies for Competitive Advantage, Edited by Alan L. Porter and Scott W. Cunningham.
ISBN 0-471-47567-X Copyright © 2005 by John Wiley & Sons, Inc.

venient! Database access mechanisms are slanted toward human users who seek a few choice records to read. Enhancement of machine access would facilitate search and retrieval of sizable search sets. Suitable intellectual property ("IP") protection for the database is needed, of course.

The number of core R&D databases depends on your tech mining purposes. For most tech mining, access to somewhere between one and five well-chosen databases should suffice. Database and gateway collaboration to facilitate integration of search results on this scale is highly desirable (e.g., Engineering Village now offers EI Compendex and INSPEC in combination). Mechanisms to allow " metasearching" of multiple databases and websites is becoming available (e.g., Scirus). Compiling web information in more depth and consistent format could boost tech mining dramatically.

Most database portals are set up to be people friendly—that is, for humans to locate and retrieve a few items. A tech mining target is to expedite computer access to these data sources. In this model, one would not extract the search results from the source database into plain text (or other) format to, then, import them into the analytical tool database. In our example, fuel cell analyses, we first export from Derwent, Science Citation Index (SCI), and INSPEC, and then import into *VantagePoint*. This transfer should become automatic. One attractive possibility is a direct machine access standard to expedite interexchanges. Databases such as SCI, MEDLINE, and EI Compendex helpfully provide a step in this direction via the option of exporting data in "ris" format (for Reference Manager and other software accepting this format). So analytical software that can import "ris" should be able to easily consolidate information from these sources.

Another appealing route forward is to integrate databases and analytical tools. We see an example of this in the provision of ChemAbstracts via the analytical tool suite of SciFinder. The catch is to provide flexibility. We seek to enable use of *VantagePoint* or ClearResearch with ChemAbstracts, or SciFinder with Derwent World Patent Index, for instance. Certainly, attaining every combination among competitive players is unrealistic, but collaboration could expand the market for all tremendously.

Special database licensing would facilitate tech mining. Customers who want to download many "10,000-record" searches for data mining can't (and won't) pay the same per-record rates as customers who want a handful of records to read. Many universities enjoy unlimited use licenses with certain data providers, but we need clarification of what constitutes legitimate uses, of what amounts of text. Present licensing uncertainties severely restrain the spread of tech mining in the universities. Yet, this is the place to build awareness and competencies in prime students to go forth and apply familiar tools.

Data access constraints and costs prevent many organizations from exploring tech mining. Creative arrangements need to be devised involving data providers, gateways (e.g., Dialog, STN), and groups of potential customers. For instance, Web of Science only became available at Georgia Tech about the year 2000 when a multistate university consortium worked out a package license arrangement with its providers (now Thomson Scientific). Another appealing arrangement involves multiple government users sharing database access with several data suppliers through a central server at Los Alamos National Laboratory. This simplifies data updating and management through centralization and also provides attractive cost savings. Similar arrangements with small- and medium-business consortia, possibly through trade associations, would open substantial new markets.

Other potential arrangements abound. For one, certain technology information providers, such as NERAC, presently provide raw search results to their customers (e.g., in response to a request on a given topic). They could enhance the usefulness of such searches by providing overviews (e.g., "one-pagers" showing the scope, leading authors, etc.) along with the data in minable form. Live links to full text resources back at the database provider would encourage further data sales. Automated accounting arrangements among the end-customers, NERAC or similar intermediary providers, and the data owners would further boost attractiveness. Another variant would be to enable technology consultancies to provide data files (e.g., in *VantagePoint* or *Derwent Analytics*) along with their reports. This would provide interactivity, enabling users to quickly dig down to specifics, as well as to explore relationships. Try out the interactive website accompanying this book to get a feel for possibilities. Again, simplified fee arrangements could track usage and share revenues between value-adding information retrieval and technology assessment firms and the data providers.

Self-documenting record formats exist. These provide "metadata" about the documents. *VantagePoint*, for example, keeps track of the search information with the resulting records. Many syndicated news sources are providing their content in "rich site summary" (rss) format. More than 200 channels and feeds exist, allowing news to be aggregated and customized according to specific audience needs. The key to this easy interchange of data is XML—a self-explanatory document format that tells you what each component of the news-feed is and how it can be used in relation to other content. XML was originally developed in recognition of shortcomings of the internet format—HTML. HTML tells you how to "lay out" information on a screen, but it doesn't tell you how to relate or structure the information you receive into usable, customizable knowledge. Now that the news is provided in a neatly shared format, it would be nice if science and technology publications and patents could also be broadly interchangeable.

Better Tools and Methods

Cross-tool information transfer is appealing. Many pairwise examples are appearing. *VantagePoint* can exchange data with BizInt SmartCharts. As another example, Reference Manager software works with OmniViz to enable "RefViz" software that generates visual overviews of research reference sets.

We see strong user demand for multifunction tool integration, but also for simplification. To some extent these aims reinforce each other, as in automated data import and consolidation and in seamless tool interfacing. Combining multiple functions, while simplifying, poses some real challenges too. "One size fits all" looks pretty unlikely when we consider complexities such as:

- Providing a single interface for multi-database searching from one's analytical tool
- Treating structured and unstructured text and numeric data
- Encapsulating knowledge from diverse domains
- Enabling entity extraction (with known search targets) as well as inductive data representations (with no imposed knowledge)
- Performing statistical analyses together with artificial intelligence applications
- Offering information visualizations with convenient export to standard office packages
- Generating reports with multiple delivery mechanism options

With U.S. National Science Foundation support, we are exploring ways to expedite tech mining processes while enabling analysts to tailor processes to particular needs.

Tech mining methods come to us from varied traditions in statistics, information science, artificial intelligence, linguistics, applied sciences, and technology futures analysis (TFA). It's a rich welter, but it suffers from a confusion of similar analytical methods bearing different names. Standards for judging the adequacy of models are lacking. Guidelines for choosing and designing models are also thin. In some ways, the sum has become less than the parts.

On the positive side, the understanding needed to support tech mining is growing rapidly (see Chapter Resources). We are gaining in techniques to systematize uncertain knowledge and even witness efforts to develop a unified framework for understanding content, document, and individual preferences. We perceive progress toward shared analytical models. Tech mining studies will benefit from tools for designing new knowledge networks and diagnosing failed networks. Understanding of how best to support technical information needs of organizations is advancing. Researchers are experimenting with ways to "activate" document repositories to anticipate and answer user questions. Tech mining is increasingly being built right into the structure of portals to the science and technology databases.

As tech mining advances, it should be continually enriched by capabilities developing in adjacent areas. For one, the discipline of probabilistic modeling—also known as graphical models, or even machine learning—offers possibilities. Probabilistic modeling draws from graph theory and probability theory to help systematize previous knowledge. It can also help solve difficult applied problems in areas such as speech recognition. These modeling approaches offer systems for recommending content and for indexing information (Popescul et al., 2001; Baldi et al., 2003). The opportunistic tech mining analyst should stay tuned and incorporate such capabilities as they become available.

Table 2-2 noted various forms of TFA, many of which tech mining supports. Most TFA methods arose in an era dominated by industrial (manufacturing) technologies and military applications (more than economic competition). Coates et al. (2001) make the case that new TFA methods are needed to address increasingly science-based technologies, especially molecular technologies. They note the increasing pace of change, warranting attention to technology transition modes (chaotic transition regimes between traditional S-curves) and continuous technology intelligence. These authors also call for fuller exploitation of information resources—namely, tech mining. As we wrote, a meeting in May, 2004, explored new TFA methods (Technology Futures Analysis Methods Working Group, 2004; see also http://www.jrc.es/home).

Better Results

We shift our considerations now to the "pull" side—How can we deliver more useful technology information products? A coming section considers technology information products (content) to address managerial needs; this section deals with basic improvements in delivering those results.

Delivering tech mining information involves elements of representation, communication, and interaction. A starting point is to orient technology analysts to focus on target users and to learn what information they need. A further step is to influence those needs by demonstrating capabilities, forming personal alliances, touting how competitors are gaining advantage from tech mining, or whatever it takes to get their attention! As covered in Chapter 14, we need to shove open doors to effective collaboration between tech mining providers and users.

Devising and carrying out effective communication of findings warrants significant effort, perhaps half of total tech mining exercise resources. Information representation, for one, must meet user preferences. Today's repertoire of information visualization and delivery options permits tailoring technology information products for different users.

In some cases attaining significantly better understanding of tech mining findings may require organizational change. We have witnessed organizations where the distance (number of intermediaries) between end users and tech mining producers was too high. The value of findings was dissipated by the indirect transfer mechanisms. Direct interaction between producers and users before, during, and after the analyses is the better approach. Fitting tech mining activities to organizational realities poses serious challenges, as explored in Chapter 14.

Tool adaptations can help. We have learned that *VantagePoint* software is most suitable for use by professionals for whom it is a "daily" tool. That has led to building a free VP Reader to allow users to interact with results but not have to learn any complicated functions. So, in addition to providing a written report, PowerPoint presentation, or whatever, an interested user can browse lists or maps and open up entities of interest to get to particulars. The key here is to foster interaction with that user to increase understanding of what underlies tech mining recommendations. We also need to make it easy to probe into the information just when he or she is actively engaged.

Another approach is to provide a range of versions of the tech mining software. Aureka's "gold, silver, and bronze" software options provide three different capabilities at different price levels. Most importantly, the cheapest option is much easier to learn and use. That opens possibilities for widespread usage. Extension of this notion could lead to shrinkwrap tech mining software priced for mass distribution.

An almost magical weapon in the drive for quick and easy is scripting. That is, sequences of software operations that are called for on a repeat basis are automated. As already mentioned, VisualBasic scripting of *VantagePoint* and MS Excel operations enables performance of quite laborious processes of data cleaning, analysis, and information representation at the click of a button. Such scripting can be built up into wizards to guide the analyst through sequences of choices and actions. These can offer both simple one-button and more complicated options. In this way, tech mining functioning can be made widely accessible.

5.2. RESEARCH PROFILING AND OTHER PERSPECTIVES ON THE DATA

The growth of science and technology databases over the past decades qualitatively changes the amount of information available to us. A technology analyst in a large multinational company told us how a particular technology search yielded 12 records in 1985 but about 120 in 1995. She anticipated an order of magnitude of 1200 in 2005. You cannot read 1200 articles, or even 1200 abstracts! So how can you exploit such quantities of R&D information? Tech mining, of course!

Tech mining offers multiple perspectives on activity in an R&D domain. Emerging technology search sets, such as those of the example "fuel cells" tech mining (Chapters 4, 12, 16), typically contain hundreds or thousands of records. Traditionally, most technologists and managers, including information professionals, think of the great R&D databases only as ways to locate a few good papers to read. Tools such as *VantagePoint* can help with this, to get to the most relevant items expeditiously. We illustrate this in Chapter 16 by zooming in to topics of interest and double-clicking to read particular abstracts. We also illustrate routines that "bucket" the search set records into clusters; one then examines the interesting buckets to read the gems, instead of dealing with the whole set.

In contrast, a "bird's eye view" seeks to "see" a research domain's overall activity. The sense of a bird flying over and looking down at the thousands of records is apt. We find high value in analyzing entire search sets—to discern emphases, trends, and links among issues and methods across an R&D domain. Chapters 16, 9, and 10 show how we can detect main patterns of domain activity, as well as fringe activity at the boundaries. We call this "R&D profiling" (Porter et al., 2002). Such research domain analysis (Borner et al., 2003) can alter the very way research is done by improving the transfer of knowledge among scientists and engineers.

Research profiling can involve many tech mining analyses and representations, including: 3-D landscape views, topic maps (Fig. 16-8), lists and maps of the main organizational players, lists and collaboration matrices of key researchers, and so forth. This information can help target prime opportunities. It can help identify potential collaborators with complementary interests. It can also help sell the proposal by helping reviewers understand how it fits in the scheme of research activities and why it is valuable and special (see Sidebar). Profiling can also help communicate research findings by literally showing how the new findings relate to other work.

Sidebar: *The Rejected Proposal—Lesson Learned the Hard Way*

In 2002 Alan Porter and colleagues submitted a proposal to the U.S. National Science Foundation to develop "research profiling." It was rejected. One of the reviewers tellingly observed that it was hard to locate where the proposed research fit. To what other activities did it relate? Where would its contributions exert influence? The reviewer was absolutely right—we had failed to hear our own message—that is, to provide a "research profile" for the proposal to develop research profiling! We hadn't situated our activities with respect to the overall research arena or specified how it fit with other efforts. If overviewing research domains doesn't come naturally for us, we can't expect it to come easily to others.

The crucial step forward is to make consideration of such overviews standard practice in R&D management and other technology management activities. Scientists, engineers, and managers don't think in these terms, so this doesn't come automatically (as per Sidebar). Our "vision" is that every research project, ranging from an undergraduate class paper to a proposed multi-million-dollar research project, should profile its research domain.

Most of this book emphasizes capabilities to identify the main research thrusts, leading organizations, etc. Additional value comes in finding the "unique." Spotting the novel concept or the stand-alone experimental finding can indicate new opportunities worth exploring. Finding a method newly applied to the R&D domain, or identifying common interests across fields, can open doors. Tech mining facilitates recognition through both simple techniques (e.g., list comparison) and more elaborate ones (e.g., a variation of PCD to identify discrete terms—see Chapter 16).

5.3. MORE INFORMATIVE PRODUCTS

Previous chapters raise prospects of

- Understanding user needs, especially those of technology managers
- Deriving answers to the questions of those users
- Drawing on available theory or modeling to devise more interpretable tech mining findings ("innovation indicators" in Chapter 13)
- Presenting tech mining findings in multiple ways to match various user personal and cognitive style preferences (numbers, words, pictures).

Tech mining as we define it is not an island unto itself. Rather, it must integrate into a range of technology management functions. We see many ways to enrich tech mining and the value it provides.

One promising extension is to widen the information resources being exploited. Our example, fuel cell tech mining, draws on research publication and patent abstracts. Mining of other resources would enrich the innovation context and market prospects indicators. Prime candidates include:

- Business information (e.g., abstract databases such as Business Index)— to obtain indications of commercial interest and activity in a target technology, to track the emergence of standards, etc.
- Market databases—to help assess market prospects by technology, product, sector, and locale
- General press—(e.g., newspaper indexes)—to capture signal strength regarding popular acceptance or resistance to a technology (e.g., environmental concerns)
- Public policy compilations—(e.g., Congressional activity, policy abstracts)—to track development of legislation and regulations, stakeholder interests

Tech mining must be more than astute data analysis. For example, we have learned that some technologists and managers favor in-depth technical treatment and explanations. Others just want the "bottom-line" message delivered succinctly. Still others like integrated treatments—weaving the technology analyses together with product development, marketing strategy, and cost analyses in one package.

We distinguish between tech mining, as driven by technological innovation process concepts, from data mining, which starts from the data. Chapter 13 provides our collection of 13 technology management issues and 39 questions, together with "innovation indicators" that we can generate to help address them. Focus tech mining analyses on answering your users' key questions but also explore the data opportunistically—you might discover some pleasant (or unpleasant) surprises.

Additionally, consider multiple ways to get your message across. Literally, don't necessarily choose one from a written report, an in-person presentation, a web-based posting, or an interactive report-data combination (e.g., providing software such as *VantagePoint Reader* to dig down into the source data). Combinations can amplify the signal.

5.4. KNOWLEDGE DISCOVERY

Tech mining is all about making more information available in new and powerful ways. A key application is discovery of knowledge implicit in text. We call attention to two variants with special promise—as a creativity stimulus and in making previously unknown links.

Tech Mining as a Creativity Tool

The creativity potential of tech mining lies in its presenting information of which people were unaware. The juxtaposition of such new information with an issue to be resolved can suggest new solutions. The Sidebar on "Tech Mining as New Idea Trigger" is adapted from a process used by a multinational consumer products company. The tech mining results were used as stimuli at a brainstorming session of technology analysts and researchers to identify which new research opportunities to pursue. Note process similarities to framing the query (Chapter 7).

Sidebar: Tech Mining as New Idea Trigger

Tech mining can be used to expand research and technology horizons. One company uses this approach with its researchers to generate new consumer products. There are many variations on this theme, but here is the idea transformed as a hypothetical case for "Bug Killers, Inc." (BKI):

1. Define an issue—We observe that ants "magically" locate food tidbits on our kitchen counters and get word to their relatives to join the party. BKI ant poisons presently kill on contact. Let's explore opportunities based on attractants (bait) that would draw ants to delayed-action poisons, to which they could recruit their entire colony.
2. Explore related research domains—start by reviewing research on ants (e.g., 8275 hits in SCI).

Continued

a. We identify ant olfactory capabilities and preferences
b. We find various types of related attractants: pheromones, juvenile ant hormones (has been used as bait), foods that ants like (vegetable oils and starches have been used as baits).
c. This exploration targets immediately related topics intersecting the central issue—What materials are ants most sensitive to? (corresponding to the "star" in Fig. 5-1).

3. The Key: Expand the search boundaries to research topics that share some, but not all, of our issue's parameters:
a. Deliberately step beyond the directly relevant material to seek novel possibilities that we might adapt to our target (pursuing the "doughnut" in Fig. 5-1 with a hole in the middle).
b. Consider other ant sensing attributes (e.g., sensitivity to proprioceptive signals, visual recognition aids, wind detection)
c. Reach out to other ant characteristics besides sensing (such as range marking, chemosensory capabilities, and behaviors—e.g., influence of the queen).
d. Compare ant species differences—e.g., research on desert ants (as opposed to dessert ants).
e. Check out olfactory attractants for other insects (e.g., SCI search on olfaction and olfactory yields 16,536 hits on Feb 23, 2003; restricting to insects yields 562)
f. Scan olfactory appeal to other animals (e.g., special human sensitivities)
g. Explore what other industries are doing as potential ideas for us to borrow (e.g., how does the perfume industry disseminate smells so effectively?)
h. Continue to pursue promising leads even further removed from the center—(e.g., olfactory binding proteins.

4. Examine information resources other than research papers.
a. Patents (pursuing particulars relating to any of the research topics)
b. Business information (e.g., new product announcements with olfactory elements)

5. Conduct a brainstorming session in which tech mining results are shared with researchers, product managers, and salespersons in a nonevaluative forum to stimulate additional ideas.

6. Screen and evaluate—identify one or more ideas with real promise and pursue (e.g., identification and formulation of cDNA for an ant chemosensory protein to produce our own attractants with special appeal).
a. Next steps—technology and new product development (e.g., bait formulations; applications, such as building perimeter treatment patterns)
b. Obtain patents to protect the target IP spaces.

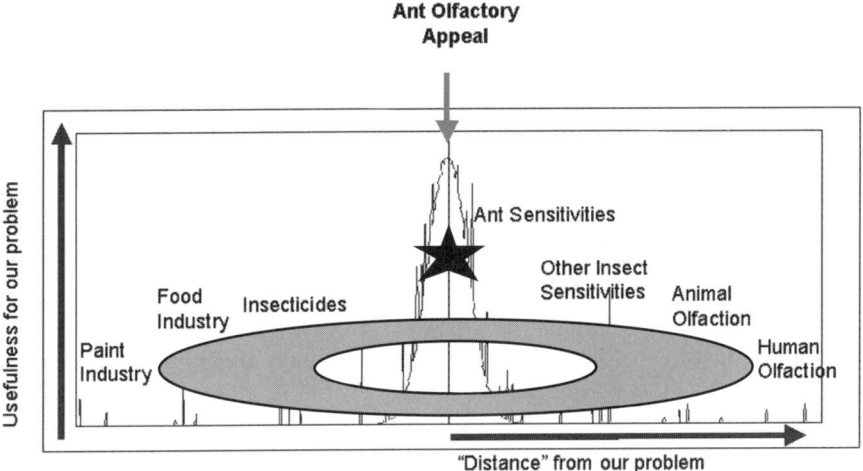

Figure 5-1. Using tech mining to stimulate creative design

This stretching of inquiry domains beyond those most directly relevant shares some notions of "analogies" with TRIZ (see Chapter 12).

Swanson Analyses

The "ants" example sets up another type of tech mining analysis—looking for new links. In our Chapter 13 example of neural network approaches within the "large dataset mining" research domain, we came upon a possible association of neural nets with statistical analyses. This can be pursued in many ways; for one, we do a new search on "neural networks" not constrained to large dataset mining. Within this new neural net search set, we explore for fresh methodological approaches. Then, we try to relate those back to our original interest area—large data sets.

Swanson and colleagues have applied this logic to generate biomedical research advances (see Chapter Resources). The gist of the approach is as follows. We search a suitable database on Concept A, uncovering interesting links to Concept B. We then execute a new search on Concept B, uncovering interesting links to Concept C. With the aid of knowledgeable researchers who know Concept A, we inquire whether A and C have been studied in conjunction with each other. If not, might this merit further investigation?

To illustrate, we (Porter and Schoeneck, 2000) have explored the worldwide decline in amphibian populations (Concept A—why are the frogs dying?). Searching in SCI, we found many factors advanced as possible contributors to the decline. Concept A is linked in the research literature to such Concept Bs as exposure to ultraviolet (UV) radiation, pH levels in streams and lakes, and fertilizer residues. Linkage is based on co-occurrence of these terms together with mentions of amphibian health problems. We then perform a new search on a Concept B to discover factors to which it is linked. For instance, a new search in SCI on "UV" exposure yields a new set of abstracts. There we find UV associated with many other concepts. One implicates UV in damaging DNA (Concept C). We then look for direct A–C links in the research literature—do any articles discuss DNA damage as a possible contributor to the amphibian health problems? In the absence of such publications, we would investigate with bioscientists whether this holds real promise for novel research. In this illustration, several aspects of DNA damage offer research promise: Populations that lay eggs more shallowly have greater response to UV-B radiation, amphibians' complex developmental sequences could be especially sensitive to disruption, and the amphibian effects include striking propensities to malformations. So our tech mining digging offers a fresh hypothesis for exploration.

If you would care to try out such knowledge discovery, Swanson and colleagues have developed an open website, called "Arrowsmith" (http://kiwi.uchicago.edu/), that facilitates A–B–C analyses on MEDLINE topics. The site interacts directly with MEDLINE to search and retrieve abstract records on Concept A and also on Concept C. It then aids one in screening candidate linkages (Concept Bs). As another example, we have recently explored putative linkage between Concept A—pesticides and Concept C—thyroid disease. We were particularly interested in mechanisms that might involve the immune system, so we screened for "Bs" accordingly. In general, we see high potential in exploring associations among chemicals (agents), organs, and disease (effects).

Another approach asks how we can enhance the Internet to actively answer our questions (Agichtein et al., 2002). The authors entertain expanding traditional search engines so that questions like "How many miles are there in an astronomical unit?" can be transformed into traditional Boolean queries (see Chapter 7) and then answered with well-targeted internet searches. Such an approach should work well with science and technology databases too. Why can't our databases help us directly answer questions?!

5.5. KNOWLEDGE MANAGEMENT

Tech mining generates valuable derived knowledge (i.e., empirically based intelligence). Here, we raise the issue of how best to cumulate such knowledge efficiently and effectively.

If an organization (e.g., a company) pays by the record for database searching, it hates to pay for the same records multiple times. A centralized information resources unit can maintain search sets for the company to be augmented and updated, rather than recreated, as needed. An alternative approach is to save the search algorithm rather than the search results, then rerun the search as warranted. The choice depends on working out mutually satisfactory arrangements with the data provider.

Similar options arise with respect to tech mining analyses. One doesn't want to keep starting such analyses from scratch, but how best to retain analyses? One solution arises with the idea of sharing tech mining information products via a corporate website with periodic updating. From another standpoint, we don't want to repeatedly redo similar cleaning, analysis, and representation steps. Easy generation of scripts (macros or wizards) can alleviate this burden by capturing and automating the repetitive elements.

As already mentioned, some companies are moving to establish systematic procedures to be followed for particular technology management tasks. Tech mining components contribute to these procedures. Standardization of these procedures promises huge gains in efficiency via scripting and in utility. Gains are achieved by establishing a limited set of technology information products for managers to assimilate. The downside to be guarded against is adoption of a standard process without experimentation with a wide range of alternative innovation indicators. Ongoing evaluation (see Chapter 14) can help ensure continuing organizational learning.

Knowledge transfer extends these notions beyond a particular entity. As explored in Chapter 14, cross-organizational exchange concerning tech mining approaches and experiences should be encouraged. Tech mining is also a powerful tool to facilitate cross-disciplinary interchange in that it fosters access to R&D outside one's own domain. Somehow, we need to share learning about tech mining itself across quite distinct communities of information professionals, technology analysts, subject matter researchers, and technology managers. This challenge is exacerbated in that each of these is really many communities whose interaction to consider auxiliary tools, such as tech mining, is uneven. For instance, we are currently grappling with how to share results on text mining of the Hazardous Substances Data Bank (HSDB) with toxicologists and environmental scientists. Most conferences and journals only consider the technical subject. Toxicology professional associations exchange information on advances in toxicology, not on information mining tools to aid toxicologists. We welcome suggestions!

5.6. NEW TECH MINING MARKETS

We anticipate *researchers* becoming major tech mining users. We have been surprised and pleased to see industrial researchers grasp the potential. Researchers at Air Products and Chemicals worked with us to try out "quick" generation of technology intelligence (Porter and Brenner, 2003). Some, not all, became quite enthused at the prospect of doing tech mining themselves. But to make tech mining analysis practical for casual users requires several necessary conditions:

- Desktop data access—unlimited use licensing of a few suitable databases affordably (one model has researchers doing early-stage, preliminary screening so that having the "best" data is not necessary)
- Desktop tool access—user-friendly software that a researcher could learn to use productively in minutes and be able to take up again after not using it for months
- Low cost per user

Note that such a dispersed tech mining model would run counter to many organizations' way of doing things. We believe that a central tech mining core of information professionals and technology analysts can support a wider constituency of occasional users (researchers, technical professionals, and managers).

The academic research market also beckons. University tech transfer and vice president for research offices can apply tech mining to help prioritize opportunities. We can imagine usage then expanding through faculty and staff researchers to graduate and undergraduate students doing funded or class research. The "Academic Market?" sidebar suggests that tapping this market will not come automatically.

Sidebar: Academic Market?

In 2003, Alan Porter participated in a U.S. National Science Foundation review panel considering small business proposals (SBIRs). At one point, a panelist challenged the commercialization plan of a proposed software development project. His concern was basic—the proposed target market was academic! The panel (largely academics) agreed on how precarious it would be to try to make a profit selling to academics. So we have a paradox—millions of potential tech mining customers, with extensive data access already paid for, but woefully short of funds to buy the tools. How can we refocus this picture?

We see hope in a scenario in which an application of research domain profiling proves distinctly successful. A major university research center proposal gains an edge or students apply tech mining effectively to ace a course. "Word" of success spreads like lightning across the campus and other campuses as students or faculty share stories. If one campus facilitates tech mining by providing a blanket software use license along with the data, word of this will spread to other schools. We see this enhancing knowledge discovery by expediting transfer of findings and methods among disciplines. And, perhaps, even offering profit potential to data and software providers.

5.7. DANGERS

As with data and text mining, tech mining poses certain concerns.* The "dark side" of tech mining depends on your perspective—one party's gain may well be another's loss. Competitive intelligence is rife with "spy vs. spy" environments. As CTI activities spread, counterintelligence efforts are apt to increase as well. Organizations may deliberately present disinformation. Some suggest that automotive manufacturers have pursued "defensive research" on electric vehicles for decades, for public relations, not commercialization, purposes. Conversely, organizations may suppress publication or patenting on a technology of high promise.

This can be carried further. Imagine drug company A determining to abandon a particular line of research as they discover serious side effects. One option is, of course, to publish their findings openly and redirect their efforts. But a more devious strategy might entail delaying the publishing on the side effect discovery while continuing to publish research as though they were still pursuing commercialization of this line of drugs, the hope being that competitor B would track A's activity and be led to waste their own R&D efforts pursuing this dead end. Were A to pursue this strategy further, they might patent in a way to entice others to pursue this dead-end line, too.

In today's global economy, as suggested in Figure 1-1, the industrializing countries exploit international R&D very effectively. Well-networked computers mean that well-trained scientists, engineers, and tech miners have full access to science and technology publication and patent databases and the Internet "everywhere." So, that small American high-tech start-up had better be alert to competition from Southeast Asia, South Africa, or Hungary, not just Western Europe and Japan. In particular, China and India have strong R&D capabilities of their own to complement CTI garnered on others' findings.

Beyond these high-tech economic adversaries, a more sinister prospect is the exploitation of advanced technologies by terrorists. Governments need to consider the implications of rapid tech transfer being facilitated by tech mining activities by terrorist organizations.

This short section should suffice to get us thinking about ethical and legal limits to tech mining. The Society of Competitive Intelligence Professionals (SCIP) considers such issues as competitive intelligence strives to establish acceptable professional behavior.

CHAPTER 5 TAKE-HOME MESSAGES

- Look to a bright future for tech mining, with a lot of work required to achieve it; track developments to quickly make use of those of value.

*This section owes its genesis to two of our helpful reviewers, Hal Linstone and Fred Rossini.

- Negotiate with data providers to secure new data access arrangements to facilitate your tech mining.
- Extend tech mining approaches to data types besides science and technology—e.g., policy, business, market, popular trends, product announcements.
- Pursue tool coordination and integration where these expedite your tech mining.
- Take advantage of scripting to expedite repetitive analyses dramatically; try new vistas for tech mining applications enabled by much quicker analyses.
- Demand research profiling on entire R&D domains at the initiation of technology development projects.
- Use multiple ways to represent technology information products and to disseminate these, including intranet sites.
- Try tech mining as a creativity tool; it can provide fresh perspectives and rich information to stimulate new lines of thinking.
- Explore using tech mining to associate disparate information, to facilitate knowledge discovery processes (Swanson A–B–C analyses).
- Actively collect and manage tech mining outputs to generate secondary payoffs at minimal cost.
- Look to engage researchers and academics as active tech miners.

CHAPTER 5 RESOURCES

We note several recent approaches to address knowledge exploitation—see the following references: Agichtein et al., 2002; Baldi et al., 2003; Barabasi et al., 2002; Popescul et al., 2001; and Power, 2002.

"Swanson analysis"—examining R&D literature to generate previously unknown A–B–C links is reflected in a number of papers—cf. Swanson and Smalheiser, 1997; see http://kiwi.uchicago.edu/references.txt. Kostoff (http://www.onr.navy.mil/sci_tech/special/technowatch/) has arrayed a host of logical relationship possibilities associated with spanning of isolated research domains.

Part **II**

Doing Tech Mining

Chapter **6**

Finding the Right Sources

This chapter guides you through the vast amount of electronic information available on emerging technologies to get to what you need. We distinguish information about R&D inputs, such as personnel and funding, and R&D outputs, such as new knowledge and technology. We point to science and technology databases as the primary gateway to R&D output information. Tech mining exploits research publication and patent information to yield valuable indicators of underlying technological progress.

CHAPTER CHALLENGE: THE DEBRAND FOUNDATION

- Imagine that you head the (hypothetical) DeBrand Foundation's 15-person research evaluation unit. Your annual research proposal solicitation deadline is approaching—you are giving out $170 million this year. You will help pick about 850 winners from some 2500 proposals you expect to receive.

- The Foundation Board has tasked you with improving the proposal review process by taking better account of other R&D. They want you to apply tech mining to ascertain how given proposals contribute to particular research domains and avoid duplication of others' efforts. They also believe that mining science and technology databases can help determine the capabilities of the proposing researchers. Over the six weeks following submission of these 2500 proposals, you are to complete this tech mining effort.

- This chapter will follow the issues faced by this fictional nonprofit as it gets up to speed with tech mining. We begin by asking: Which on-line databases should

Tech Mining: Exploiting New Technologies for Competitive Advantage, Edited by Alan L. Porter and Scott W. Cunningham.
ISBN 0-471-47567-X   Copyright © 2005 by John Wiley & Sons, Inc.

the Foundation use to further its mission? Has it chosen an optimum means for accessing the data? What will it cost to get these data?

6.1. R&D ACTIVITY

Scientists and engineers conduct a tremendous amount of research—how do we find out who's doing what? Most of our attention is directed at their "outputs." Luckily, they publish hundreds of thousands of research articles and file for hundreds of thousands of patents every year. This number is growing some 20% every decade. There are thus millions of public research publications and patents, ripe for investigation. So, how can you get to them?

Focus on two main sources on R&D outputs—the Internet and the publicly accessible (syndicated) databases. The Internet is a vital source of technology information. Its importance grows because of the increased popularity of on-line publication and the low transaction costs for retrieving on-line publications (Lawrence, 2001). Complementing the Internet, a number of databases compile a wealth of science and technology information. We want to make the most of these sources to monitor R&D activity and its implications for impending technological progress.

Before embedding ourselves in R&D publication and patent considerations, we more briefly pose R&D input and other output measurement possibilities.

Input and Output Measures

Input measures represent investment in a given science and technology area. One key data type concerns *personnel*—for example, the number of scientists and engineers engaged in R&D. Another addresses *financial* commitment—for example, investments in R&D. Both are compiled at national and international levels (e.g., U.S. National Science Board, 2002; UNESCO and UNCTD) and can be pursued at organizational levels as well. Appropriate sources vary widely. For instance, PriceWaterhouseCoopers compiles a database on how much venture capital is going into particular technological areas. For another, the "Community of Science" compiles a Funded Research database that tabulates which businesses have received Small Business Innovation Research (SBIR) grants.

Promising sources for tracking R&D inputs, policies, and performance by nation are the "Community Innovation Surveys" being conducted by European countries and a growing number of others. These contribute to the "Innovation Scoreboard" (http://www.cordis.lu/innovation-smes/scoreboard/home.html). The Organisation for Economic Cooperation and Development (OECD) publishes a variety of benchmark figures for national and industry R&D (http://www1.oecd.org/publications/e-book/92-2001-04-1-2987/A.4.1.htm). See their "Science & Technology Scoreboard."

Besides direct R&D outputs such as publications and patents, we can consider many other measures. This book addresses *citations* to those outputs in several places. *Royalties and license fees* constitute another science and technology output of real interest from corporate to national levels (e.g., the World Bank's *World Development Report*—http://econ.worldbank.org/wdr/). We expect technological innovation to result in new commercial products and services, and interesting data pertain (e.g., number of new product announcements; portion of sales from products introduced in the past three years). Further downstream, we look to economic growth. Treatment of these less direct measures, potentially ranging from single organizations to multinational regions, is beyond this book. Brainstorm measures that could serve your technology analysis needs and play detective to locate them!

This short treatment has "accordioned" from thinking about quite specific technological innovation measures—for example, for one target technology as being developed by one company—to incredibly broad ones—for example, European innovation writ large. Beware that anytime we treat the overall science and technology effort of a substantial enterprise, not restricted to a particular technological domain, those data will be very heterogeneous.

Data Volume and Its Challenges

Tech mining studies range dramatically in scope. You might be profiling R&D activity *across Western Europe* to look for potential collaborators. "Edison House" (a tri-services U.S. military office) does so to "make marriages" between American and European researchers. They mine science and technology databases in support of human networking to identify exciting research activity and bring potential cross-Atlantic collaborators together. Tech mining helps identify productive conferences on particular topics and to identify highly active researchers to seek out. It amplifies the effectiveness and efficiency of "good ole boy" networking. As a different example, you might be focusing on research activity in *a particular city* to determine its "cluster" potential. Or competitive technological intelligence ("CTI") often keys on *one company's* work—either in one technology or across technologies. Sometimes, we focus on *a particular research center* or one investigator's recent activity. At another level, scientometric studies may do a retrospective over *decades of R&D in a given domain.* Obviously, tech mining studies vary greatly as to which data best tap into the R&D networks for particular purposes.

Scientific research is organized in hierarchies of discipline, field, and specialty. Disciplines (e.g., chemistry) are the largest units of scientific organization, encompassing many fields. Fields (e.g., molecular chemistry) are intermediate areas of research. Specialties, or research domains (e.g., nanotubes), are often the most fruitful level to address in tech mining. Patent studies are often organized by technology and industrial applicability. You might be asked to profile the competition in R&D on adhesives—for biomedical applications.

TABLE 6-1. The Scope of Scientific Publication

Science and Technology Output	Annual Magnitude
Publication: World production Patents: World awards	1,000,000+
Publication: Output of leading nations Patents: Awards of leading nations	100,000+
Publication: Disciplinary totals Patents: Industry awards	10,000+
Publication: "Field" annual totals Publication: "Specialty" lifetime Patents: Corporate lifetime totals Patents: Sector annual totals	1,000+
Publication: "Specialty" totals Patent: Technology totals	100+

Tech mining thus varies tremendously by topical breadth, temporal horizon, organizations included, and geographic locus. The amounts of information involved correspondingly range enormously. Table 6-1 represents the rough orders of magnitude involved.

Reading "hundreds" of publications and patents would be a challenge—even to digest that many abstracts. But way beyond this level, some studies require analyzing hundreds of thousands of articles. One of us (SWC) did his doctoral research profiling all of British science, processing millions of publication titles. Regardless of the specific scope, tech mining uses computers and software to analyze large amounts of science and technology information.

But there's a major conceptual hurdle here. What do you think of when we talk about abstract databases and finding thousands of research articles or patents? Nearly all of us consider articles as something to read. Nearly all of us cringe at the very thought of reams of printouts—"Here's 3500 abstracts on fuel cells; have a great weekend!" But if you really had to—imagine you're the poor grad student assigned to "review the literature" and handed those 3500 abstracts—what to do? Winnow, winnow, winnow! Start with the most recent ones; maybe scan for ones from the leading journals; possibly key on words like "breakthrough" or "the future of"—anything to reduce the number to a "few good abstracts." Then, like an effective information professional, you would give your advisor a handful of papers to read. If you have successfully read his or her mind as to what is most salient, you are thanked for finding a gem or two.

Tech mining can help mine for those special nuggets, as per this illustration. However, tech mining has much more to offer. In the next two sections we discuss how databases can be leveraged to better understand R&D. We also discuss how to select the right database for your needs. In later chapters, we will show how to profile entire bodies of R&D information to see the big picture, spot trends, and discover new associations.

The Internet

The Internet offers a vibrant and rapidly growing resource, but it is a frustrating source to mine. Peer-reviewed S&T articles are not routinely available on the Internet. Chapter Resources mentions a few exceptions, and the Internet often serves as a conduit to access syndicated databases. The Internet contrasts sharply with the S&T abstract databases. Most significantly, much of the Internet consists of unstructured text, whereas the databases provide field-structured information. In addition, most Internet sources lack quality assurance.

Despite this, there is increasing evidence that the Internet is becoming an important repository of scientific information, much of which is freely available to scientists, engineers, and the interested public. This stems from its low overhead costs as well as the high visibility it affords scientists who distribute their research over the internet. It is a vital source of technology information.

Nonetheless, we recommend letting others compile the internet information for us (see Chapter Resources and Appendix A). A good tech mining motto is *Don't compile data yourself if someone else is doing it for you*! Then access the prime records from consolidated sources—namely, databases. We use the Internet to augment tech mining of database information. In the tech mining example (Chapter 16), we show how to pursue leads generated through analysis of search sets (from databases), using the Internet to get at current research work.

The coming chapters focus on exploiting syndicated S&T databases. Mining the scientific information (freely) available on the Internet poses significant hurdles in gathering and structuring those data. Overcoming these hurdles requires additional technical capabilities and is best left for more advanced discussion. Such efforts get into applied artificial intelligence (AI) methods, such as the use of "bots" (intelligent agents, crawlers) to automatically search for particular information. Entity extraction approaches do good work in finding associated information, if you know ahead of time what you're looking for to seed the searches. Search Technology and others are exploring ways to effectively combine such approaches with the statistically based tools emphasized here. Combinations of free-text and field-structured text analyses offer particular promise for tech mining. In the future, expect strides in utilizing S&T databases together with the Internet in powerful ways. For now, we advocate "picking the low-hanging fruit"—that is, exploit the great compilations of S&T information in databases and follow up on specific leads over the Internet.

The next section looks at the data, specifically the S&T databases and the sort of records they contain.

6.2. R&D OUTPUT DATABASES

Text as Data

A familiar, overarching categorization of data is the distinction among data, information, and knowledge:

- Data: raw, unprocessed content and discrete facts
- Information: data organized into meaningful patterns
- Knowledge: insight; information interpreted into a useful framework

Tech mining treats text as data–that is, we manipulate, count, and analyze text—much as data miners do with numerical data. It's been charged that this is an engineer's approach—to escape having to read. (But, alas, selective reading is still mandatory!) The big payoffs in tech mining lie in finding relationships across entire sets of text records retrieved on a topic of interest.

Tech mining transforms textual data—thousands or even millions of chunks of data—into actionable knowledge about emerging technologies. Let's examine what kinds of data to use to gain the most valuable insights.

Publication and Patent Databases

Two of the most accessible sources of technology information are publication and patent abstract databases. Full-text, as contrasted to abstract, sources are increasingly available, and obviously highly valuable. But for our detective work, the abstract databases provide a better starting point because they concentrate tremendous quantities of raw information into well-structured records. They are ideal for tech mining.

Let's first consider scientific publication abstract databases. The "premium" information consists of peer-reviewed articles from scientific and technical journals. Some databases, such as Science Citation Index, restrict themselves to abstracting articles from such journals. Most S&T databases include peer-reviewed, primary research articles along with reviews, non-peer-reviewed articles, letters, conference papers, and so forth. For some tech mining purposes, you won't care about these source quality distinctions; for others you will. Many databases include a field denoting publication type (e.g., the "PT" field in the MEDLINE abstract shown just ahead).

Patent databases provide the other main tech mining resource. Patents are the key public disclosures of invention. The U.S. Patent and Trademark Office (USPTO) is the largest database of patents in the world (http://www.uspto.gov/; also see Appendix A). Because of the extent of American invention and the appeal of the U.S. market, if you were to pick one resource for measuring patent activity, this would be a good one. However, USPTO presents problems as well. It has historically made available the patents granted, but lags average several years from submission, with some patents not appearing for as long as ten years. Only beginning in 2001 has USPTO made available information on patent applications (18 months after submission, with exceptions). Tech mining analysis of U.S. patents alone biases toward American inventors; this is increasingly unacceptable in our global economy. Moreover, retrieval of large search sets is essentially unworkable.

Virtually all industrialized nations have their own patent systems. The Japanese Patent Office (JPO) is the second most prominent, with English language patent availability on its website (http://www.jpo.go.jp/). The European Patent Office (EPO) helpfully consolidates much European patenting (http://www.european-patent-office.org/), but national patenting remains legally important. The World Intellectual Property Organization (WIPO) (http://www.wipo.org/) provides resources such as the Intellectual Property Digital Library with international registrations and patent cooperation treaty filings (these provide international protection for a year, after filing in one country to decide in which other countries to file). We will say more about picking patent databases later and pursue such matters in Chapter 12.

We want access to these publication and patent databases electronically. This may well be through an Internet connection, but that's just the medium to tap into the database. Note that most S&T databases charge to retrieve documents. Exceptions include the national patent office databases and other government databases, such as MEDLINE. Certain database providers take publicly available information (such as U.S. and other patents) and structure and organize the information for better access (e.g., Derwent World Patents Index, MicroPatent)—for a price.

The information contained in such databases is rich and suitable for "mining." For instance, the tech mining example we present in Chapter 16 profiles development of fuel cells. R&D relating to this electrochemical technology generates many papers. A Dialindex scan of 34 science and technology databases included in Dialog (a gateway service to over 400 databases), finds almost 28,000 chemistry-related article abstracts in Chemical Abstracts and much engineering also—over 7500 abstracts in INSPEC and 6500 in EI Compendex. Energy Science & Technology yields over 25,000 abstracts (coverage ranges from about 1970 to 2002 for these tallies). These searches overlap to a fair degree, depending on the topical coverage. Additionally, one might explore business interest in fuel cells via 3600 items found in the Business Index since 1988. Often, we search and analyze such volumes of R&D information to identify leading research groups and then visit their websites to find out about their latest activities. Tech mining also profiles the totality of this R&D to discern relationships and trends (upcoming chapters).

Database Organization

Simply stated, an S&T database is a compendium of individual records of data. Each record usually reflects one publication or patent, whether as an abstract or a full-text record. The records are stored in a central location and structured so that users can fetch specific kinds of information on request.

Note that most syndicated S&T databases are set up for use by people who want to retrieve a very few records—perhaps downloading one or two key items. Technology miners, however, typically grab thousands of records. Convenient downloading in quantity is a critical enabling attribute for tech mining. Fortunately, many gateways facilitate this type of electronic retrieval. One needs to be aware of and respect database restrictions. These aim to protect the database provider's intellectual property ("IP")—for example, to dissuade users from setting up a copy database.

Individual Records

To ensure that we are on common ground, let's examine the content of a representative database, down to an individual record. In this chapter we will discuss the arrangements necessary to access, search, and retrieve records from this, and other, S&T databases.

MEDLINE exemplifies one such structured abstract records database freely accessible via the internet (http://www.ncbi.nlm.nih.gov/PubMed). One of the 12 million+ records found on MEDLINE is summarized as a sidebar. This record tells us that Kamoun and his coauthors did an empirical study relating to a particular genetic change associated with diabetes. Their write-up of this research appeared in a French pathology journal in 2001. It has been indexed for MEDLINE and assigned an identification number.

Sidebar: Example MEDLINE Record Summary

Kamoun Abid H, Hmida S, Smaoui N, Kaabi H, Abid A, Chaabouni H, Boukef K, Nagati K.

Association between type 1 diabetes and polymorphism of the CTLA-4 gene in a Tunisian population. Pathol Biol (Paris). 2001 Dec;49(10):794–8. French.

PMID: 11776689 [PubMed—indexed for MEDLINE]

MEDLINE offers a number of choices to go beyond this limited Record Summary. In many cases it offers the opportunity to order the full article while you are on-line. If you'd like to find more articles, it points to another 364 articles that relate to the topics of this article (as of Jan. 2004). We emphasize abstract records, noting that some disdain these (sidebar).

Sidebar: Abstracts

Text mining purists demean abstracts. They tout full text as the "whole truth," and anything short of that is suspect. We take a different perspective. Abstracts constrain authors (or indexers) to tell what's essential about a piece of work, briefly and clearly (to be understood by a wider audience than the full text targets). R&D publication abstracts make for rich mining—they generally try to incorporate all of the key terms and they entail less noise from marginal text discussions. They are analogous to the executive summary of corporate reports. Collections of abstracts on a given topic provide excellent coverage of the content of the field—hence, a fine source on which to profile an emerging technology.

For tech mining purposes, we prefer to work with an intermediate level of detail—abstract records. By selecting the "MEDLINE" formatting option on the website, we obtain these. Furthermore, we elect to retrieve not just one or a few abstracts of special interest, but the entire set on the topic of our search. In seconds, this yields some thousand or so abstract records right on our computer. We will discuss this transfer of articles between computers in Chapter 7. Now, however, let's examine the full abstract record of the article by Kamoun et al. in Exhibit 5-1.

EXHIBIT 5-1 *A Sample MEDLINE Abstract*

UI	21633248
PMID	11776689
DA	20020104
DCOM	20020117
IS	0369-8113
VI	49
IP	10
DP	2001 Dec
TI	Association between type 1 diabetes and polymorphism of the CTLA-4 gene in a Tunisian population
PG	794-8
AB	Susceptibility to type 1 diabetes mellitus is strongly associated with particular HLA class II alleles. However, non HLA genetic factors are likely to be required for the development of disease. The candidate genes include the cytotoxic T lymphocyte associated 4 (CTLA-4) located on chromosome 2q33 and designated (IDDM12), which encodes a cell surface negative signal T molecule providing for activation. We investigated CTLA-4 exon 1 dimorphism in 74 type 1 patients and a control group of 48 healthy subjects from Tunisia using two methods PCR (polymerase chain reaction) allele specific and polymerase chain reaction restriction fragment length polymorphism (PCR RFLP). The CTLA-4/G allele was found on 68.9% in type 1 patients as compared to 51.02% in controls (p = 0.002), mostly in homozygous from 43.24% versus 22.45% (p = 0.0058). These results indicate that CTLA-4/G allele was significantly associated with predisposition to type 1 diabetes in our group from Tunisian population.
AD	Institut national de nutrition et de technologie alimentaire, 11, rue Jebal Lakhdhar Bab Saadoun 1006, Tunis, Tunisie.
FAU	Kamoun Abid, H
AU	Kamoun Abid H
FAU	Hmida, S
AU	Hmida S
FAU	Smaoui, N
AU	Smaoui N
FAU	Kaabi, H
AU	Kaabi H
FAU	Abid, A

Exhibited 5-1 *Continued*

AU	Abid A
FAU	Chaabouni, H
AU	Chaabouni H
FAU	Boukef, K
AU	Boukef K
FAU	Nagati, K
AU	Nagati K
LA	fre
PT	Journal Article
TT	Etude de l'association entre diabete type 1 et polymorphisme du gene CTLA-4 dans une population tunisienne.
CY	France
TA	Pathol Biol (Paris)
JID	0265365
RN	0 (Antigens, Differentiation)
RN	0 (CTLA-4)
SB	IM
MH	Adolescence
MH	Alleles
MH	Antigens, Differentiation/*genetics
MH	Child

In this abstract record, we see many fields typical of abstract databases. Separate fields of potential tech mining interest appear for authors, authors' organization, title, date, journal name, publication type (article), and keywords. Other fields shown are usually of less tech mining interest. Patent abstracts are roughly analogous, but with differences (e.g., those preparing patents may not want to communicate their intents fully). Patent data are considered later

TABLE 6-2. Answering Tech Mining Questions by Using R&D Publication Abstract Record Content

- How recent is this research? [Use Date of publication (DP) or the date on which the record was added to the database (DA)]
- How might you describe the key ideas in this research? [Could use: Title (TI); Keywords (MH); and/or Abstract terms (AB)]
- Who is working on topics that concern me? [Look at Authors (AU, FAU)]
- Which organizations are publishing the most on particular subtopics of interest? [Check First Author's Affiliation (AD)]
- What nations lead in this research? (Extract the "country"—embedded in the AD field (First Author's Affiliation), not CY, which is the publication place of the journal!)
- How did the authors make their discoveries? [Examine clues in the Abstract (AB)]
- How would others classify this research and identify related topics? [Try Codes or Keywords (MH for MeSH—medical subject headings)]

in this chapter and in Chapter 12. Table 6-2 shows the basic strategy of posing a tech mining question and finding record fields that could be tallied to answer it, with the aid of tech mining software.

Some repositories provide other valuable types of information as well. These include useful "input" information in the form of sponsorship data and additional "outcome" information in the form of citation data (in Science Citation Index). Research may be sponsored by one institution but conducted by another. Tech mining can use this information to track research funding emphases and research participation patterns.

Individual authors almost universally provide citations (references) in their publications. The role of citations is to acknowledge an author's intellectual debt to others. Those can be compiled as a valuable form of output information. We can mine this information to identify communities of shared interest. Additionally, this information can suggest which authors, research groups, and institutions are held in the highest repute by their peers—potentially quite helpful in research evaluation tasks.

We will come back to consider the information available in typical fields of an S&T database and how to exploit this. First, however, we step back to consider the nature of the tech mining enterprise—measuring technology.

6.3. DETERMINING THE BEST SOURCES

What are key characteristics to consider in evaluating potential S&T databases? We also note a number of auxiliary sources. Ultimately, selecting the specific sources involves consideration of the tech mining priorities, available information resources, and attendant costs.

Database Characteristics

A few key qualities of the data make MEDLINE, and other databases, particularly suitable for tech mining. Let's examine these (Table 6-3).

Scientific repositories are not all based on text. We also have databases containing chemicals and their properties, genetic information, geographic

TABLE 6-3. Database Characteristics Conducive to Tech Mining

- Each record describes specific developments in S&T
  (i.e., primary research findings being reported).
- Each record is attributed to specific researchers and identifies their organization
  (at least for the lead author).
- The repository is rich in pertinent text records.
- It comprehensively covers an important S&T domain.
- The repository provides suitable quality assurance.
- Each record is structured with pertinent data fields.

positioning information, and imagery. These are all interesting sources of information—however, this book focuses on R&D abstract resources.

Tech mining strives to assess the knowledge of whole communities of scientists and engineers. Complementary "expert opinion" techniques seek the views of individuals. In contrast, one of tech mining's aims is to analyze the "whole"—the body of knowledge pertaining to a technology of interest. The goal is therefore to obtain as many relevant records as possible—not those few that we would traditionally have read.

Records for a tech mining study may derive from a single source (database). A "dense" repository contains many relevant records in a single place. This is certainly convenient and provides advantages in consistency (e.g., "keywords" compiled across databases may merge very different hierarchies). However, other tech mining work may warrant your blending searches performed in multiple databases.

Science and technology are quality-assured processes. Scientists and engineers work in communities and have their work peer-reviewed by others. This helps check that new knowledge can be assimilated and, most importantly, reproduced by others. The history of S&T is full of maverick ideas and researchers that rechart the course of scientific history. Nonetheless, the majority of R&D being conducted is "normal" science. It works by using established principles, with accepted forms of scientific proof and argument. The tech miner should favor information that has been quality assured. (Exceptions do arise—for early warning purposes you may want to analyze "fringe" literatures to identify concerns that may later spread.) Quality can be assured, to a degree, by using data from peer-reviewed scientific journals or patents that have been examined to determine novelty. Some scientific content, such as articles, book reviews, many conference proceedings, and editorial commentary, is not peer-reviewed. Consider whether or not you want to include these in your analyses. You can often restrict your search to research articles by selecting the appropriate record type (e.g., in the MEDLINE abstract illustration, by checking the value of "PT").

A whole "ecology" of information exists about S&T. Sources of information may be directly or indirectly attributed to those actually conducting R&D. Technology information miners would often favor primary research articles and patents because these more directly reflect R&D activities.

Records come in "structured" and "unstructured" formats. A general internet search exemplifies "unstructured" format materials. Tech mining uses computerized tools to assess vast quantities of information. Although these tools are increasingly capable of parsing data in complex formats, the miner should select structured or "machine-readable" file formats if available. Structured records consolidate important information more accessibly than free text, and these records are therefore much easier to mine. The qualities of data we note here offer solid guidelines for evaluating suitable databases.

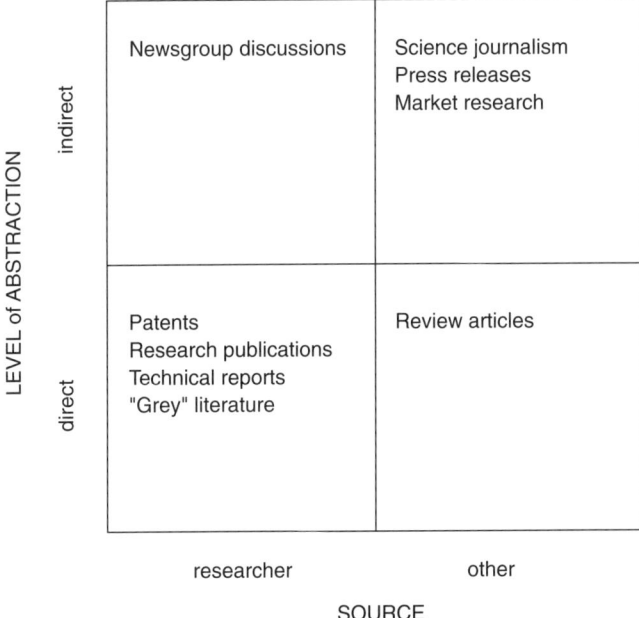

Figure 6-1. Alternative sources of S&T information

Other Sources

Not all sources are of equal quality or validity. Figure 6-1 characterizes the wide range of S&T databases. We can variously locate direct descriptions of "raw" R&D work, up through interpretations of the implications of entire bodies of such work. So information about "stem cell research" ranges from highly technical reports on specific research studies through papal pronouncements on the meaning of life.

You can tap directly into the description of R&D findings by the researchers themselves. These tend to be rich in technical detail and scientific specifics. Or, alternatively, you could examine sources written by outside observers, analysts or journalists, to gain an independent perspective on the research.

Which sources of information are right? The answer "depends"—on the issue you are trying to resolve and which types of information speak to that resolution. Generally, we want to track R&D advances and the prospects for successful innovation (commercialization) of a certain technology. Had we such interests for "stem cell" development, we might very well want to profile both the outpouring of technical gains and the resulting societal reactions.

Figure 6-1 distinguishes four sorts of information based on the source and level of abstraction. For technical depiction, one would usually prefer to operate in the lower left quadrant. Most S&T databases include review type articles (lower right quadrant) together with primary research reports. This usually poses no particular concern for tech mining purposes.

Indirect, but relatively specific, content is illustrated by on-line newsgroups (upper left quadrant). It may be difficult to establish the real identities or affiliations of those posting therein. When tech mining is being used to track the actual conduct, ideas, and activities of scientists and engineers, you generally prefer to use the most direct depictions of these activities that are available. Market research reports reflect secondhand interpretation of a given industrially relevant S&T area. These may indeed be useful sources of information; however, using these metastudies is another topic. For now, we suggest you augment your tech mining work with these additional perspectives.

Indirect discussion of S&T work appears in news articles and press releases. These sources have been filtered, often by third parties, to make the information approachable and of interest to some larger community. Such sources can help you as a tech miner understand the basics. They can also shed light on popular understandability and acceptance, and on the likelihood of "show-stopper" issues blocking the development of the technology in question.

One can extend the kinds of information, and databases, used for tech mining. There are sources to track public (e.g., newspaper article compilations), legislative (e.g., Congressional Record), and economic interests (e.g., databases such as Lexus-Nexus) attendant to S&T developments. You can extend tech mining approaches to both enrich your business intelligence and create "metastudies" of S&T development. However, full treatment is beyond our scope. We concentrate on the "raw" R&D content that is closest to the researchers themselves—abstract records of conference and journal papers and patents.

Evaluating Databases

Regardless of the specifics, you apply similar steps to evaluate the suitability of a given database.

First, consider the characteristics of the database as a whole, ensuring that it meets a set of minimum standards in line with your information objectives. Considerations include:

- Suitable coverage (e.g., covers the topics of concern; inclusion of classified technical development)
- Comprehensiveness of coverage
- Biases (e.g., toward English language publication; capturing industrial R&D efforts)
- Content quality
- Record structure (e.g., inclusion of essential information for your purposes; consistency)

• Keyword availability (e.g., inclusion of subject index terms; consistency of these)

For some purposes, analyzing the entire repository in limited ways can help calibrate one's specific analyses. In sections on "innovation indicators," we note how "term frequency" in a particular search set can be benchmarked against the frequency of those same terms across the entire repository (database) to characterize the R&D emphases associated with that topic.

Consider two dimensions in selecting S&T databases. First, determine where your intelligence needs lie along the "vertical" maturation axis (Fig. 6-2). For instance, if your analyses serve R&D management in selecting which proposed projects to support, move upstream (toward Fundamental Research). Consider using databases like Science Citation Index (SCI). In contrast, if you're managing IP, shift downstream. Include one or more patent databases and consider enriching with an engineering or industrially-relevant source as well.

A second dimension distinguishes "horizontal"—topical—emphases (not shown). For instance, if you are interested in the practice of medicine, start with MEDLINE. If your interests lie in medical research, augment with SCI. If your interests lean toward medical devices, access EI Compendex as well.

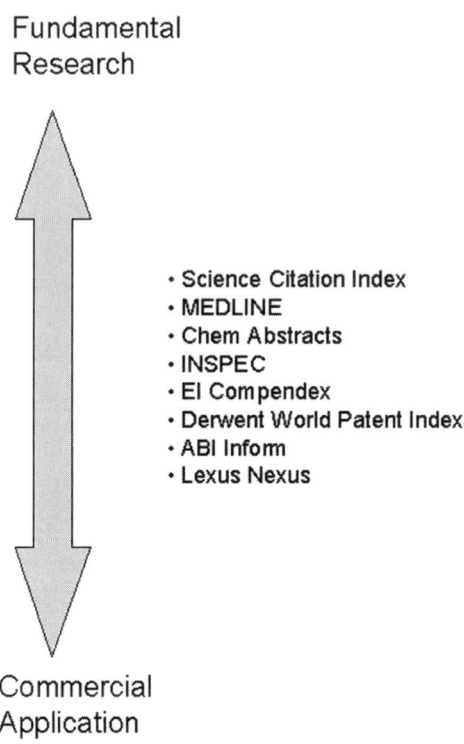

Figure 6-2. S&T maturation axis

On the other hand, if you are working with chemicals, the starting point is most certainly Chem Abstracts.

Stretch to include other sources as warranted. Database providers will help you recognize topical affinities. For example, Engineering Information (EI; www.ei.org) offers subsections of the website: "Engineering Village," "Chem Village," and "Paper Village."

Second, examine the content available in the records to determine which features serve your tech mining needs. In terms of software such as *Vantage-Point*, this means determining which fields to mine. For example, the full repository records may contain over a hundred fields (e.g., the Hazardous Substances Data Bank maintained by the U.S. National Library of Medicine) from which to select those relevant to your tech mining exercise.

Selecting Databases

How do you decide which S&T databases to mine? Our three pointers:

- Be clear on your objectives and which types of information support them. Separate what you "really must have" from what "would be interesting."
- Know what's at hand. Check throughout your organization to find out which databases you have already licensed for unlimited use. Second tier, identify additional databases whose value is clear, even though you have to pay for each record you retrieve. Third tier, identify additional databases that seem promising to investigate.
- Balance against your available resources. How much can you spend for data for this study? How much for tech mining work this year all told? How much time do you have to learn an unfamiliar database? Will a cheaper source eat up more of your time and resources trying to make do with inferior-quality data?

Weighing these three factors, pick one or a few databases that offer the richest information resource for your needs. In other words, weigh both benefits and costs. Suppose you are pursuing CTI about certain information technologies. We suggest you consider INSPEC, EI Compendex, and a patent database, such as Derwent. The "cost" section to follow will make the case that your organization wants to license a few key databases for unlimited searching to avoid paying by the record. Tech mining calls for "carefree" (or nearly so) access to tens of thousands of records.

Refer to Appendix A for a listing of some leading S&T databases.

6.4. ARRANGING ACCESS TO DATABASES

We now turn to a key enabling process—how do you gain access to, then search and retrieve the information to mine? We will also address cost because these resources usually come quite dearly. Cost considerations should strongly influence your organization's arrangements to access syndicated databases. We begin by considering two sources of information—internal and external databases.

Internal Databases

Your organization maintains internal data. These data are precious because they directly address your business, but data turn into valuable information only if you explicitly determine to exploit them. These data can range enormously—for instance, from credit card records of purchases, to cumulations of maintenance record notes, to design histories, to you-name-it.

Exploiting internal data requires careful data management—perhaps even a data warehouse. Data warehousing is not our focus, but we offer three observations. First, fight against the sacred cow called "security." The pendulum of fear swings wildly; we assert that it generally sways too far toward protection at the cost of access. Second, in determining how to structure internal data repositories, we recommend flexibility. Rapidly changing information technologies and rapidly evolving information needs imply that today's optimum structure has little chance of being optimal in the future. Third, as we turn our attention to external data resources, we favor minimizing compilation within your organization. Anything you can do to take advantage of someone else managing and updating information resources, rather than taking on these tasks yourself, the better.

However, beyond this bit of cheerleading to use the data your organization collects (see the sidebar, too), we will largely emphasize external data. Two reasons—we can all use these external sources (within bounds), and tech mining is heavily into intelligence about what's happening outside our organization.

Sidebar: Exploit Internal Databases

We bet that your organization compiles some potent data and information. Too often, internal databases are scattered around the organization, difficult to keep track of, and awkward to access (particularly in conjunction with each other). Knowledge management wrings its hands over such "black holes"—resource sinks where a lot goes in but nothing comes out. We've just been working to access a fantastic internal projects database at Georgia Tech. Different people had to authorize ALP to access the database, lead him through the process to get to it, and provide a digital ID to use it. That's enough small hurdles to sap initiative. But four months after initiating the process, he had access to the database. Step aside for the holidays and some travel. It's now early February—and here's an E-mail that his digital ID is expiring and we need $13.95 funding to extend it for a year. Yes, this database really merits high security, but access hurdles sure discourage usage.

The lesson in this is not to do the things the Georgia Tech way—instead, centralize internal databases, and make authorized access as painless as possible.

External Databases

Few questions can be answered well by using only internal databases. Consulting external databases, however, requires specific institutional and technological arrangements. Actually, exploiting both internal and external information resources together is often best—for example, profiling external R&D on a topic of concern along with review of related internal projects.

To understand the external alternatives, we first consider the institutions that may be involved in accessing them. Next we discuss the mechanics of information transfer. Finally, we examine the economic arrangements involved in accessing S&T databases. Join our DeBrand Foundation in the sidebar to kick us off.

Sidebar: Selecting Databases at the DeBrand Foundation

What databases should the Foundation acquire? Three things are dictated by the Foundation's mission. First, a strong interest in medicine. Second, a concern with agricultural technology. Third, a broad emphasis on global issues, particularly demographics and population in the developing world. Of particular interest to the Foundation therefore are the following: MEDLINE (the database of the U.S. National Library of Medicine), Agricola (a leading agricultural science database), and the Science Citation Index (the foremost database for science).

Organizational Relationships

The diagram in Figure 6-3 shows five groups—the user, the host, the distributor, the provider, and the researcher. We consider relationships among these groups and their roles in creating or distributing technology information. First a brief definition, and then a more detailed discussion of their roles:

- *Users* access science and technology information.
- *Hosts* provide the necessary network and licenses to access the information.
- *Distributors* provide a gateway to a variety of databases.
- *Providers* (also called aggregators) create electronic, and online forms of content.
- *Researchers* write scientific and technical articles and submit them to journals.

First and foremost there is a user. The user may have a local host that provides access to one or more technology repositories (databases). For instance,

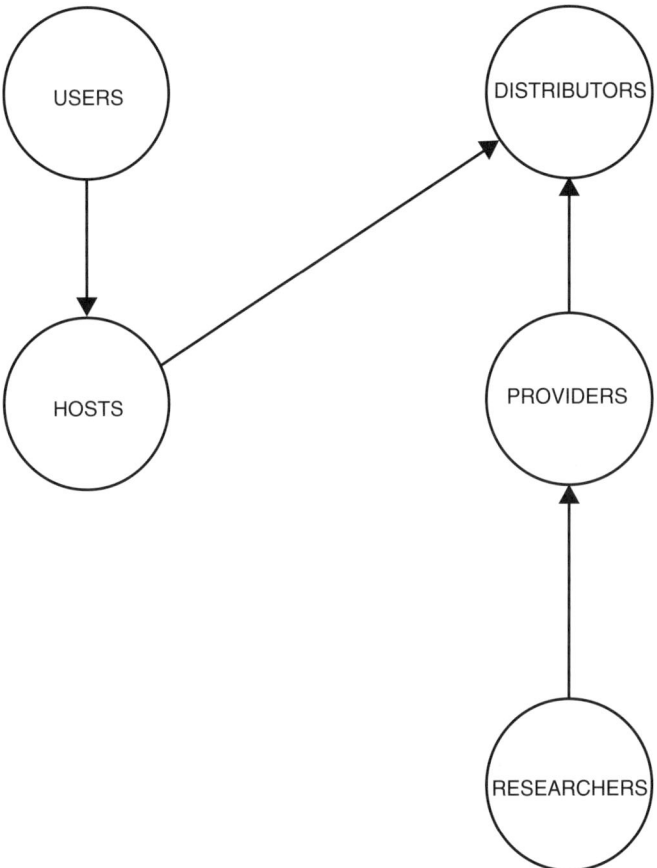

Figure 6-3. Five user groups

our Technology Policy and Assessment Center (TPAC) at Georgia Tech accesses several key sources via the Georgia Tech Electronic Library, others through the statewide Galileo network, and some directly over the internet (e.g., MEDLINE). See sidebar.

Some database providers work directly with customers and/or local host units, whereas others utilize intermediaries to assist in the distribution of the information. Many offer multiple avenues of access. Providers may be part of a syndicate with specific rights to published research, such as a publication company (e.g., Reed Elsevier or Thomson). Database providers compile data from multiple sources—for example, INSPEC abstracts papers from thousands of journals and conference proceedings.

Intermediary distributors are also known as "gateways." Gateway services can take you to multiple database providers through a single common inter-

Sidebar: Institutional Access at Two Technological Universities

Many universities these days provide their students and faculty with unlimited use access to many databases.

For instance, at Georgia Tech, we have had a four-tier system. First, loaded directly on Georgia Tech servers for immediate access have been INSPEC, Engineering Index (EI Compendex), NTIS (National Technical Information Service), U.S. Patents, Computer Database, Company Intelligence, Business Index, and others (e.g., Georgia Tech Library Catalog).* Second, Georgia has a statewide system providing access to about 150 databases, through various providers such as EBSCOhost and ProQuest. These include Agricola, Applied Science & Technology Index, Environmental Sciences & Pollution Management Set, National Newspapers, and Readers Guide Abstracts. Third, the State has obtained access to over 100 databases via the web, including Chem Abstracts, Community of Science, Gale Business Resources, and Science Citation Index. Fourth, additional databases are available on CD-ROM. This availability implies that most graduates will enter corporations and agencies familiar with these sources and how to use them.

At the Delft University of Technology (TU Delft), in The Netherlands, information is available to students and faculty through a single library website that serves as a useful gateway. Many databases, accessible through the web, are suitable for tech mining. There are three tiers. The first tier comprises publicly available databases that anyone can access freely. This includes, for instance, MEDLINE. The second tier includes syndicated databases that require institutional subscriptions—these are available only from campus. This includes, for instance, SCI. The third tier of databases entail premium access but provide full-text access to articles online (in lieu of the hard copy of the journal). This is made in arrangement with publishers and is also available only on campus.

The two universities share some characteristics in their access to data. Both provide information on-line through campus gateways. Both provide Internet access to much of the data but limit access to members of their respective institutions. This model of controlled access via the Internet seems likely to become the dominant model that most institutions will adopt (e.g., in preference to mounting of the databases on local computers, as Georgia Tech used to do). We also note that license agreements vary in terms of their openness to data mining. Some are highly restrictive; others are quite liberal.

---

*In late 2003 this system was heavily revamped to provide better linking to full text sources at the expense of other features including proximity searching and quantity direct downloads of records.

face with the convenience of single-point billing. We note some leading gateways later in the chapter. At times, TPAC has contracted with Dialog, one leading distributor, to access over 400 databases, at one flat rate.

Keep in mind that the content of the abstract databases traces back to multitudes of individual scientists and engineers—the researchers themselves. Without the participation of researchers, technical discoveries would not be made or documented. Ultimately, we consult S&T databases to learn about those discoveries.

Costing

There are two basic costs involved in acquiring data—time and money. Ideally, the tech miner has all the data needed for the analysis right at the desktop at no additional charge (unlimited downloading). More often, you'll need to work with your organization's information services to obtain the needed data.

Time-based costs accrue when searches are difficult, records are diffusely distributed, and the download times are lengthy. If the records are not well structured, analyses consume more time as well. Trade-offs enter between search and analysis, too. Better searches may take longer to craft but will probably save more analysis time in trying to clean and interpret dirty data sets.

Monetary costs occur when users access syndicated data sources. You may want to begin with free sources (e.g., MEDLINE, ResearchIndex, and others). As noted earlier, organizations may also have internal repositories of data, ready for use without charge.

How can you properly weigh the various costs involved? Evaluating whether or not particular data costs are acceptable to the organization requires a cost model. Table 6-4 introduces variables to consider.

Tech mining units usually begin with a budget, typically on an annual basis. That budget may or may not reflect income from performing tech mining for others. Sometimes, a set amount is allocated to technology information access and analysis. Income refers to funds received for tech mining efforts, either from external or organizational sources.

By value, we explicitly mean the contribution of tech mining to decision-making. This includes the knowledge added by integrating with other information and in performing technology analyses (Table 2-2). Improved

TABLE 6-4. Cost and Benefit Considerations

- Budget (available funds)
- Income
- Value
- Information overhead
- Acquisition costs
- Discounting
- Licensing fees

decision-making affords value to your organization, but measuring that value may prove very challenging.

Information overhead expressly pertains to acquiring information. Other elements of overhead, which are not considered here, include facility costs, salaries, and utility bills.

Acquisition costs include salaries paid to information specialists or software engineers to restructure the acquired data. Opportunity costs should be considered when the data are time consuming to acquire and process. Software, such as analysis tools and crawlers, can permanently reduce acquisition costs, often with a single, one-time fee. Data acquisition may therefore involve both fixed and variable costs.

Discounting may be possible if multiple units or organizations are willing to share access, and cost, for the data. For instance, Portugal negotiated access to Web of Science (including Science Citation Index) for all its public universities. Likewise, discounting may effectively occur if the data can be reused in later studies (but some database licenses preclude this).

We are concerned with both costs and benefits. The goal is to gain the maximum benefit at the lowest cost. Benefits derive from direct value gained from the information, plus derived knowledge from integrating that information in various ways. Exhibit 6-2 presents our resulting model. You can adapt this basic model to consider whether benefits of tapping a source of data outweigh its costs for your unit.

EXHIBIT 6-2 *Data Costing Model*

> Costs = Net Licensing Fees + Setup Expenses + Analysis Costs
> Benefits = Income + Value

Then the resulting licensing fees should not exceed the following:

> Licensing < (Benefits – Costs)

The model is not neutral to the nature of the licensing fees. Fixed and variable licensing fees may be evaluated very differently. This model is deliberately generic—tailor it to fit your considerations. For instance, if you must have certain data no matter what, your modeling should compare alternative sources for those data and consider other data as additional incremental costs (e.g., if a package deal might be arranged).

For certain patent analyses, for instance, you might:

- Pay nothing for the data, but spend many hours manually downloading individual records from the U.S. Patent website
- Pay nothing, but spend even more time by manually downloading from multiple websites (e.g., USPTO, European Patent Office, Japanese Patent Office)

- Pay modestly for a database that has compiled patents granted by the U.S.
- Pay more for a database (i.e., Derwent) that rewrites abstracts for added value from multiple sources (USPTO, EPO, JPO)
- Explore additional options (e.g., patents issued, patent applications, patent citations)

Which is the best way to go? Although that choice certainly depends on particulars, in general we suggest you strongly consider the enhanced quality and expedited access options.

The model points toward ways in which greater benefits can be attained at reduced costs. These include seeking cleaner data, sharing your costs, and reusing data once acquired. Data reuse poses knowledge management challenges of coordinating access across business units (which may be spread out internationally) and the retention of purchased information. Nonetheless, it affords considerable benefit to the organization. The specifics of this cost-benefit analysis can be clarified through a numerical example (sidebar).

Sidebar: Costing Database Access at the DeBrand Trust

Let's sketch an example with imagined figures faced by the DeBrand Trust's Research Evaluation Unit (REU). The goals will be to 1) see whether anticipated value justifies tech mining and, if so, 2) determine an upper limit licensing fee for acquiring desired external databases. Start with the 7-factor costing model of Exhibit 6-2.

Value: Let's imagine that DeBrand's REU pilot tests tech mining value in assessing proposals received. Suppose they determine that tech mining findings alter decisions regarding 10% of sample proposals. The yearly R&D investment of the Trust is $170 million. Spending that 10% more effectively accrues benefits of $17 million (that can be applied to further the mission of the Trust). No additional Income is gained for DeBrand by this tech mining effort.

Also, assume that this pilot effort has determined that unlimited access to 3 particular databases is the best way to go.

Costs: Local hosting and management of the databases has been assessed as a better option than having individual analysts prepare data on a case-by-case basis. This is interrelated with the decision to commit to a multiyear, large-scope (e.g., 2500 proposal topics to address this year) research intelligence program. Initial setup will entail significant costs for a workstation server for the data and powerful personal computers for each of several analysts. It also entails data management design and initial data preparation costs (staff time). Assume that a data specialist will be hired. Let's take total setup costs as $500,000.

Amortization of those costs over a multiyear period is vital to justify this investment. Let's superficially divide charges evenly over 5 years, so the $500,000 becomes a charge of $100,000 this year. Annual tech mining

Continued

analysis costs, beyond what REU would have done anyway, are estimated at $100,000.

Results: The annual tech mining benefits are estimated at $17 million. Costs are ($100,000 amortized setup + $100,000 additional analyst time + data licensing fees). This strongly validates pursuing the data and tech mining. Conversely, should DeBrand choose not to purchase syndicated data, they could only access 1 of the 3 databases directly (MEDLINE via the U.S. National Library of Medicine) and would still have to pay to automate access and clean the data.

Imagine that REU's annual budget is $2 million and DeBrand policy anticipates information overhead (cost of data) at 10% of that, or $200,000. As a small cadre of users within a nonprofit foundation, REU should be able to negotiate unlimited access to the 3 databases at far under that ceiling.

Retention (warehousing) of downloaded information is a concern if you purchase by the record. We can relate "Believe It Or Not" tales of companies buying the same patent several times. We turn next to the contractual arrangements necessary for acquiring data.

Contractual Arrangements

As noted, sometimes it is more cost-effective to pay a monetary premium than to expend a lot of time in cleaning and downloading equivalent data via cheaper sources. This is because ofttimes costs of time and money are interchangeable. Individuals have salaries, and organizations have budgets.

Technology information is rarely fixed in price. Negotiate! Prices charged by distributors and providers may depend on the size of your organization, whether it is an academic, industrial, or governmental institution, and the use to which you put the data.

Routinely, academic institutions pay less. Smaller work units usually pay less. So, negotiating a license for unlimited access for a year will be somewhat proportional to the expected usage level. If you work for a large multinational, but only two people at your site intend to use the data source plus a couple of people at other company sites, price the alternatives with care. On the one hand, company-wide site license is appealing to standardize access, minimize learning curves, and share information. On the other hand, it may be more cost-effective for each corporate site to strike a separate deal.

Licensing fees are monies paid to distributors or providers for access to data. A number of different kinds of licensing fees are common. These include annual fees (e.g., subscription, per-seat fees) and variable charges based on the number of records downloaded and, possibly, the length of time spent connected. Initially, your organization should probably not commit to expensive licenses but favor free and then pay-per-record arrangements. Gain experience to help assess whether you want to pursue tech mining actively and which sources are best for you. Then give strong consideration to a fixed fee (unlimited use license), because active tech mining can consume astounding amounts of data.

Sidebar: Licensing and Accessing Databases at the DeBrand Trust

What licensing arrangements should the Trust make regarding these data? It should eventually pursue unlimited access to its three key databases. However, it should start with MEDLINE as it is the most relevant for medical research, and also free. Two of the three databases (Agricola and MEDLINE) are U.S. government information products. In contrast, SCI is provided and distributed by the Institute for Scientific Information, a Thomson subsidiary. DeBrand weighs licensing these databases directly from the producers versus obtaining all three through a single distributor (e.g., Dialog).

What access arrangements should the Trust make regarding these data? All three databases are available both on-line and via CD-ROM. As noted, the tech mining searches performed in support of the DeBrand mission are expected to be extensive. Hosting the data locally (as assumed in the "Costing" sidebar) minimizes bandwidth issues and data cleaning energies. CD-ROM is an attractive solution. However, as part of its information technology budget, DeBrand will also have to supply the additional hardware and networking to ensure that its Research Evaluation Unit has full and easy access (to save time).

Imagine that the Trust negotiates with a couple of distributors and database providers to secure the best deal. They make their case as a non-profit with relatively few users. Working with the providers/distributors, they estimate annual demand and strike a bargain.

The next step in the tech mining process is, after identifying your sources, to create and evaluate useful queries and search strategies for retrieving information. We discuss this topic in detail in Chapter 7.

CHAPTER 6 TAKE-HOME MESSAGES

- A vast amount of information is generated by and about S&T; much is compiled in fee and free databases; consider these as primary tech mining sources to be augmented by the Internet.
- Publication and patent databases provide some of the best available output data on S&T.

- Consider sources covering a range from research to commercialization emphases, plus direct and indirect levels of abstraction.
- Do not create your own database if you can possibly mine someone else's.
- Tech mining favors records that are both structured and quality assured (i.e., from major R&D databases).
- Decide what S&T content you need and weigh alternative data sources in terms of data quality, usability, and cost.
- Perform benefit-cost analysis to help decide on data sources and access modes.
- Think small—be highly selective in picking databases to mine (e.g., 1–3 sources will probably suffice initially), but then think big—retrieve lots of records. (Think of text records as data to be collectively mined, not just to be selectively read.)
- Explore multiple avenues to negotiate the most favorable data access you can for your organization.

CHAPTER 6 RESOURCES

A number of repositories compile scientific information (research paper abstracts, full-text links, and, in the case of ResearchIndex, citations) on the internet. Consider:

- ResearchIndex, or CiteSeer, emphasizing computer and information sciences (http://citeseer.nj.nec.com)
- Scirus: a broad blend of web and database resources from Elsevier (http://www.scirus.com/)
- arXiv, covering physics, with mathematics, nonlinear sciences, computer science, and quantitative biology as well (http://arxiv.org/)

Dialog provides a handy description of the content of its 400+ S&T and other databases (http://library.dialog.com/bluesheets/html/bloS.html). These "bluesheets" can give you a quick sense of which databases cover a topic you are pursuing. Also, its "Dialindex" searches allow subscribers to see how many hits a particular search yields in each of a selected set of databases.

Chapter **7**

Forming the Right Query

Chapter 6 addressed database selection. In Chapters 7–9, we guide you through the process of exploiting these databases by constructing a query, downloading records, and performing basic analyses. To initiate tech mining, you must translate your topic into a search strategy—a query—for selecting articles from the database. This chapter discusses how to form an effective search strategy.

CHAPTER CHALLENGE

- In Chapter 6 we learned about the DeBrand Foundation, a fictional nonprofit dedicated to improving human (and particularly children's) health, nutrition, and welfare. In pursuit of this mission, the Foundation supplements public funding for scientific research. However, the Foundation wants to sharpen its grant-giving process. What fields of science and technology should it monitor given its objectives? Let's formulate a database query to help the Foundation track important developments.

7.1. AN ITERATIVE PROCESS

Tech mining is an iterative process. At each step, you learn a little more about the topic under investigation. Sometimes the right approach is to press ahead and complete the original study, and then consider a follow-up. At other times,

Tech Mining: Exploiting New Technologies for Competitive Advantage, Edited by Alan L. Porter and Scott W. Cunningham.
ISBN 0-471-47567-X   Copyright © 2005 by John Wiley & Sons, Inc.

it is best to restart the process with a new query that more effectively captures the topic you "really" wish to investigate.

Regardless of the aim of the study, it is wise to invest substantial time up front to consider the question being asked and to discover the best query to retrieve the available information. Some queries are based on key terms. Others look for a specific person, place, or article. Still others utilize database index structures. Combinations of these approaches often work best.

We first discuss how to frame a query and how the choice of a query affects the results you get. We also discuss how to measure the quality of the articles retrieved—both the articles you do see and the articles you fail to retrieve. The chapter then treats changing a query to better meet your study needs. We defer focus on patent searching to Chapter 12. Patent search nuances include the importance of combining index-based searching (e.g., using patent classes) with term searching.

An experienced technical librarian suggested the following general process:

- Describe the information you seek in general terms—that is, list the subject areas.
- Nominate terms (words or phrases) that you think would capture that subject information.
- Translate into search logic (i.e., Boolean phrasing—see Sidebar).
- Determine the types of sources desired (e.g., patents granted or applications; journal articles or conference presentations; books or popular articles).
- Consider which sources to search (which databases, websites, etc.).
- Try the search on a small scale (e.g., the most recent year or so); assess the results; refine.

7.2. QUERIES BASED ON SUBSTANTIVE TERMS

Information in most databases is accessed through Boolean algebra—a special language, but a pretty intuitive one. This section turns its attention to the logic of inquiry, rather than emphasizing these mechanics (but see Sidebar).

Consider two aspects in formulating an effective query. First is the breadth of the search—are you looking for broad inclusion (strong *recall*, missing minimal relevant items) or a sharp focus (strong *precision*, getting minimal irrelevant items)? Second is to consider the nature of the language being used. Are you using scientific language in your selection of queries? Or, alternatively, is it more appropriate for the search to use the "natural" language common to scientists and nonscientists alike?

Sidebar: Boolean Searching

Boolean search applies a select set of operators—including AND, OR, & NOT—that allow the user to precisely specify a query. Boolean algebra uses set theory to select and delimit articles. If you are comfortable with Venn diagrams, they depict the logic well (if not, don't worry).

The search—"FUEL AND CELL"—fetches articles that contain both terms. In set theory, this is the class of articles at the intersection of both "fuel" and "cell" search terms. In contrast, the search—"FUEL OR CELL"—fetches articles that contain either the term "fuel" or "cell" (or both). It is therefore a more encompassing search than "FUEL AND CELL." In set theory terms, this is the union of the "fuel" and "cell" search terms.

Specific operators, handling of multiple term phrases, delimiters, and field options vary among search engines. For example, SCI supports a proximity operator—SAME—that restricts to finding both Term A and Term B in the same sentence or section.

We can also restrict retrieval to terms occurring in certain fields (e.g., authors). A query might also limit searches to certain SCI records, say those with 2004 in the "date" field. Other options help you capture term variations. Common truncation markers are "*" or "$." CELL* would retrieve both "cell" and "cells," for instance.

In sum, don't be put off by Boolean searching; a little experimentation will quickly get you the hang of it. Trying variations on the "FUEL CELL" search to observe the change in number of hits and how the captured terms change will readily convey the nature of the process. (For more on Boolean, see the Sidebar later in this chapter.)

Breadth of the Query

Consider the trade-offs between broad and narrow searches (Table 7-1). In Chapter 6 we considered the organization of science and technology. We discussed disciplines, fields, and specialties, recognizing that they differ by orders of magnitude in levels of yearly output. Which is the correct unit of analysis? It depends on your topic and target tech mining uses. If an aim is to spot unusual, nonmainstream R&D, you want to capture items with the barest threads of association to your topic—reach out broadly. Conversely, the more you know just what you're looking for—say, what research group X is publishing on subtopic Y—the narrower you prefer.

Suitable breadth also depends on database portals (see Chapter 6). If it is easy to retrieve the large numbers of records (that well suit tech mining), you

TABLE 7-1. Breadth of Queries

Nature of the Query	Consequences
Broad	Broad queries capture many articles across multiple disciplines or fields.
	Scientific progress in the form of transfer of ideas is best captured in these queries.
Narrow	Narrow queries capture fewer articles of a more specific nature.
	Scientific progress in the form of research concentrations is best captured in these queries.

can confidently broaden the search. Tech mining software will later let you focus to whatever degrees you want.

A paradox of scientific language involves the phenomenon of "self-reference." Scientists and technologists operate in insular communities that have little need to refer to themselves directly. Scientific groups are, in a sense, self-selected. Specific topics of investigation are often far more productive queries than a query using the name of the field of science from which they originate. As an example, searching Web of Science (January, 2004) on "transgenic plants" finds a vigorous topic of research with 6371 articles. In contrast, searching on its parent discipline's name, "botany," yields only 2665 articles. This cautions against using naïve queries as an indicator of the publication "health" of a discipline, field, or specialty. Another illustration—we will later explore aspects of "nanotechnology," an incredibly hot topic as we write. Often as a research domain grows, researchers abandon use of its nominal name in their papers—terminology fractures while activity builds. (See "nano" discussion just ahead.)

Queries: Use Scientific or Natural Language?

A second consideration in framing the query is to consider the role of "natural" versus "scientific" language. What terms should you use when searching? Natural languages are spoken by scientists and laypeople alike. Natural languages are rich, expressive, deeply and often deliberately ambiguous. Computer scientists contrast natural with "machine" languages—languages constructed for use by computers. Machine language is precise, structured, unambiguous, but rather limited in its expressive power. Scientific language exists because scientists and engineers desire precision. To express scientific or technical ideas in an exclusive way, scientists either appropriate words or create entirely new words (often rooted in Latin or Greek). These new scientific words occur within a specific theoretical (and often disciplinary) framework.

For this reason scientists often resist adopting the "common sense" language used by society when discussing S&T. In the upcoming DeBrand Foundation query example, the commonly used term "jasmine rice" fetches 8 articles. In contrast, the scientific term for rice—Oryza sativa—fetches over 6600 articles.

The natural language of S&T is increasingly English; this is not to say that extensive publication does not occur in other languages, only that the highest-profile publication outlets tend to be in English. This facilitates tech mining as most of the S&T databases appear in English. For instance, JICST, the database covering Japanese S&T, is available in English as well as Japanese. PASCAL, emphasizing French language S&T, blends English and French titles, abstracts, and keywords. The prevalence of English also facilitates compilation of global S&T by databases. Although many of them do abstract papers from languages other than English, most favor English language publications. This indexing influences researchers to publish in journals or conferences that are abstracted by the major S&T databases, providing ever-increasing impetus toward English as the dominant language of science.

This also suggests caution in evaluating publication performance for non-native speakers of English. There is an English language bias that can cause non-English-speaking countries or institutions to appear underrepresented or undercounted in scientific output. The extent of this bias is not fully known.

A concrete example of the differences between popular and scientific language occurred when the authors sought to assess developments in the area of "nanotechnology" in the mid-1990s (Porter and Cunningham, 1995). This field involves the ability to manipulate processes at the molecular level. Researchers in many disparate specialties conduct nanoscience and nanotechnology. But many of them avoid "nano" terminology. In our tech mining study, searching on "nano" came upon a small amount of "supramolecular chemistry" work. Knowledgeable colleagues explained that this was closely related, so we expanded our search to include "supramolecular chemistry," finding a huge body of activity with molecular level emphases. The message is to try multiple terms to obtain the coverage you need.

Table 7-2 summarizes some of the consequences of framing a question with natural or scientific language. For many purposes, combining terms of both types will yield richest results.

A devilish tech mining pitfall occurs with homonyms—the same word or phrase meaning different things. Take the word "environment"; it may mean the natural environment, the built environment, or even an informational environment. Working around synonymy requires developing careful queries, and developing an iterative approach to get the articles you need. Tech mining techniques can help disambiguate terms and identify the different communities of meaning resident in a single term.

A nasty result of these language aspects is that tech mining works much better in technical research domains. With careful query formulation, we can

TABLE 7-2. Queries in Scientific and Natural Languages

Nature of the Query	Query Consequences
Using scientific language	More likely to capture the conduct of science
	More likely to capture theoretically rich papers
	More likely to capture specified areas of research interest
	May uncover stagnant fields of research as theoretical ideas are abandoned
	May uncover rich sources of endogenous change
Using natural language	More likely to capture articles of wide interest
	More likely to capture multidisciplinary research
	More likely to engage societal interests or problems
	Less likely to uncover abandoned lines of research
	May uncover rich sources of exogenous change

obtain highly satisfactory collections of R&D outputs in domains such as fuel cells or nanotechnology. In contrast, we struggle to search and to analyze activity in domains such as "management of technology" (MOT) or "knowledge management" because the terminology is so heavily and ambiguously "natural language." Alternatives to term-based searching (see upcoming sections) help in these domains, but tech mining analysis remains challenging.

A widespread need among research performers and research managers is that of information retrieval. For these same reasons of ambiguity in language, prioritization and management of research outputs present serious challenges. Our colleague Ron Kostoff commends an iterative approach, using relevance and feedback from queries, to improve information retrieval. Advanced tech mining approaches, complemented by more "traditional" literature surveys, can provide complementary perspectives.

Queries to Capture New S&T Developments

Tech mining assists individuals in monitoring S&T developments. The selection of a search strategy can strongly impact which kinds of new developments are found, or slip by unnoticed.

Changes in S&T often occur in a "bottom-up" fashion. Scientists and engineers in frontier areas of science make new discoveries. These discoveries are tested, further explored, and then eventually adopted and spread by others. As a result, "pinpoint" (narrowly focused) queries have the best chance of uncovering these processes of "normal" scientific discovery and diffusion. A second major source of change occurs as existing ideas are recombined. Theories, methods, and technologies are disseminated and then used in new and often unexpected ways. Barriers to interdisciplinary communication are typically high—tech mining helps transcend these barriers. Here, "broad-brush" inquiries often fare best.

S&T developments also are reflected in the choices of language used by writers. Here we distinguish between exogenous and endogenous sources of change. Scientists reference, discuss, and modify each others' work. These processes of scientific change are "endogenous," largely internal, to scientific communities. In addition scientists are affected by societal events and driven by economic motivation. Such external, or exogenous, inputs also affect S&T. Terminology ebbs and flows with funding fashion (e.g., a "war on cancer," everything "nano").

So, in formulating queries, balance these dimensions. Examine records retrieved to identify additional elements to enrich your search. Look for terms that tie into adjacent literatures. Have several knowledgeable persons review the content (e.g., list of leading keywords) of the records you retrieve to suggest additional search elements.

7.3. NOMINAL QUERIES

The previous section discussed using substantive (topical) terms and phrases—language—to select subsets of articles from an S&T database. We now discuss queries based on specific people, places, or articles. We call these "nominal" queries because the tech miner begins the process with specific names in mind.

Citation-based queries track references to specific articles or authors. These have a long history of tech mining investigation. We first examine a generic strategy for querying based on specific people or places, and then consider citations.

Queries Based on People, Institutions, Regions, and Countries

Many tech mining studies concern a specific person, institution, region, or country. For these studies it is entirely appropriate to form a query that searches appropriate fields (such as the "Author" or "Institution") for the names being investigated. Such studies can be particularly important for competitive technological intelligence ("CTI"). For instance, imagine General Motors wanting to track R&D emphases of Ford, Toyota, etc. In addition, searches focused on particular organizations (or persons or countries) can contribute to technology monitoring, technology foresight, and technology process management (see Chapter 2).

Such queries face specific challenges, which we call "fuzzy boundaries." These occur when data input errors, variations, or uncertainties accumulate in publication or patent databases, preventing users from getting exactly the information needed. Consider author names. Does the database provide an author's first name (unlikely), first initial, or first and second initials? Even if you have both initials, can you be certain you got the intended person? Difficulties compound as authors change names (marriage) or have names with multiple segments. In a study of Iraqi engineering (Porter, 2003), we came upon the work of Abdul-Halim A.K. Mohammed. We found his name with every conceivable ordering and abbreviation it seemed. (*VantagePoint* "author clean-up" did an excellent job at "finding him' in multiple forms!) With substantial data cleaning, and combining evidence from multiple fields (including affiliation and year—e.g., A. Mohammed of the University of Baghdad in 1990), the diligent tech miner can achieve the target. In the meantime, however, it becomes necessary to download all potentially relevant articles to later cull those not wanted. Fuzzy query boundaries also occur with institutions and place names. Even simple names can be spelled with a bewildering variety of abbreviations, variations, and outright errors. In some cases, we suspect firms deliberately use institutional variations and encourage patenting and publication under affiliated organizations.

Another challenge for nominal queries occurs when you do not initially know all the ones significant for your study. Creating a prioritized list for monitoring is a worthy tech mining investigation in its own right, often requiring considerable iteration. Tech mining software, such as *VantagePoint* (or the version optimized for use with Delphion, Derwent, and Web of Knowledge data, *Derwent Analytics*), provides a vehicle to build, then keep improving, specialized thesauri to facilitate continual query improvement (e.g., a thesaurus of Iraqi engineering organizations).

Queries Based on Indexes

Another searching option takes advantage of the database's categorizations. These might consist of:

- Classification codes (e.g., INSPEC codes such as "A4255N" for fiber lasers and amplifiers)
- Database controlled terms (keywords) (e.g., EI Compendex indexes articles using "Fuel cells" or "Fuel cells—electrolytes")
- "CODENs" (designators of specific journals; likewise, Conference Codes in EI Compendex)
- Document type (e.g., restrict to journal articles)

MEDLINE's "MeSH" index is a prominent example of a multilayer, hierarchical index. This helps associate variations; for instance, "asthma" nests under lung diseases.

Use of indexes is particularly critical in patent searches. We elaborate on various patent class codes in Chapter 12.

Existing, top-down classifications also present challenges. One of us (SWC) considered an existing R&D classification used for policy purposes (Cunningham, 1996). When the classification used by science funders was matched with how scientists classified their own research, severe discrepancies resulted. Roughly one in every three publications could not be fit into a compatible framework acceptable to both parties. A wide gap in accounting for financed research has long been noted. The issue involves different definitions of research among funders and performers of research (Office of Technology Assessment, 1986). Fortunately, tech mining can help in creating classifications of research, development, and knowledge that are based on empirical evidence drawn from large bodies of research.

Queries Based on Citations

Scientists investigate the literature and reference related research, as part of the scientific discovery process. Likewise, inventors (or patent attorneys) cite other patents to delineate their intellectual property ("IP") from prior art. Patents also increasingly cite scientific work (more on particulars in Chapter 12). We can map the sometimes elaborate network of citations among papers and/or patents in an R&D domain to ascertain intellectual and social ties.

Scientific authors exist in a competitive marketplace where the reputation of ideas and research is at stake. Most papers are rarely read; few are heavily cited (the most common number of citations to a paper is zero). A few papers and authors in any specialty are cited repeatedly. Those papers already cited become easier to find, and more attractive to scientists looking for key references. As a result "the rich get richer." This is known as "the Matthew Effect." Tech miners can exploit this "reputational market." For instance, in our 1995 probe of nanotechnology, we examined who cited its instigator, Eric Drexler, and who did not. You may want to examine the body of literature that references a specific scientific paper, author, or institution.

Queries can begin with a specific scientific paper or patent in mind. Unfortunately, few databases include citation information. The Web of Knowledge (from Thomson Scientific)—including the Science Citation Index (SCI) and Social Science Citation Index (SSCI)—is the notable exception. Research Index also provides citation information. Section 12.6 discusses sources of patent citation data.

Users of citation analysis should beware—citations are a rich, but ambiguous, measure of article relatedness and worth. Authors reference work for multiple motives, including self-aggrandizement. Some citations reflect disagreement with the perspective of a rival community or even criticism of particular work. Citation-based searches may yield a heterogeneous collection of articles as a result. The meaning of a citation also varies according to the S&T community involved. Factors affecting the worth of a citation include the amount of publication involved in a specific discipline, the frequency of citation, the frequency of self-citation, the presence of "hub" research leaders, and the connectedness of research within and outside the domain. Citation comparisons among disparate research communities require suitable normalization and expert review.

Citations are valuable, for instance, in CTI, where it is important to determine the experts in a field. Citations serve as a measure of scientific expertise that augments sheer publication activity. Other tech mining studies may seek to identify communities of practice (as reflected in mutual citation). Citations occur in a network of papers and authors. Depiction of a full and representative network often requires repeated data collection and iterative analysis. And finally, tech mining should usually combine citation-based queries with other queries.

7.4. TACTICS AND STRATEGIES FOR QUERY DESIGN

Types of Queries

Having recognized distinct kinds of queries, we now suggest ways to develop the right query for your tech mining study. We note specific questions that tech mining seeks to answer (Chapters 2 and 13). In Chapter 9 we illustrate how specific data record fields might help answer these questions. This and the next section consider using various features that you can manipulate to generate effective database queries to get the information you need for tech mining.

A key dimension concerns broadening or limiting your search (Table 7-3). For example, think about "fuel cells." Although our query underlying the example analyses (Chapter 16) is quite straightforward, it could be tailored many ways. One might narrow to a particular type of fuel cell, relying heavily on technical language. Or key on a particular application, which might draw mainly on natural language. Conversely, we could expand to related electrochemical energy technologies beyond fuel cells per se. We suggest trying multiple search strategies to determine the best blend. Table 7-3 suggests you take advantage of most search engines' ability to restrict to occurrence of terms in particular fields (recall the sample record and its many fields of Exhibit 6-1).

TABLE 7-3. Query Tactics

	Broad Query	Narrow Query
Scientific language	Try names of specialties or R&D domains. Try journal titles, classifications, or keywords.	Try specific phrases that might occur in titles. Try citations to specific on-target or key scientific publications. Try specific, prominent institutions and authors. Try limiting queries by year.
Natural language	Frame the query in terms of broad societal or economic issues. Look for words in abstracts.	Frame the query in nonscientific terms of known policy or popular interest. Look for words in abstracts.

TABLE 7-4. Query by Tech Mining Study Type

Technology Analysis Being Supported	Queries Suggested
Technology monitoring—cataloguing technology development activities	Usually term based—broad queries on a target technology
Competitive technological intelligence —finding out who is doing what	Usually nominal, or a mix of nominal and topic specific
Technology forecasting—anticipating future development	Usually term based—often quite specific explorations of related technologies
Technology roadmapping—tracking evolutionary change in families of related technology	Often a composite—broad coverage of a main technology augmented by explorations on maturation of component and competing technologies
Technology assessment—anticipating the possible unintended consequences and new uses of a new or altered technology	Often broad to pick up contextual influences and effects
Technology foresight—national strategic planning for technology	Various broad-sweep technology and contextual treatments
Technology process management— getting people involved to make decisions about technology	Often broad, seeking new directions, intersecting S&T, and contextual factors to stimulate creative thinking
S&T indicators—tracking advances in national (or other) technological capabilities	Usually broad technology interests

Integrating Study Requirements into the Query

Specific tech mining study goals affect query formulation. In Chapter 2 we noted specific types of technology analyses that tech mining serves. Table 7-4 suggests kinds of queries (either term or nominal) that you might use in starting a search. The Sidebar imagines our DeBrand Foundation initiating searching in support of a broad monitoring program in support of its S&T emphases.

Sidebar: S&T Monitoring at the DeBrand Foundation

The DeBrand Foundation might profitably start by monitoring exogenous, issue-driven changes driving science and technology. Recall their interest in issues of human welfare, food safety, and nutrition. Broad queries might address: pesticides, herbicides, malnutrition, hunger, child health, salt tolerance, salinity tolerance, food safety, food security. These queries range from 1000 hits to more than 15,000. More specific, narrow issues might include: desertification, child mortality, food crops, food policy, manufactured foods, nutriceuticals. The narrowest of these queries resulted in only 20 hits.

Continued

The Foundation has strong prior knowledge and expertise in food processing, biotechnology, and agriculture. It might also want to monitor changes in technologies with which it already has in-house expertise. An interesting example of a biotechnological query is "Oryza sativa," known popularly as "plain old rice." Over 5000 articles have been written in the last decade on this one plant. Examples of broad scientific or technological topics include: agriculture, biotechnology, genetic engineering, botany, and food processing. More specific queries of S&T might include: iron fortification, bioprocessing, and crop science.

They might also use citation-based querying. Norman Borlaug won the Nobel Peace Prize in 1970 for his work in reversing chronic food shortages in India and Pakistan. His interests and career make him an iconic figure— a contributor to a body of thought well worth monitoring by the Foundation. We find 173 articles that cite Borlaug's work. These cover diverse topics, written by diverse authors, in multiple institutions, across multiple fields.

The Research Evaluation Unit might also develop proposal-specific queries. For instance, they might try to devise an automated query formulation process. They could require proposers to suggest search terms to capture the essence of their topic, or just request keywords. They might script the searches in their three databases licensed for unlimited search, based on these nominated terms. Imagine having 2500 such search sets! Again, scripting would be needed to automatically provide initial screening analyses on such aspects as:

- Does the Principal Investigator show up as an active publisher? Is (s)he widely cited?
- How hot is the topic area?
- Who are experts in the area (potential reviewers)?

Precision and Recall Metric for Queries

Imagine that your query yields exactly the right publications for the question you seek to answer. Consider the ways that a real study might depart from this ideal. First, you might find only a fraction of the "ideal" articles that are out there. The fraction of "ideal" articles captured by a query is known as "recall." Second, although you might collect most, or even all, of your ideal articles you might also include some irrelevant articles. The fraction of relevant articles is known as "precision."

The ideal query thus has high precision and high recall. You collect a lot of relevant articles with little extraneous noise. Unfortunately, there is often a trade-off. Some queries offer more precision at the expense of a loss of recall. Others offer recall at the expense of precision. Which is worse—a lack of precision or a lack of recall?

Experience suggests it is a good idea to choose queries with high recall. Tech mining techniques can help you identify and eliminate extraneous hits. In addition, a comprehensive query can be focused later once you better understand the research domain. Little can be done, however, if new developments of interest are omitted by a badly chosen query. This recommendation should be tempered to fit your database access arrangements and costs.

Techniques for Evaluating Queries

How do you know if your query is working well? We suggest five approaches for assessing queries:

1. Use multiple, redundant search terms and search sources and compare results.
2. Use indicators to determine how well your query is faring.
3. Read a small fraction of the articles yourself.
4. Ask a subject-matter expert to review results.
5. Utilize "queries by example."

We will discuss each of these approaches in turn.

There are numerous specific tactics to consider. For one, preselect a group of articles known to be of interest. Design your query. Then determine how many of these articles were found by your query. There are limitations to this method. First, creating a sample of ideal articles can be difficult. Second, this does not assess precision. However, it does get at recall, which is often tougher to do.

Use of multiple, possibly redundant, search terms and sources (databases) invariably increases recall. Unfortunately, a certain loss of precision is also inevitable—more extraneous articles are bound to be included as a result. However, it is better to err on the side of a comprehensive (but somewhat noisy) search.

Indicators might include comparison against statistical distributions that describe the highly skewed distribution of words in articles. Use of these distributions can help you estimate how homogeneous your articles are, and whether you have reached a complete sample of the universe of possible articles (see Chapter 9).

Reading some abstracts yourself is always a good idea. You gain familiarity with the research activities and get an idea of the relative number of relevant and irrelevant articles in the sample. Additional techniques, such as sampling theory, can help you rigorously assess the potential quality of the query. This technique also helps improve the precision, but not the recall, of a query.

Asking knowledgeable persons to peruse a list of leading keywords from your initial search is an easy way to check quality. Ask them to flag additional candidate search terms and noise terms that should be excluded. You might also ask what sources (journals, conferences) they consider to be most relevant to check whether your search results center on these. Possibly, have the experts suggest a search strategy. Be wary that the expert may have specific interests, and therefore may not be aware of broader developments in related fields, and may not be an expert at formulating queries.

Another task concerns "funneling down" from a diffuse search to get on target. Title scanning, plus review and marking of selected abstracts, allows players to indicate their "favorites." Some tech mining software can apply relevance scoring to identify publications or patents most similar to the favorites. This can be pursued to refine search terms. In *VantagePoint*, one can also examine title or abstract phrase lists to flag particularly attractive, or irrelevant, terms. Under Advanced Analyses (Chapter 10), we note ways to map keyword clusters to help perceive knowledge structure in the R&D domain.

Query by example is an interesting technique enabled by some database portals to help searchers find additional articles of interest. Such techniques assume the user has found one article of interest and would like to find other articles of a similar character. Similarities between articles are assessed based on items such as semantic similarity or common patterns of citation. These techniques are applied directly to the database to help users find additional interesting articles. We note three drawbacks to query by example. First, the queries are rarely "well structured." That is, the user has to figure out how and why the query was effective. Second, the query by example is out of the control of the tech miner; you cannot structure the query to account for the kinds of similarity most significant for your own study. Third, query by example is usually implemented with the scientific researcher, not the tech miner, in mind. This means that using query by example to download and save large numbers of articles may be awkward.

7.5. CHANGING THE QUERY

What happens when the chosen query doesn't meet your needs? Sometimes this occurs relatively early in the study, sometimes much later. We consider whether cleaning and augmenting existing data is preferable to downloading entirely new data sets. Sometimes compounding several discrete queries can help to filter your results so they more closely meet your needs. Finally, we discuss strategies for entirely reframing your query. As noted at the start of this chapter, tech mining is an iterative process. Sometimes the study itself changes, necessitating completely refocusing the query.

Reasons for the Change

When using a query you may discover that it yields too many, or too few, hits from the data sources. (See query variation results in Step 6 of Chapter 8's

Search & Retrieval illustration.) In these situations the query is not at the right level of generality, and it must be reframed. Narrowing the focus results in queries with a higher level of precision; broadening the focus improves the recall. In searching multiple databases, you will probably discover that "one query does not fit all." Vary the search parameters, but also try to maintain logical consistency across the sources.

Secondly, queries may be altered later in the process, once articles have already been downloaded. A preliminary examination of the articles may reveal that the query is not what you had intended; the collection may be "corrupted" by the unintentional download of irrelevant articles and abstracts. Modifications can improve the precision of the selected query. Most times, tech mining software can separate out unwanted records without the need for redoing search and retrieval.

We repeat—the tech mining process is iterative. A tech mining analysis having been completed, new areas of investigation will likely be suggested. In addition, in support of routine technology monitoring tasks it may be necessary to supplement existing collections of articles with updates or other new materials. This poses knowledge management issues to most efficiently build current, well-targeted data sets.

Creating New Queries Versus Cleaning an Existing Query

If you have already downloaded articles (Chapter 8) that are not fully suitable for your study, you may face the choice of starting afresh. The download process can be laborious, and, even if you have an unlimited use database license, downloading voluminous waste records could encourage the provider to press for higher charges on license renewal. So, it is often best to choose a query carefully before commencing full download.

In Chapter 9 we discuss how tech mining tools can help you clean and filter an existing download. The goal in such filtering is to eliminate known sources of error, thereby improving the precision of the articles collected. Weigh the option of adding new articles to an existing collection to improve recall.

Compounding Queries

We discussed how homonyms threaten effective queries by introducing unintended articles and content. We now turn to synonyms; these offer an opportunity to strengthen queries and enrich data. Synonymy occurs whenever two distinct words show strong overlap in meaning. This occurs in science, for instance, when two different research communities share common interests but have a different language for describing their interests (e.g., the aforementioned "nano" and "supramolecular chemistry" communities).

We have also tried tech mining on "cognitive linguistics" and "natural language processing," to be amazed at the commonality of terminology but the distinct communities—different journals and little overlap in researcher participation. Check with knowledgeable researchers to be sure you are tapping the right S&T information.

Using multiple, related terms in a query helps broaden the query and ensure that you get the articles that you intended. Two strategies for compounding include using "truncation operations" and Boolean operators. (see Sidebar). A concrete example of this process was seen in the Foundation example—articles used both "salt tolerance" and "salinity tolerance" to reflect an interest in breeding plants able to withstand growth in brackish water.

Sidebar: More Tips on Boolean Searching

Truncation operators are a useful way to broaden your search. For instance, the search term CELL* would capture occurrences of all words beginning with the letters C, E, L, L This is useful because it can find plural or adverbial forms of words (often what we want), but it may also capture entirely different words (not what we want—e.g., cellulose). The specifics of truncation vary by search engine (e.g., $ for unlimited strings, other symbols for individual character or internal character strings).

The use of **proximity** in searches is a good way to narrow or broaden a search. The search term ADJ, for instance means "adjacent" in some search engines. So the search string "FUEL ADJ CELL" would capture only the phrase "fuel cell." Some search engines enable operators such as "NEAR#," where # says within how many terms of each other, in either order. What about the search phrase in quotes—"FUEL CELL"? It depends on the individual search engine. The ISI Web of Science reads this query as "FUEL ADJ CELL."

A third tip is to **exclude** words from your search. So for instance "FUEL CELL NOT HYDROGEN" eliminates a substantial number of articles about a specific facet of fuel cells. This capability to limit or exclude articles in searching is often exactly what is needed. And our best bit of advice—check the search engine's "Help" instructions to be sure of what capabilities are available. Experienced searchers develop an intriguing variety of ways to accomplish similar ends—you can surely gain insights from your organization's information pros.

Reframing the Query

Sometimes it is necessary to elaborate on an existing query. Do you still have that query? Be sure to save your search phrasing. Some search engines provide for this internally, but it is still a good idea to copy into your own text file.

As you iterate through a tech mining study, you learn more about the topic, and new ideas for getting additional related articles may occur to you. Reframing may occur as a result of broadening or narrowing the query. Or you discover the need to transition between natural and scientific language. Very commonly in a tech mining study you begin with a general query and decide it is necessary to explore certain aspects of the topic in much greater detail. Sometimes you need specific articles that might have been omitted by the original query. As a tech mining exercise progresses, it often moves from everyday terminology into the jargon of the area in question. Having access to domain expertise greatly facilitates sharpening the query.

In Chapter 8 we take a specific query and data source and discuss how articles are selected and downloaded from on-line databases. This is preparation for the basic and advanced analyses to be discussed in Chapters 9 and 10.

CHAPTER 7 TAKE-HOME MESSAGES

- Your tech mining questions determine the right query formulation.
- Effort spent up front in designing a good query will be amply repaid.
- Queries may be based on substantive terms or phrases, on specific people, organizations, and places, and/or on indexing structures.
- Consider using both scientific and natural language terms for your search.
- Tech mining is an iterative process; plan on performing initial "quick and dirty" searches, reviewing results, and then reformulating the search strategy.
- Recruit someone who knows the research domain to help you refine your query and assess the quality of search results (e.g., by scanning a list of prominent keywords returned).
- Learn Boolean searching; it's easy. But also work with information professionals to learn ways to improve your searching.
- Research community norms provide useful information, too; use these by taking advantage of specialist terminology and by tracking citations.
- Document your search strategy, steps, and results; this can prove essential in interpreting findings and refining searches.
- Assess your query effectiveness in terms of precision and recall.
- Try the five approaches to help evaluate queries.
- Take advantage of search engine power by using tools such as field delimiting and truncation.
- Watch for homonyms; use synonyms; find out how researchers and patent attorneys phrase the issues of interest.

Chapter **8**

Getting the Data

This chapter shows you, step by step, how to access and download records from an on-line science and technology database.

8.1. ACCESSING DATABASES

The bottom-line message of this chapter is that getting the data is easy. Arrangements must be made (see Chapter 6), but then the actual access and transfer of records is usually quite seamless. For instance, a Georgia Tech student or professor, from his or her laptop—at the office, at home, or on the road—can connect through the Georgia Tech Electronic Library to the databases. He or she can search hundreds of millions of research publication or patent abstracts; then download selected electronic records to that computer, typically *within seconds*. The world's science and technology is at our fingertips.

Before stepping through how to get data electronically, let's look at the system elements involved (Table 8-1). We consider each of seven main items: of hardware, software, or protocol, plus "computers" as the general processors. (These elements could be ordered many ways.) The table indicates the several items, gives an instance of each, and points to who is most likely to be concerned about each.

We'll examine each of the elements in turn, starting with the transfer protocols.

Tech Mining: Exploiting New Technologies for Competitive Advantage, Edited by Alan L. Porter and Scott W. Cunningham.
ISBN 0-471-47567-X Copyright © 2005 by John Wiley & Sons, Inc.

TABLE 8-1. Hardware, Software, and Protocols for Data Access

Hardware, Software, or Protocol Elements	Example	Of Interest To
Computer	Personal computer	Users, hosts, distributors, and providers
Storage	CD-ROM drive	Users, hosts, distributors, and providers
Transfer protocol	http	Users, hosts, distributors, and providers
Analysis tool	*VantagePoint*™	Users
Crawler	IBM's "Aglets Software Development Kit"	Users, distributors
Client (application)	Netscape browser	Users, hosts, and distributors
Server (application)	Apache web server	Hosts, distributors, and providers

Computers have *transfer protocols* used in exchanging information. Multiple protocols evolve over time—for instance, "http" (hypertext transfer protocol), "ftp" (file transfer protocol), and "mime" (used in exchanging E-mail). These facilitate interactions between computers using web pages, direct file downloading, or E-mail. These protocols use existing networks such as the Internet.

Ideally, you would directly tap into the data resource, and signs point toward movement in this direction (i.e., integration of analytical tools with databases). However, in most cases today, data are retrieved from one data structure (the source database), transferred, and then reconverted into another data structure (in our examples, *VantagePoint* files).

Computers also need *storage* (memory) to retain the full data repository itself (in some cases) or just the selected records (search results). Storage includes, but is not limited to, a hard drive. Portable storage devices, such as CD-ROM, can also transfer and save data. If your organization determines to license the database on site (as Georgia Tech did with several main S&T databases through 2003), storage, networking, and computing resources become significant considerations.

Several types of software are vital to the data access process. The *client application* enables exchange of data between computers. A familiar example is a web browser (e.g., Netscape). Clients for exchanges of E-mail or files provide other key capabilities. For instance, if you perform a search at ISI's Web of Knowledge, you can download your search results to your computer using a "save" option or an "e-mail" option. Users and hosts need client applications.

Analysis tools include text mining programs that assist in handling and analysis of the records retrieved. Some such tools (e.g., *VantagePoint* or *Derwent Analytics*) act on data stored in memory at the desktop, others as central server tools acting on data stored at the server (e.g., web-based Aurigin, now integrated with MicroPatent as Aureka).

Servers distribute information. Tech miners today most often deal with web servers. However, some providers and gateways still maintain a "telnet" interface option that may offer more end user control over the search process. Hosts and distributors typically use servers to assist in disseminating raw data to others.

Additionally, there is a piece of software called a *crawler*. Other names for a crawler include "spider," "intelligent agent," or "bot." This software automates the finding and downloading processes, particularly when multiple repositories are involved. Search engines deploy web crawlers to check millions of websites for new information to index. Users who have the choice of many repositories may wish to do the same in collecting S&T information. At least one database distributor, NEC with its ResearchIndex, compiles S&T information with Internet crawlers. Its crawlers seek out research websites, downloading scientific publications when they are found.

You may be wondering why anyone needs databases and statistical analyses when intelligent crawlers can go and get data "everywhere." Databases compile particular, defined classes of information. They offer advantages of specified coverage and quality control (e.g., MEDLINE and SCI draw only on peer-reviewed journal articles). They also provide consistent formatting of records that enormously facilitates tech mining. The Internet offers diversity and currency—very complementary resources to the databases.

We likewise see a fundamental complementarity between artificial intelligence (AI) capabilities, reflected in crawlers, and statistical analysis approaches, emphasized in this book. The AI approach works well when you can specify exactly what to look for. Your instructions can be very encompassing—as in "find any scientific article pdf files posted on the web." Or, they may be tightly constrained—"find mentions of Vladimir Putin (Russian president) in news feeds over the past month." On the other hand, statistical techniques do well at exploration and discovery—as in "profile that information on Putin" so we can see patterns and associations. These approaches (often "disciplinarily dissociated") can work well together to find and understand information.

Accessing databases requires some real care in assessing your organization's support resources to determine the best way to go. The DeBrand example (Chapter 6) illustrates an extreme in starting from scratch to mount and maintain selected databases locally, tailoring them to the target uses. At the opposite extreme, most academic users can pretty much assume it's "all taken care of"—just check with the campus library to find out how to access hundreds of databases from your PC. If you are the path-breaker in an organization lacking much information services support, strongly consider working through a gateway. For instance, Dialog and STN provide access to hundreds of databases direct to your PC, with various licensing arrangements.

The following section steps through how to access a syndicated database.

8.2. SEARCH AND RETRIEVAL FROM A DATABASE

This example illustrates use of a gateway that gains you access to multiple databases. We'll go through each of eleven steps for one case to illustrate what's involved. With a little guidance to figure out what's required the first time, it's really easy! Your specifics will vary, but this should provide perspective on what's involved.

In this example, we'll access the Science Citation Index (SCI) through the Georgia Tech Electronic Library (GTEL). The GTEL website serves as a gateway to many information providers—some of which are limited to Georgia Tech faculty or staff. Accessing web-GTEL requires a computer with an Internet connection as well as a browser. Incidentally, this example was generated remotely (off-campus).

1. Browse the home page of the gateway.

In Figure 8-1 we've typed in the URL into an Internet browser (say, Internet Explorer), and arrived at the front page of Georgia Tech's Electronic Library (as of 2003; it has since been modified).

2. Select a database.

Navigating to Georgia Tech's selection of on-line science databases (see Fig. 8-2), we note the availability of Science Citation Index (SCI). SCI is a funda-

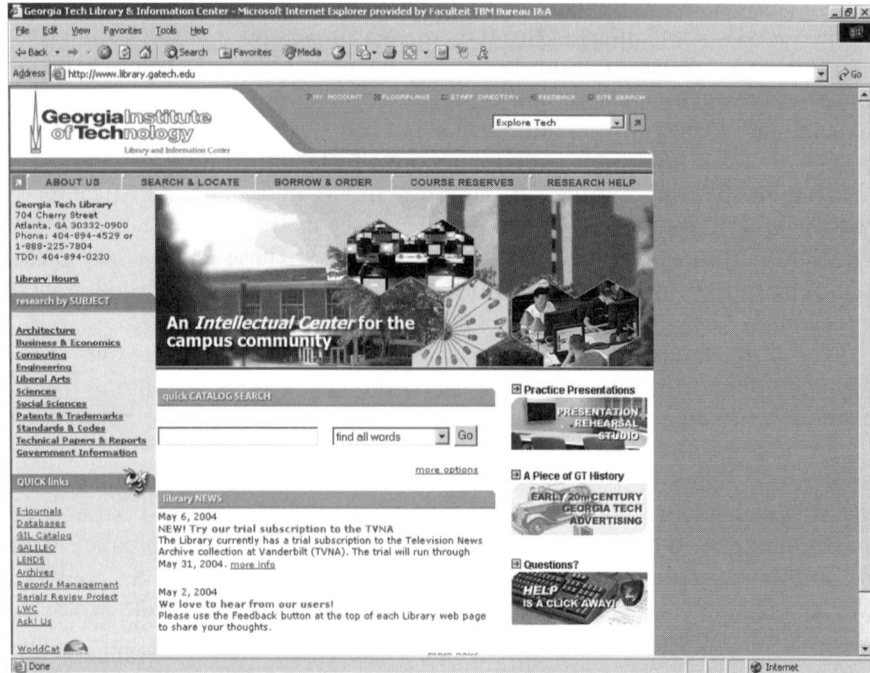

Figure 8-1. The GTEL home page

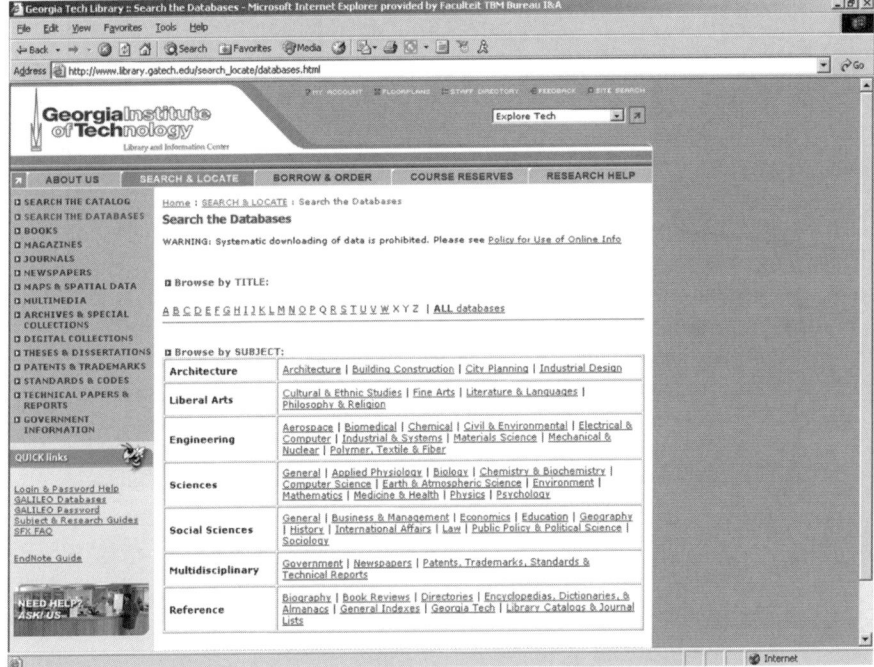

Figure 8-2. Available databases from the GTEL gateway

mental science database that is suitable for many tech mining studies, and one of three databases we use in the fuel cell example analysis (Chapter 16). SCI also incorporates a lot of applied research, as well as literature more "managerial" in character.

3. Log in to the portal.

The SCI database is a licensed property of Thomson Scientific & Healthcare. Georgia Tech has a specific licensing agreement that limits access to this database to its students and faculty. We must therefore use a password and user name supplied by Georgia Tech (see Fig. 8-3).

Now that we have been authorized through the Georgia Tech website, we have been redirected to a web portal (Thomson's Web of Knowledge) containing the specific database in which we are interested.

4. Begin searching the database.

We select the "full-search" option to locate relevant articles (see Fig. 8-4).

5. Address the size of the search.

Our running examples involve a tech mining study of "fuel cells"—a compact energy source (to power vehicles such as automobiles, for stationary power sources, etc.). Our search strategy is basic. So we enter "fuel cells" under the topic section, and hit SEARCH (Fig. 8-5).

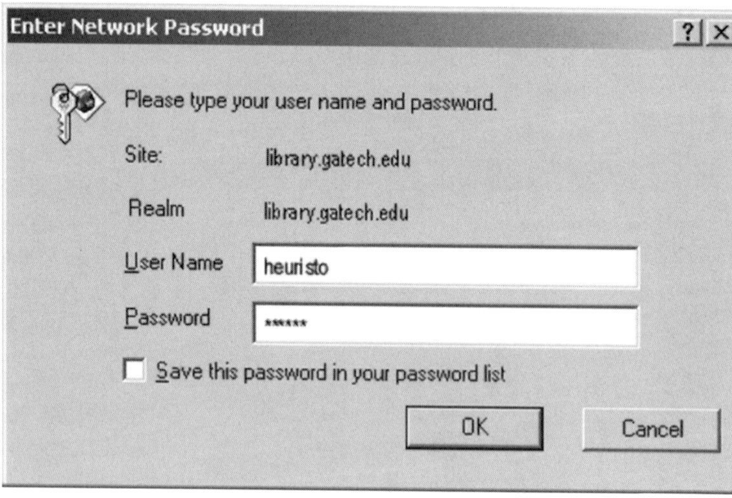

Figure 8-3. Log-in pop-up

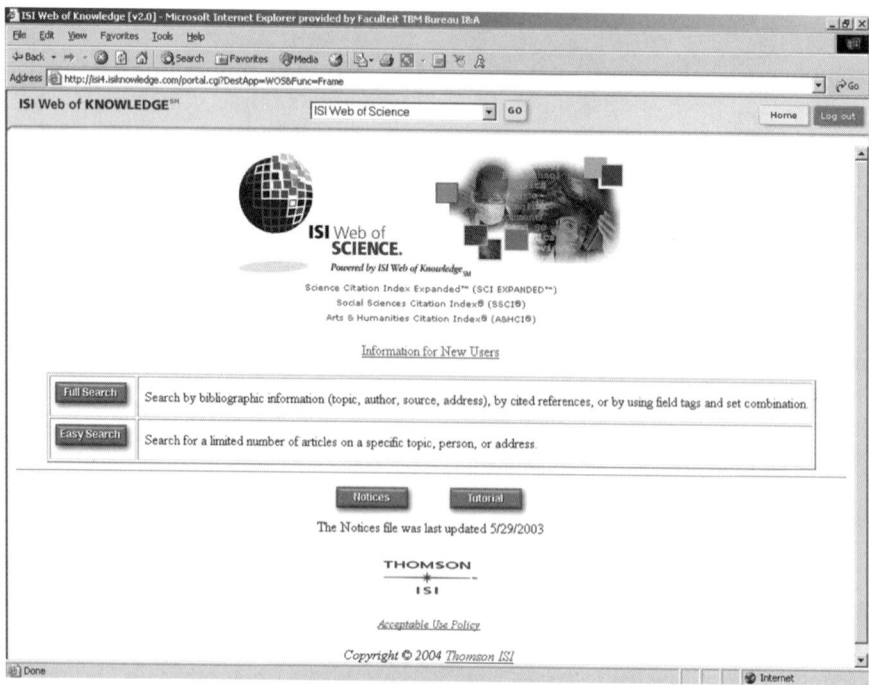

Figure 8-4. SCI start page

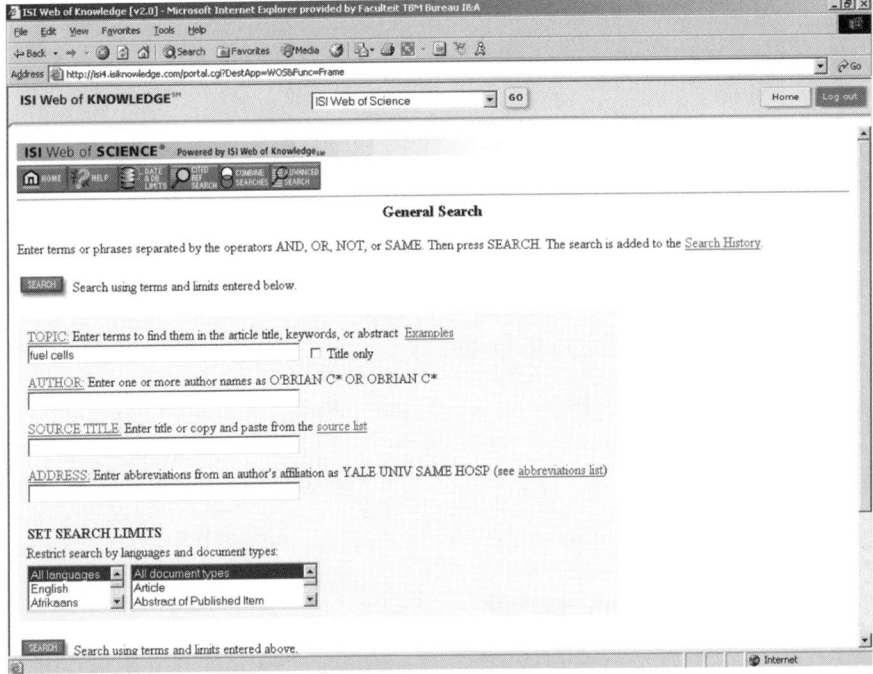

Figure 8-5. Query page

There are many choices we might make when conducting this search, including changing the article type and language. We'll justify our choice of all documents and English language later in the chapter.

The raw results of the search reveal 3332 articles containing "fuel cells" (as of July, 2002) (see Fig. 8-6). This is a fairly typical order of magnitude, considering we are investigating research across the lifetime of an emerging technology.

Note that the search was from years 1988–2003, a standard "window" of dates for ISI databases. We often want as many years as possible of data to better understand growth and change in a research domain. This window of years is increasing of late: Providers such as ISI seem to be incorporating increasingly older articles in electronic format as well as the newest publications.

6. Explore alternative search strategies.

Chapter 7 introduced many considerations in selecting your search strategy. Table 8-2 gives some empirical results showing how much difference "minor" changes can make!

In this example, inclusion of both plural and singular forms of "fuel cell" yields an additional 2400 articles to the plural form alone (row 1 vs. row 2).

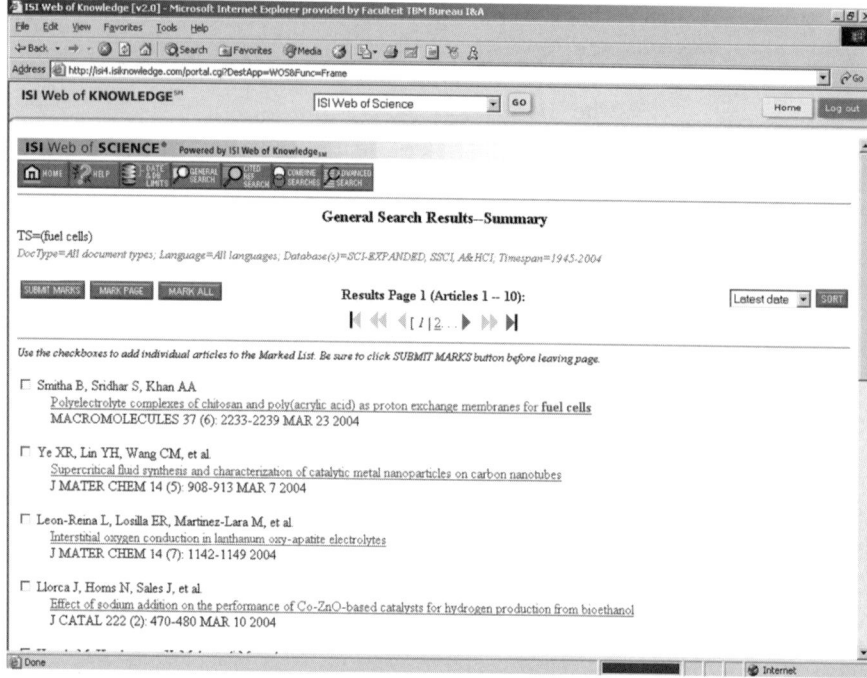

Figure 8-6. Search page

TABLE 8-2. Alternative Search Strategies

Search Term	Hits (Number of Articles)
fuel adj cells	3809
fuel adj cells or fuel adj cell	6220[†]
fuel cells and fuel cell	1573
fuel adj cell*	6220
fuel and cell	7533
fuel and cell*	7876

[†]Checking again in late October, 2003, we found 7975 documents from 1970 through the present, suggesting the value in periodic updating of TM studies.

This is definitely desirable; both terms capture valuable articles—neither alone is sufficient. Row 3 is of incidental interest; some records actually contain both forms. Using all alternative endings for the word cell could yield a lot of irrelevant articles (row 4), because it might include words such as "cellular," "cellophane," and "cellulite." In this instance, we neither expected nor wanted alternatives beyond the terms cell or cells. The search tallies confirm our hopes that we were getting only the articles we wanted in this respect.

Changing from fuel adjacent to cell to a general association between words ("fuel and cell") yields another 1300 articles. Broadening still further to "fuel and cell*" nets another 500. At this point, however we've cast the search too widely. Articles such as the following title are being collected by this search:

"Correlation between in vivo and in vitro pulmonary responses to jet propulsion fuel-8 using precision-cut lung slices and a dynamic organ culture system"

This article is about toxicology, not electrochemistry! The costs of cleaning out these articles on toxicology (and other noise) to gain an incremental addition to our knowledge base on fuel cells is not worth the trouble. We therefore continue with our search using the strategy "fuel cell or fuel cells." Chapter 7 discusses techniques for reviewing your search and reformulating it if you have not retrieved the kinds of articles you need.

7. Count the number of articles by year.

Counting the number of "fuel cell or fuel cells" articles published in each year from 1987 to 2002 provides a useful check. This can be done at the website by limiting the search to one year, then redoing for the next year, etc. (see Fig. 8-7). Alternatively, it could be done all at once at the desktop in *VantagePoint*. The results of the search, by year, appear in Table 8-3.

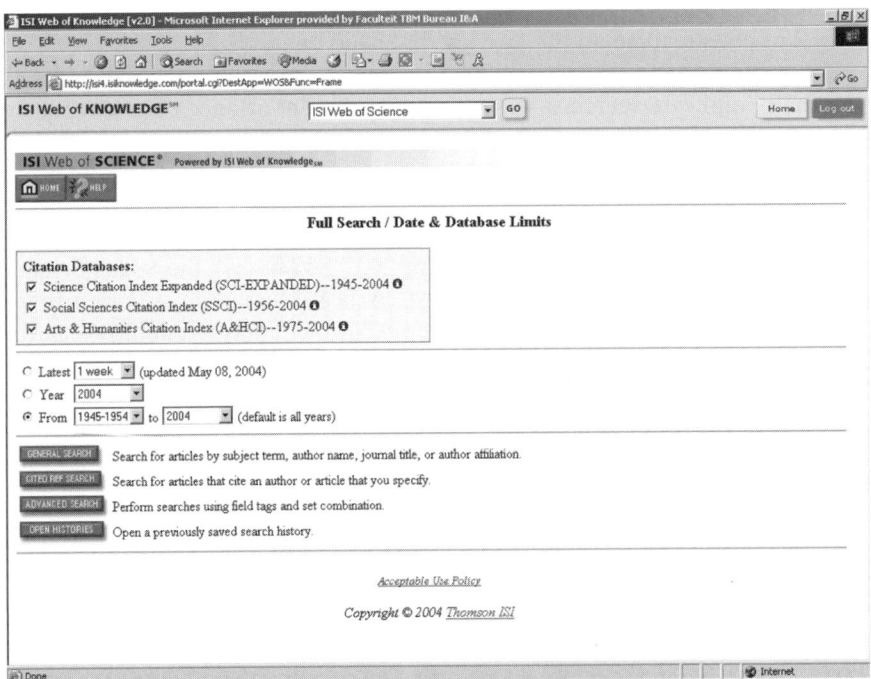

Figure 8-7. Date limit screen

TABLE 8-3. Fuel Cell Articles in SCI by Year

Year	Fuel Cell or Fuel Cells	English Articles
1987	n/a	33
1988	108	43
1989	58	18
1990	128	55
1991	112	63
1992	222	135
1993	195	110
1994	275	165
1995	293	167
1996	423	277
1997	382	241
1998	534	344
1999	536	355
2000	754	458 (461)
2001	869	553 (557)
2002	1046	304 (645)
2003	285	171

These results verify that most research activity in "fuel cells" has occurred after 1987. The 1987 starting year, as imposed by ISI, will not hinder our study. The field continues to grow strongly—so strongly that we will have some challenges in downloading all the articles, discussed next.

On a "final check" in January, 2004, we revisited these searches. It is tough to exactly replicate searches—a reminder of the importance of precise documentation on search phrasing and parameters, as well as dates conducted. Anyway, column 2 actually tallies "any documents" with the terms, fuel cell and/or fuel cells, present in title, abstract, or keywords. Column 3 restricts to articles in English. These Web of Knowledge searches actually include results from the Science Citation Index (e.g., for 2001, 865 items), the Social Science Citation Index (4 items for 2001), and the Arts & Humanities Citation Index (0 items for 2001). The values in parentheses show counts on redoing the initial search at a later date. Note that the data for 2002 have increased tremendously, whereas a few year 2000 and 2001 articles have been added to the database considerably later. Specifics aren't important for our purposes here, but do be aware that database entry lags publication and that the latest year is often incomplete data. In analyses, we will suggest you apply some form of normalization to incomplete year data.

8. Mark the articles for download.

This could be a time-consuming step, depending on the search engine interface. Some websites allow marking only of single records one at a time. Others allow marking of small sets, such as a page of records. Some, such as

MEDLINE, allow you to mark the entire search set at once. The speed of the Internet connection can affect large downloads, but these days it's usually not a serious matter.

It's best to develop a strategy to systematically download all articles with the minimum number of steps. In this case, we face a 500-article downloading limit. (Limits, if any, vary by database interface and institutional agreements.) We can break the search set into sets by time periods, following Table 8-3. For instance, we might separately download the 2003 records at one time, with the 1987–1993 records as another, separate search and download. Step 11 addresses ways to handle the 2001 record set, which exceeds the 500 limit. For the others, one, first, marks those records using the button provided (see Fig. 8-8).

9. Request download of the articles.

SCI provides several options for downloading articles (see Fig. 8-9). Among the choices are "format for print," "save to file," "export to reference software," and "e-mail." For this example, we choose the button labeled "save to file."

Note that the web portal allows users to customize the download by specifying specific portions of the record to include or exclude. Recall Exhibit 6-1 illustrating available fields in a MEDLINE record. For our tech mining

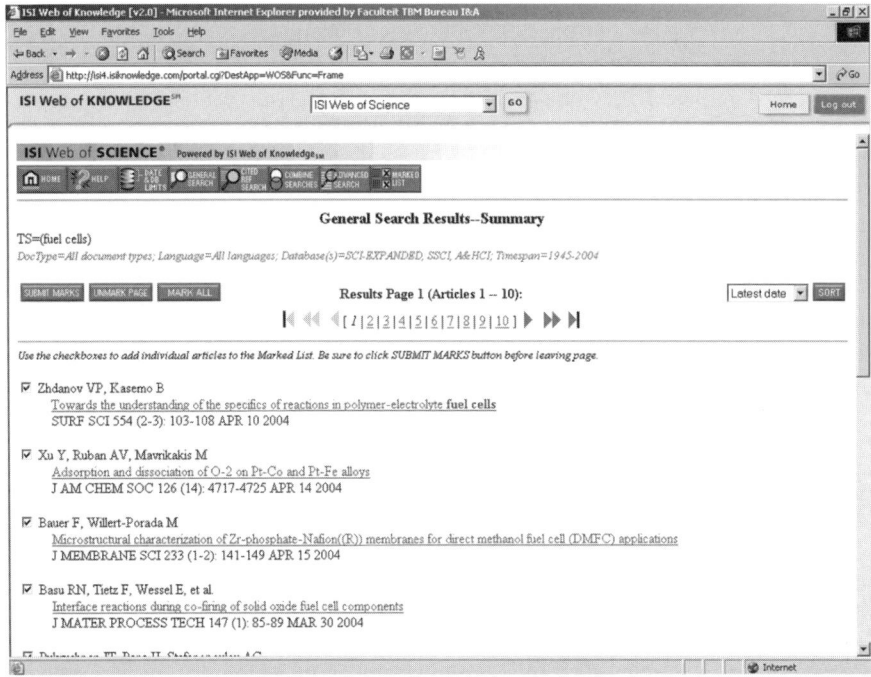

Figure 8-8. Selecting articles for download

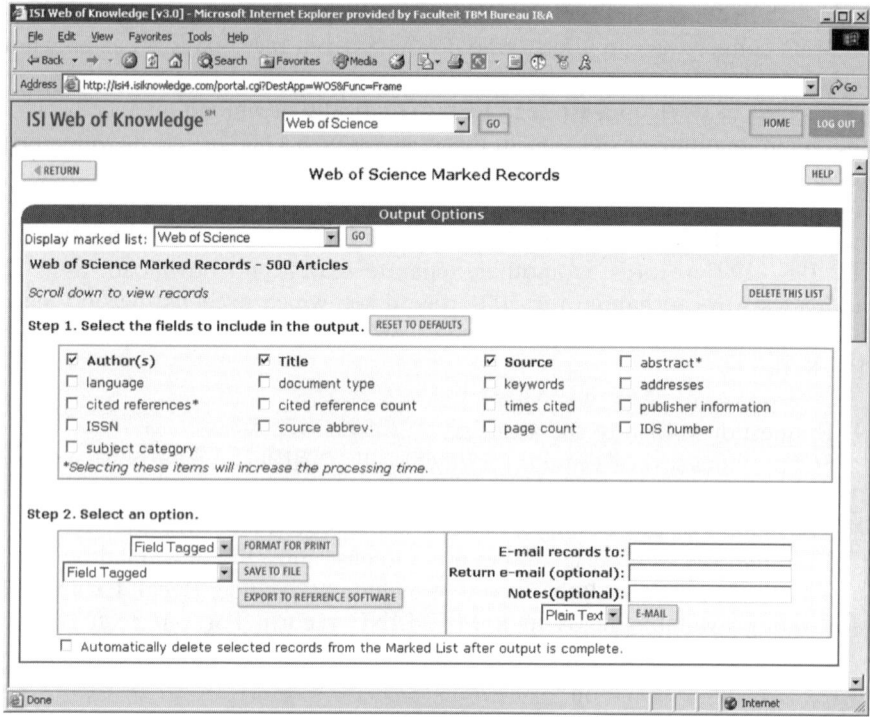

Figure 8-9. Options for download

example (Chapter 16), we have taken: title, author, source, abstract, document type, keywords, addresses, and times cited.

10. Save the file on your local computer.

We have now queried the database for an appropriate subset of articles, marked the list of articles for download, and requested that the file containing the articles be saved. At this point a pop-up window appears, allowing you to choose where the file will be saved locally on your computer (Fig. 8-10). Hit "save" and save the file in a convenient location on your PC. (In this example, the local computer is an Intel-compatible machine running Windows XP. The exact steps in saving the download may vary a little according to your operating system.)

11. Break the download into smaller files, where necessary.

In this illustration, we have 553 abstract article records for 2001. Retrieving this batch of articles involves introducing some additional terms that are irrelevant to the search, to allow us to break the year 2001 articles into two smaller groups for downloading. Most effective for this purpose would be common words widely used in these abstracts. Unfortunately, many common English words are "reserved" and not used by database search engines. The

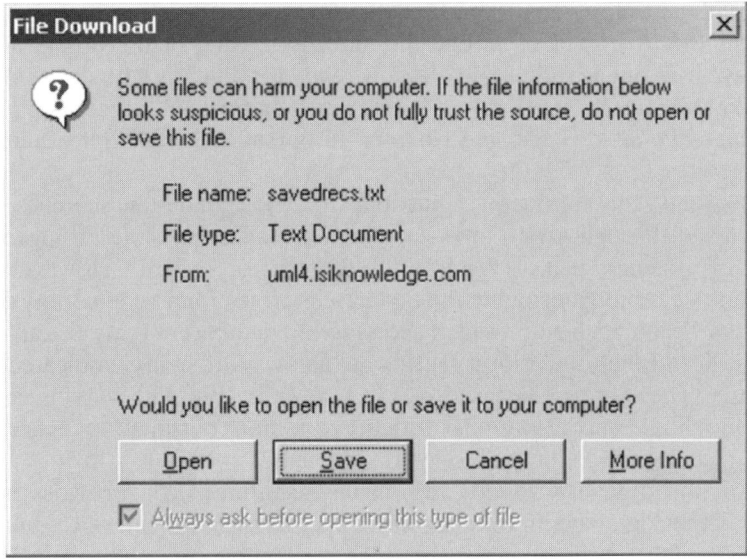

Figure 8-10. Download options pop-up

next tier contains common S&T words such as "system," "source," "method," "design," and "algorithm." Use of any of these can serve to split a data set satisfactorily.

We construct two queries that together enable us to get all the 2001 "fuel cell articles":

- ("fuel cell" OR "fuel cells") AND "2001" AND "system"
- ("fuel cell" OR "fuel cells") AND "2001" NOT "system"

The queries effectively reduce the search sets to 196 and 357 record subsets of the 553 article set for 2001. (Watch out—"2001" could occasionally appear as a number, not the year, in a document.) An efficient naming and storing strategy for such files helps later when you wish to reassemble them into a single unified collection. At a minimum such a strategy should consist of separate files with a name for the topic, year, and portion (e.g., FuelCell2001a and FuelCell2001b).

8.3. WHAT TO DO, AND NOT TO DO

General Principles

The search is complete. All the SCI "fuel cell(s)" records are now saved on our personal computer. However, they reside in a number of files. These can be recombined in various ways. For one, use any text editor or word processor (e.g., MS Word) to open one file and insert each of the others into it, then save it as one text file. You are now ready for analysis with a text mining tool like *VantagePoint*. Chapter 9 sets forth basic analysis elements with the *VantagePoint* software, and Chapter 16 pursues the fuel cell example in detail.

Note some general principles that apply to almost all on-line searches. First, we accessed through a web browser. There are other possible arrangements (using approaches such as file transfer protocol or E-mail). However, web browsers are rapidly becoming the de facto interface for a wide variety of on-line tasks. Second, a host provided access to the database. In this example, the Georgia Tech Electronic Library serves as "gateway" to many syndicated S&T databases. These are all licensed by the University, but with varying restrictions; one should become familiar with the conditions pertinent to text mining. This is a standard sort of academic arrangement.

Third, the database vendor (Thomson Scientific's ISI) provides access to the actual database in various ways. We used a web interface, but SCI is also available through gateways such as Dialog, and they provide the database itself for local mounting by certain licensees. Most databases limit access to registered users, so you need to register, usually as an organizational unit.

Finally, the search itself typically requires using Boolean logic to specify which records you want. The specifics may vary, particularly by fields (e.g., how date ranges are to be specified). Regardless, the general process is much the same, and really pretty intuitive. When you first try it out, it is wise to ask an information professional or experienced user to guide you.

Pitfalls and Obstacles

We mention three obstacles when accessing on-line S&T databases. First, interfaces are usually designed for researchers—not tech miners. As a result, web portals favor the downloading of small numbers of abstracts (or articles), with an emphasis on finding a very few "nuggets" and, possibly, linking to full text sources. Tech miners, in contrast, seek to download many abstract records spanning entire research domains. We have stepped through related issues such as marking records. Sometimes, alternative access modes may greatly improve performance (e.g., for years telnet access held advantages over the web interface at Georgia Tech).

This brings up a second obstacle tech miners encounter—download limits. Providers may want to manage load on the server and to protect their intellectual property. Bibliometric applications, such as tech mining, are within the "fair use" clause of nearly all on-line databases. What is prohibited, however, is resale or repackaging of the raw information. Despite this "fair use," large downloads are not consistently and easily supported by database vendors. Steps 8 and 11 above discuss tactics to work with download limits.

The third difficulty is encountered at times by all who use the web—congestion. Possible remedies include securing a faster connection, arranging for a local copy of the database, or working during off-hours.

To conclude this "getting the data" chapter, we mention some underlying data pitfalls to avoid. Keep in mind that publication and patent data are S&T "output" indicators. You may want to augment output indicators with "input" indicators (e.g., R&D funding) also.

As output measures, publication and patent activity can distort realities. For one, company policies on publishing and patenting vary greatly; many companies limit publishing. Patents are less suitable indicators in certain sectors. Many of the S&T databases abstract from leading English language and non-English language journals and conferences, but favor English language publications. R&D that makes its way into the databases may not be fully representative of a nation's overall R&D efforts (e.g., for Germany).

Publication and patent (see Chapter 12) practices differ across areas. A publication in mathematics is not necessarily as indicative of technological progress as one in physics or medicine. Authorship patterns vary, with some areas typically having huge numbers of authors whereas others favor individual work. These differences in publication activity stem from intrinsic differences in the conduct of science itself. Certain fields emphasize conference presentation, whereas others lean toward journal publication. The career rewards for publishing and patenting vary by sector (academic, government, industry, nonprofits), as well as by scientific discipline. The difficulty in publishing an article, or patenting an invention, also differs by area, and by particular outlet (e.g., journal). So, be wary in making cross-field comparisons.

We suggest that you use a variety of empirical information sources. In addition, expert opinion strongly complements empirical tech mining.

CHAPTER 8 TAKE-HOME MESSAGES

To get your tech mining data:

- Learn and exploit the information resources available within your organization.
- Start small—pick one (or a very few) databases of greatest relevance to license for tech mining; augment with additional information as needs arise.
- Choose the best access method for you.
- Try it out; explore an emerging technology of current interest.
- Begin with an information professional or experienced user to "learn the ropes" in accessing target databases and in performing effective searches.
- Gain experience with basic Boolean searching.
- In doing tech mining, get a person knowledgeable about the subject to check your search results and suggest improvements.

Chapter **9**

Basic Analyses

This chapter guides you into tech mining of science and technology records. Chapter 8 demonstrated the process of downloading articles selected from suitable databases. This chapter offers pointers on cleaning the data and in performing basic analyses. It suggests how even simple analyses can usefully inform technology management.

CHAPTER CHALLENGE

- The DeBrand Foundation wants to analyze articles citing the Nobel Prize-winning agronomist Norman Borlaug's work. This is "exploratory" tech mining, monitoring for interesting and novel research opportunities. (In Chapter 6 we introduced this hypothetical research funding organization. Elsewhere we pose other tech mining tasks for DeBrand, including support of proposal evaluation in Chapter 6.) We have 173 records from Science Citation Index (SCI)—what do we do with them? The chapter will trace our steps as we examine these data, do necessary cleaning, and initiate analyses.

9.1. IN THE BEGINNING

The rationale for, and organization of, this chapter deserve explanation. The chapter sets forth to explain how to begin to analyze the S&T publication and patent data we have been working to obtain. On the one hand, we could take

Tech Mining: Exploiting New Technologies for Competitive Advantage, Edited by Alan L. Porter and Scott W. Cunningham.
ISBN 0-471-47567-X   Copyright © 2005 by John Wiley & Sons, Inc.

a down-to-earth approach and just show how to do this. On the other hand, there are important underlying analytical principles worth noting. We also confront tensions between showing lots of analyses and explaining how to use these in management of technology ("MOT") decision support.

Here is our resolution. Sections 9.1 and 9.2 lay out the basics; they can have you going and doing tech mining in a few minutes. (Chapter 16 provides more example analyses on fuel cells to further "show you how.") Section 9.3 goes deeper to provide the conceptual framework that underlies these basic analyses. Section 9.4 develops this framework to show how you can investigate a bountiful set of relationships in S&T data (see especially Table 9-5). Section 9.5 then implements corresponding analyses, pointing toward MOT applications. Finally, Appendix C notes options if you do not have access to tech mining software, and Appendix D discusses several characteristically skewed distributional patterns seen in these bibliographic data.

In this prelude we first note the relationship between "basic" and more "advanced" forms of analysis. One way of considering basic analyses are "reports" on the data. Reports provide simple, pragmatic information drawn directly from the data. These may be counts of authors, institutions, or publications. Some reports "roll up" or aggregate data; others provide slices across time. An example of slices includes numbers of publications by year.

Examining and Assessing the Data

Getting right down to business, the first step in any analysis is to explore the data. Browse through the records obtained to get a feel for coverage and contents. In the Chapter Challenge, an analysis of Borlaug's work was requested, so the DeBrand Foundation selected and downloaded 173 SCI article abstracts citing his work. Figure 9-1 shows one abstract record to illustrate the nature of the content at hand.

Often, casual browsing will give ideas on expanding the search. In this case, the Foundation's interest is to identify promising approaches to remedy world hunger through improved agriculture. The article shown is a sociological analysis of technology, but the article set as a whole is highly varied. The download includes technical agronomy and biotechnology articles, as well as issue-oriented articles that address the status of genetically modified crops. You might augment this search via additional term-based queries to pursue certain topics suggested by these articles (Chapter 7 discusses database querying).

"Wallowing in the data" can also uncover data problems. By reading over abstracts you may spot items that do not belong. A brief review of 30 articles reveals only one apparent outlier: "Marine mammals in the next one hundred years: Twilight for a Pleistocene megafauna?" This appears to be about animal extinction—an important topic, no doubt, but not directly relevant to the DeBrand Foundation's hunger interests. Depending on the sensitivity of your tech mining objectives to the data, you may want to assess the quality of your search (see also Chapter 7). We illustrate an approach for data quality assessment.

```
FN ISI Export Format
VR 1.0
PT J
AU Kroma, MM
   Flora, CB
TI Greening pesticides: A historical analysis of the social
   construction of farm chemical advertisements
SO AGRICULTURE AND HUMAN VALUES
LA English
DT Article
DE agricultural media, environmental sustainability, ideology,
   pesticide advertisements, social risk
AB Ideology is maintained and driven by powerful symbols.
   Agricultural media such as farm magazines achieve this by
   appropriating societal values of currency
   them in imagery that accompany ad
   ~oducts, including pesti
```

Figure 9-1. Sample "Borlaug" article content

Remember the distinction between precision ("getting only the right articles") and recall ("getting all the pertinent articles") from Chapter 7. The sample with 29 of 30 articles deemed relevant estimates the precision of the query. The "binomial test of proportions" can help gauge overall search set precision (Chapter 7 discusses this). This sample with 1 "flawed" article among 30 (3%) suggests a true rate of error between 1% and 15%. In short—don't assume too much confidence in a small sample of articles!

Estimating recall is much tougher. How can analysts determine whether they have collected all the relevant articles? One step is to identify certain known entities such as a set of relevant articles or a few researchers whom you know author on the topic at hand. Then test whether a search that ought to capture these pieces does so. Another step is to request subject experts to review your search results to identify missing work.

Tech Mining Software Helps Clean and Filter Text Records

This book presumes use of tech mining software to facilitate analyses. Appendix B notes several software packages. Appendix C offers some suggestions on what you can do without such software. We use the package that we know best, *VantagePoint*, to illustrate capabilities. Other tech mining software will achieve most of the same ends, but not identically. Specialized tech mining software, like *VantagePoint**, offers many advantages. One, we can clean downloaded text datasets. Two, such software aids in discovering

---

*You may come across other versions of this software. *Derwent Analytics* is optimized for use with Delphion, Derwent, and Web of Knowledge (including Science Citation Index) data. *TechOASIS* is a version developed for U.S. Government use.

previously unrecognized relationships. Three, a range of flexible and sophisticated analyses are enabled. Fourth, repetitive analytical processes can be automated to a degree. This chapter and Chapter 10 set out how to achieve these advantages.

Tech mining software helps clean collections of text records in several ways. In the Chapter Challenge, we discovered at least one unrelated article in the collection by inspection. On the basis of sampling theory, we suspect that there are others we probably should find and remove. Using a software filter could eliminate these. Our filter might include the term "human impact" or "extinction" in the keyword or abstract fields. Conversely, *VantagePoint* can select subsets of the records for special analyses (e.g., to focus on those mentioning genetic modification).

Tech mining software is particularly strong at *combining multiple sources* of data and making a repository of data for continued analysis. This enables you to examine the data with a single, consistent interface regardless of where the records may have originated. For instance, if we performed another SCI search, we could combine results with the 173 Borlaug-citing abstracts. More challenging, we could search and retrieve from another database (e.g., MEDLINE, Agricola) and consolidate with the SCI records. To the extent the fields are the same, this kind of data combination is straightforward. If the fields are distinct, then some of the records would have blanks (e.g., the Agricola records would lack the "total citation" counts that SCI provides). It is trickier when both sources provide similar, but not identical data. One must then judge whether to combine these or not. For instance, "authors" may have different formats. Even trickier, "keywords" (or descriptors or subject index terms) take on many guises. SCI, for example, offers "Keywords" (supplied by authors), as well as "Keywords Plus" (words or phrases appearing frequently in the titles of an article's references).

MEDLINE, in contrast, offers its own hierarchical subject index terms called "MeSH" (medical subject headings). Depending on the tech mining purposes, you may want to mix various types of keywords or not. This is not an issue in our "exploratory" Borlaug analysis, but it may be critical in situations where precise nuances are important.

After fusion of data sets, the next step is to identify, and eliminate, duplicate documents. When performing tech mining analyses, it is useful to uniquely identify entities in the system. In Figure 9-2, we call attention to *documents* (or *records*) and to their information *fields* (or *terms*—e.g., authors, institutions, locations, or citations). Noise may be introduced because of any number of reasons—variations in typing, formatting, and spelling are most common. These may be introduced by the author, the journal editor, or the database company. Or, as we combine multiple queries, duplicate articles may be introduced into the set. Duplicates can produce false counts and give results that are unrepresentative of the underlying data. For these reasons it is important for tech mining tools to identify and remove duplicates.

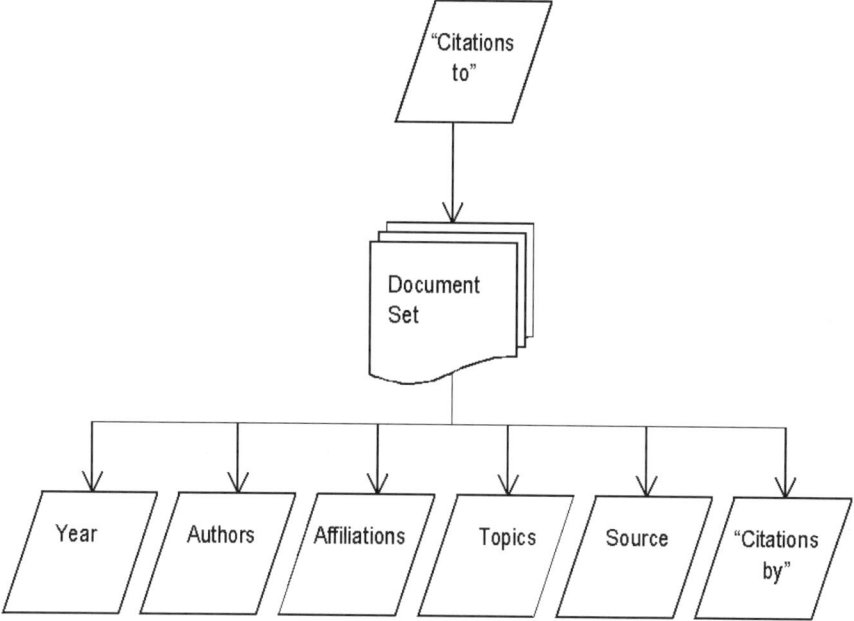

Figure 9-2. Document-term relations

Considering *documents*, we want to determine whether we have captured the same item (i.e., publication or patent abstract record) more than once. Particularly when fusing records from different sources, some of the fields may not be identical, even though they reflect the same document. In our fuel cells case (Chapter 16) we address issues in recognizing whether one record obtained from SCI is the same as another from INSPEC. Considering the content of the *fields*, we again face challenges in deciding what is really the same and what is different. We illustrate just ahead.

VantagePoint provides *fuzzy matching* capabilities to help clean. The analyst has a range of choices including:

- Remove exact duplicate records (documents).
- Match and remove those records with certain fields identical (e.g., 2 papers by the same author published in the same journal).
- Perform fuzzy "near matches" to various degrees on desired fields (e.g., special fuzzy match on author giving higher weight on probable last name matches than on first names and initials).
- Combine exact and fuzzy matching on multiple fields to maximize removal of true duplicates with minimal loss of actually distinct records.

Your sensitivity to error determines what strategy best suits a situation.

To exemplify some of these issues, let's check our 173 article abstracts. Two have the same title—"Feeding a world of 10 billion people: The miracle ahead." Borlaug authored both—one that appeared in *Biotechnology & Biotechnological Equipment* in 1997, and the other appearing in *In Vitro Cellular & Developmental Biology—Plant* in 2002. (Note that these Borlaug articles appear in our search set because they cite other of his work.) So, despite having the same title, these are two different articles. Our list of authors appears quite clean. *VantagePoint*'s "author clean-up" suggests combining only: "Pretty, JN" with "Pretty, J." This makes good sense. However, such consolidation would be riskier were this a more common name in a larger data set—for example, "Smith, JN" with "Smith, J."

Other cleaning tools include dictionaries of "stop words" (extremely common words such as "the"), techniques for discovering and counting phrases, and methods for identifying synonymous or otherwise variant words. *Thesauri* are vital. They collect variations on the same entity. Use of thesauri can consolidate alternative spellings (e.g., English and American variations) in a standardized fashion. You can build your own thesauri and keep enhancing these over repeated analyses in your area of interest. We have done this with American universities and companies, accruing variations (e.g., Massachusetts Institute of Technology and MIT). We also have been building a thesaurus that groups organizations as academic, industry, government, or other. It does so by accumulating both general cues (e.g., "univ" or "college" implies academic) and specifics (e.g., MIT is academic). Tech mining software facilitates repeated analyses by applying dictionaries and thesauri to consolidate terms that reflect the same entity.

The DeBrand Foundation's commitment to ongoing tech mining to support proposal review and other ends suggests that they develop specialized thesauri. These would attune to their continuing focus on medical and agricultural R&D. We might start by grouping variations on frequent keywords, institutions, and researcher-authors to initiate three thesauri. For instance, the author thesaurus would facilitate consolidation of forms of the names of active researchers and proposers in their R&D domain (e.g., the Mexican authors discussed in Section 9.3). In following years, this would continue to build upon the thesauri previously developed. So, over time, tech mining would get better and easier. We recommend that you strongly consider such a continuing "learning" strategy—develop thesauri! As tech mining matures, user groups might well share particular thesauri. To date, our VP experience is that customers performing competitive technological intelligence ("CTI") prefer not to share.

Other software algorithms (e.g., regular expressions) recognize strings of letters. This may help group terms. A simple example of this is grouping the plural and singular versions of a word for further analysis. Judicious combination of these capabilities can even approximate "understanding" the syntax of an article. Such syntactical analyses can help identify all the verbs, or proper nouns, in an article collection. *VantagePoint*, for example, routinely parses noun phrases from titles or abstracts for further analyses. This software combines various cleaning and data manipulation capabilities. Data manipulation may be used to piece together composite records from multiple sources, or remove or combine duplicate records. The software also expedites more advanced tasks such as exporting subsets of records of interest, defining criteria for matching records, and the creation of thesauri.

The tech mining analyst often confronts quite messy data. By rapidly analyzing data, and removing less significant items from consideration, tech mining software allows the analyst to drill down and discover the most significant articles in a collection. Automatic listing and tallying of the records comes easily for text mining software. List processing functions include sorting, finding, and grouping. These are conceptually simple but amazingly powerful in making sense of thousands of records.

Furthermore, analyses can be duplicated, repeated, and even "canned." If you want to know the yearly publication rates for some term for every nation in your sample, tech mining software can easily generate these. "Canning" of software manipulations is called scripting or writing macros. This can automate repetitive operations, providing a quantum leap in ease and speed of data processing. Additionally, automating cleaning manipulations can markedly increase tech mining accessibility to nonexpert users. However, this requires solid understanding of the data and the intended uses. An organization's information resources center may be able to help. If you do automate certain cleaning operations, be sure to spot-check cleaned against raw data to be sure it's doing what you want.

9.2. WHAT YOU CAN DO WITH THE DATA

Now we're ready to have fun! Let's discover "who's doing what" in the R&D domain we want to know about. Tony van Raan (1988), a world leader in bibliometrics, usefully distinguishes:

- 1st Level Analyses—lists
- 2nd Level Analyses—combining two lists to make a matrix

Lists help you comprehend the data set. For example, our 173 SCI records contain articles published in 104 different sources (mainly journals). We could list those 104 and sort alphabetically to see whether a particular journal appears. Or we might prefer a handier list of the "Top 10" most frequent authors in this collection. We could go a bit further to provide frequency information like the summary shown in Table 9-1.

VantagePoint provides a handy macro to create a custom profile. Table 9-2 presents a trivial author profile for these four. The software has queried the 173-record file for each author, retrieving the three most frequently occurring title phrases parsed via natural language processing (NLP). It has then added

TABLE 9-1. Top 4 Authors Citing Borlaug

Author	Articles
Pfeiffer	10
Mergoum	8
Rajaram	8
Borlaug	6

TABLE 9-2. Author Profile

Author Top 4	Title (NLP) (Phrases) Top Phrases	Publication Year Top Year
Pfeiffer, WH[10]	registration [9]; "Cerrillo-TCL99" triticale [1]; "Supremo TCL-2000" triticale [1]	2001 [7]
Mergoum, M[8]	registration [8]; "Cerrillo-TCL99" triticale [1]; "Supremo TCL-2000" triticale [1]	2001 [6]
Rajaram, S[8]	registration [6]; "Cerrillo-TCL99" triticale [1]; "Supremo TCL-2000" triticale [1]	2001 [7]
Borlaug, NE[6]	miracle ahead [2]; 10 billion people [2]; WORLD FOOD PRIZE SEEKS SUPPORT [1]	1992 [2]

EXHIBIT 9-1 *Author Co-Occurrence Matrix*

	Pfeiffer	Mergoum	Rajaram	Borlaug
Pfeiffer, WH	10	8	6	0
Mergoum, M	8	8	6	0
Rajaram, S	6	6	8	0
Borlaug, NE	0	0	0	6

another field, showing the one leading publication year for each title phrase. For Borlaug, the title phrases correspond to the duplicate title previously noted. For the other three authors, note that the title phrases are suspiciously similar. Indeed, Exhibit 9-1 confirms that they work closely together. For instance, six of Rajaram's eight articles are joint with Mergoum and Pfeiffer, all eight of Mergoum's articles were authored with Pfeiffer, and so on.

We could build up lots of examples, and we will do so throughout the chapters. For now, just note that we can easily add a third dimension to these analyses.

We can take any one field's information (e.g., year, topic, author) and focus on one or more values. We might want to create time slices, say, dividing the 173 articles into an early period (through 1994), a middle period (1995–1999), and a recent period (2000–2003). For each period we could combine any two variables. For instance, we might examine the leading Borlaug-citing authors for 2000–2003, and tally the topics they address. This is one way to address three entities at once—that is, 3-D analysis.

In one respect, this is it for basic analyses—you can go forth and explore your data. However, we find it helpful to step back at this point to reflect on the logic of inquiry into text records. Section 9.3 offers a conceptual framework to help you consider the possibilities more fully. Section 9.4 returns to practical matters to offer suggestions on types of basic analyses that may help answer particular technology management questions.

9.3. RELATIONS AMONG DOCUMENTS AND TERMS OCCURRING IN THEIR INFORMATION FIELDS

Document by Term Occurrence

Each document (record, publication) downloaded for analysis presents a wealth of potential information contained in fields such as Author, Title, and Date. At the most basic level, we have documents and terms that constitute a given field of information. For instance, one set of "terms" would be all the author names appearing in the document set.

The data set file could be represented as shown in Table 9-3. We show the presence of a term in a document by a "1," its absence by a "0." (For other purposes we might favor actual counts so that a term appearing three times in a document would have a "3.") In the Borlaug example, each document abstracts certain information from a scientific article. As per Figure 9-1, terms constitute the content of any one of several interesting fields—type of publication, author, abstract phrases, etc. In our analyses, "document" (which record) is one dimension. One information field is a second dimension. We almost always focus on one field at a time. So, "term" in a table like Table

TABLE 9-3. Document by Term Occurrence—a "Burt" Matrix

Document\Term	A	B	C	D	E	F	(. . .) m
1	1	1	0	1	0	1	
2	1	1	1	0	1	0	
3	0	0	0	0	1	1	
4	1	1	1	1	0	1	
5	0	0	1	0	0	1	
6	1	1	0	0	0	0	
(. . .) n							

9-3 might be each of the thousands of keywords appearing somewhere in the document set. Chapter 10 builds upon Table 9-3 to discover relationships in the data. We call these "Burt matrices" in honor of Ronald Burt, the pioneering sociologist who used such matrices in performing network analyses of social relationships (cf. Burt, 1983).

Document Fields ("Terms")

Let's consider interesting terms for these sorts of documents. These vary somewhat by type of document (e.g., publication vs. patent abstract) and by the content provided by the database (e.g., science vs. business). We won't attempt an exhaustive list, but some of the prominent S&T fields include:

- *Author(s)* or Inventor(s)
- *Citations* and references to prior researches or patents
- *Topics* or other measures of content—including keywords, title phrases or words, abstract phrases or words, classification codes, international patent classifications, database class codes
- *Institutions*, including but not limited to the affiliation of the author or inventor, patent assignee, or sponsor of the research
- *Location*, and in particular the geographic site where the research was conducted, such as the country, state or province, city
- *Source* for R&D, particularly the journal or conference where the research was presented
- *Year*, and in particular the date of publication or indexing into the database; other dates applicable for patenting include the date of invention, patent application, or patent issued

Note that some of these imply a single value for a given document (e.g., date of publication, source, location). Others can be multivalued (e.g., authors, citations, topics). We explore term-term relationships in the next subsection.

It is slightly esoteric, but important, to note that we are dealing with the information contained in the document (i.e., the patent or publication abstract record), as opposed to the underlying reality. Some of the records in our Borlaug set lack abstracts, for instance, so an "abstract phrases" field would be empty. Obviously, this does not mean the published article lacks content. Also, most S&T publication databases do not report certain information that could be of interest—especially citations (SCI is a notable exception) and sponsorship.

A particularly vexing lack is explicit information on *author affiliation* in S&T publication databases. Tech mining wants to know "who is doing what," and the institution (organization, company, university, agency, etc.) is vital. Beware the problems caused by this missing link. Most S&T databases provide the names of all the authors of a document, but only the institutional affiliation of the first author. SCI does better; it gives all the institutions of all the authors, but not unambiguously. The tech miner is usually left guessing on several important issues:

- Which institutions (companies, agencies, universities) collaborate with which others?
- Which institutions contributed most to the particular research?
- Which institution is a particular author associated with?

You can "play detective" to come up with reasonable estimates. For instance, suppose the DeBrand Foundation wants to locate Pfeiffer and ascertain whether the research team is still together (Exhibit 9-1). We can generate a matrix of author by institution to see which institutions are indicated on publications by Pfeiffer, Mergoum, and Rajaram. We find: UAAAN, Coahuila, Mexico; CIMMYT (different subunits), Mexico City; and UAEM, Toluca, Mexico. We check further on one of four papers on which Mergoum is first author to see that Mergoum is specifically associated with CIMMYT, International Maize & Wheat Improvement Center. But we don't find Pfeiffer as first author on any papers. We can then try a Google web search on Pfeiffer. This turns up an article in *Crop Science* with a full text link that explicitly indicates the same affiliation as Mergoum. Locating an individual's home page also helps confirm that this person has not since relocated. So, DeBrand might contact these CIMMYT researchers to pursue a line of inquiry. When tracking down corporate links, this sleuthing methodology becomes increasingly convoluted. Hicks and Katz (1996) pursued automated techniques for assigning institutional addresses for all of Britain's researchers!

We single out the field "Year" (usually the preferred tech mining simplification of "Date") for special treatment because of our interest in tracking developments over time (the focus of Chapter 11). Date is also a rare numeric variable in this primarily textual realm. Consider key combinations of Documents, Terms, and Year (Table 9-4). For example, consider "authors" as the term of interest at the moment. "Documents by Terms" would tell us how many of the documents each authored. "Documents by Years" ignores authorship; it just tells us how many papers were published each year. "Terms by Years" would show each author's publishing by year. "Term A by Term B," in

TABLE 9-4. Combinations of Document, Terms, and Year

Combination	Tech Mining Interest
Documents by Terms	Basic occurrence
Documents by Years	Overall R&D domain activity trend
Terms by Years	Trends in topical emphases
Term "A" by Term "B"	Co-occurrence (basis for term relationships)
Term "A" by Term "B" by Years	Trends in relationships

this case would show which authors have coauthored papers with which other authors. "Term A by Term B by Years" would let us see which coauthorships occurred in particular years (e.g., to see whether a research team is still actively collaborating).

Combining Documents and Terms

The following section will explore our primary interest—relating terms based on their patterns of occurrence across documents. Here, we step back to note that the basic data (Table 9-3) can be analyzed several ways. Consider our main tech mining example on fuel cell publications (Chapter 16). The "n" of documents is 11,764. In a typical principal components analysis (PCA), we might cluster the top 200 or so keywords. (Chapter 10 gets into PCA and other advanced analyses.) So, our Burt matrix (Table 9-3) would contain 200 columns and 11,764 rows—a sizable matrix. To map *term-by-term relationships*, we create a new type of "co-occurrence" matrix. It consists of the 200 keywords by the same 200 keywords. Cells contain counts of how many times keyword A and keyword B occur together in any documents. Values could thus range from 0 to 11,764. PCA actually draws upon a variant of this, in which the matrix cells contain a similarity measure, such as correlation or cosine between the respective terms. Note that computations on a 200×200 matrix are easier than on a $200 \times 11,764$ matrix.

Conversely, we could examine *relationships among the documents* based on the terms they share. This can be derived from a document co-occurrence matrix. It is big—11,764 by 11,764. It yields document maps such as the famous "Mt. OJ" showing the "mountain" of concentration of U.S. news articles on the OJ Simpson trial compared with smaller peaks for other topics during that time period. This approach is especially good at "query by example." If a user identifies a document of special interest and wants to know about similar R&D activity, such landscape maps indicate similar documents. You may have experienced this, say, at the Amazon website. You order a book and they suggest similar books you might like (based on relevance determined by shared content or by what other purchasers of your book also buy).

The close relationships of these three alternative data representations are notable. Analyses of term relationships based on the "documents by terms" matrix (Burt matrix like Table 9-3) should be essentially equivalent to analyses based on the "term by term" matrix (self-join). It is true, however, that the data about the documents has been lost in this representation. Similarly, analyses of document-to-document relationships based on the "documents by terms" matrix (Burt matrix) should be essentially equivalent to those based on the "documents by documents" matrix. Technically speaking, this concept of structure being unaltered by self-joins is known as "idempotency." Comparing analytical results for idempotent matrices provides a test of reasonableness. We pick up on this in Chapter 10.

9.4. RELATIONSHIPS

The essential tech mining payoff is to ascertain relationships from the S&T data. Section 9.2 dove right in to do so directly. Section 9.3 has laid out a framework to help you consider which relationships you wish to pursue, in what ways. We continue by delineating some direct term-by-term associations of interest, and then some extended ones.

We distinguish the following essential terms: authors, citations, topics, institution, location, source, and year. Recall that these "terms" are the content found in the information fields that appear in our documents (Figs. 9-1 and 9-2). Table 9-5 refines these slightly and suggests some of the interesting intersections that can be pursued among types of terms.

The relationships posed are neither definitive nor exclusive, merely suggestive. In brief, cascading down the columns of Table 9-5 from left to right:

- Teaming—Coauthorship patterns can help ascertain development teams and core individuals (valuable in discerning competitor "knowledge networks").
- Research Community—Profiling the collective body of knowledge *cited by* our document set, in conjunction with author (researcher) publication activity, can help figure out the extent of the community and the prominence of certain individuals therein (useful in identifying core expertise).
- Co-citation Analysis—Henry Small and colleagues at ISI have generated a stream of compelling depictions of research fronts based on which documents (with their attendant term characteristics—e.g., topics) are cited together by others.
- Esteem—Citation to our work (e.g., as authors, institutions, or countries) is one metric of value imputed to our research.
- Knowledge Transfer—Citations to a body of research can be profiled to see what fields are picking up on the results or using the methods.
- Expertise—Which authors (or institutions or countries or other units) address which topics?
- Cluster Analysis—Statistical procedures can help identify which topics tend to co-occur, implying relationship.
- Self-reliance—The extent to which actors cite their own work (especially patents, but also articles) is important in determining interdependencies (especially important for intellectual property ("IP")).
- Collaboration—the extent to which institutions (or countries) work together.
- Core Sources—which journals or conferences are most cited by our documents overall, implying heavy reliance on them; this can be refined to identify core sources for particular topics.

TABLE 9-5. Term by Term Relationships

	Authors	Citations by Our Documents	Citations to Our Documents	Topics (e.g., Keywords)	Institution	Location	Source	Year
Authors	Teaming	Research community	Esteem	Expertise	Expertise	Expertise		Currency
Citations By		Cocitation analysis (research fronts)			Self-reliance	Self-reliance	Core sources	Research pace
Citations To			Knowledge transfer		Esteem	Esteem	Impact factor	Research pace
Topics (e.g., Keywords)				Cluster analysis (research thrusts)	Expertise	Expertise	Core sources	Currency
Institution					Collaboration			Engagement
Location						Collaboration		Engagement
Source								Currency

- Impact Factor—which journals are most cited, implying that their papers exert greatest influence.
- Currency—What's hot? Inspecting trends shows temporal activity patterns.
- Research Pace—Noting what proportion of citations, by and to, are recent tells about how fast the R&D domain is moving (this holds for patents and for publications).
- Engagement—We often want to know whether an organization or country is still actively involved with this R&D domain.

This list is somewhat messy as it mixes outputs and analyses. We have tried to label the entries by the most-used notion. Note that these entries relate to one another. You might think of the relations as the verbs in the system. Some of the underlying actions are: research, publish, and cite.

These data are rich! Many S&T development issues are reflected in these outputs in ways that can be usefully measured. Which measures to use depends on the questions being addressed. We suspect that tech mining veterans will find some of Table 9-5 familiar and some strange. This reflects differences in emphases. In particular, one cadre focuses largely on CTI for companies. They key on patenting by key competitor companies. Another cadre concentrates on R&D policy, often at the country level. They lean toward national or global publication measures. This section aims to enrich everyone's considerations of possible measures. Later in the book we will go further to generate "innovation indicators" from the direct measures.

So, tech mining data allow for the investigation of many two-way relationships. Examples include, for a given search set, combinations of "terms" from particular fields (1, 2):

- Who are all the (1) researchers active at (2) Georgia Tech (publishing or patenting)?
- In what (1) years can you find (how much) (2) information on a specific subtopic (say, keywords)?
- How many (1) articles can you find in that (2) journal?

Three-way (and higher) relationships within this search set are also possible. These can concern terms from the same or different fields, as well as across documents (1, 2, 3). We might ask:

- Over what (1) time period did those (2, 3) two authors collaborate?
- To what degree (1—how many records) did each of the (2) leading institutions publish on (3) certain subtopics?
- Are (1) citations to (2—a country) Chinese research increasing or decreasing over (3) time?

Less direct, term-to-term relationships may also prove valuable. For instance, which authors share interests? In contrast to teaming (a socio-logical measure—coauthoring), we can discern which authors address related topics (i.e., they use closely related words or phrases), whether or not they actually collaborate. This can suggest potential value in sharing knowledge.

Tech mining is especially valuable in identifying potential common interests among researchers who do not presently interact—that is, across disciplines. This is similar to how the military research intelligence user Edison House works. Their mission is to locate European researchers who work on topics related to American Army, Air Force, or Navy research interests and act as "matchmakers," bringing them together. Tech mining application helps Edison House identify key researchers to look for at particular conferences or prominent research labs to target.

We offer one more illustration of less direct relationships. We have information on the "core sources" for a given document set; we also have authors with the most publications therein. We could now ask which authors published most in the highly cited journals. Also, we might do new searches in SCI to determine which of our authors have been most highly cited by others. We could compile these multiple measures to help identify prominent figures in the field under study.

Examining these three "indirect" relationships we see:

- By comparing (1) content across (2) documents, what do we learn about shared interests of (3) researchers?
- Using (1) citations to (2) journals to establish prestigious journals and (3) publication counts by (4) authors to establish the major research contributors, we can now ask, Which authors contribute to the most prestigious sources?

One way of thinking about the wealth of data available for analysis is to create a diagram of entities and relations (Fig. 9-2). Figures like Figure 9-2 are known in the database world as "entity-relationship diagrams," often depicted with detailed formalisms. That is not our intent here; rather, it is to visualize our main tech mining information resources.

In Figure 9-2, the terms are shown as boxes. Excepting "citations to," all the terms derive from within the document set. That is, what we know about authors, topics, etc., comes from the downloaded set of records (usually patent and/or publication abstracts). In this book, we will not treat "citations to" our document set heavily. These potentially useful queries are difficult or impossible to make with many database portals available today.

Most of our tech mining analyses thus derive solely from information resident in the documents. We set "Year" slightly aside because, as already discussed, we often want to treat this information specially. Note that relations between terms link through the documents. In most cases this is unambiguous, but recall that the link between authors and affiliations is usually somewhat ambiguous.

We've seen entities—the people, institution, and topics—that can be investigated using tech mining. We've seen relations—ideas such as collaboration, sponsorship, or research—that can be investigated using tech mining. Now, let's look in a little more detail at how to conduct analyses using entities and relations.

9.5. HELPFUL BASIC ANALYSES

Analytical Perspectives

Figure 2-1 loosely associated the "4 Ps" that tech mining can provide—product, process, predict, and prescribe—to types of technology analyses. Figure 9-3 does the same for some of the terms (field information) we have available in our document set. Again, the ties are weak, merely hinting at uses you may find for these information resources. Tech mining *products* make use of all the information, but especially topical coverage (content). We often see author, source, and citation information as helpful in stimulating discussion of ways to address the R&D domain in question. In contrast, *process*-oriented studies may be more interested in understanding and characterizing the research actors in the system—the collectives of researchers represented by

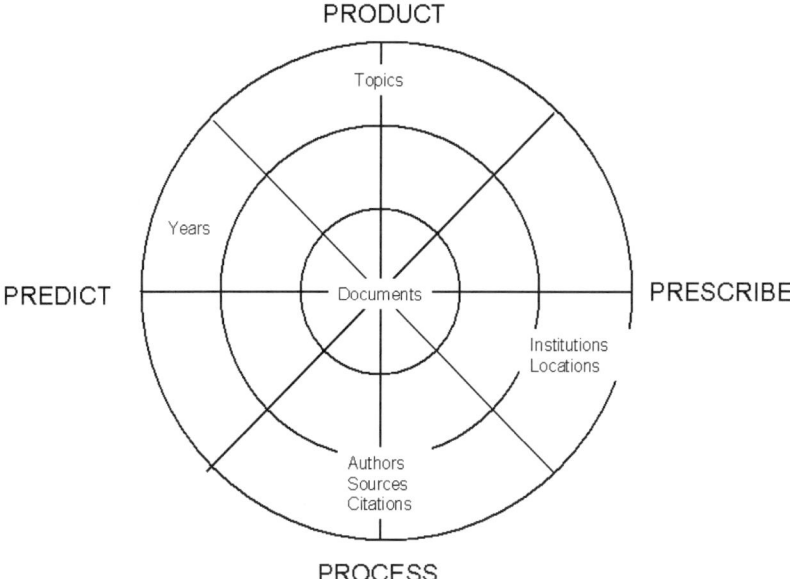

Figure 9-3. Relating the 4 Ps of tech mining to the terms

journals and by institutions. Most obviously, we associate years (dates) with *prediction*; Chapter 11 investigates use of time series data such as these to project into the future. Institution and location information are salient to benchmarking efforts that often lead toward recommendations (*prescription*).

Let's consider ways to associate and represent the data. Our starting point is the "list." Tech mining software readily combines list information, a process sometimes known as "joining" tables. One issue is how to handle long lists and, worse, their combination into matrices. Don't allow yourself to be overwhelmed by the volumes of detail! Our modest sample of 173 Borlaug-citing articles yields some 6700 cited publications. There are over 3200 journals cited by or publishing our 173 articles. This suggests a need to determine "cutoffs," for analyses and for presentation of findings. How many journals do we really need to know about? We might choose the "Top 20" journals (in number of our 173 articles published) and neglect the rest. Alternatively, we could set a level of significance—list all journals that have at least 10 citations from this collection of articles.

Be guided by what your target users need to know to make specific decisions. For instance, you may focus on a subset of the documents, setting aside the rest. Or you may key on certain key players, ignoring the others. Chapter 13 offers a decision-support framework to guide tech mining efforts toward answering MOT questions.

Crossing Lists (Fields) with Themselves

An interesting class of relational analyses occurs when lists are combined with themselves. (Technically this is known as a "self-join.") These forms of analyses are very versatile. Table 9-6 distinguishes four key analyses. All of them rely on forms of co-occurrence. That is, if certain terms occur together in multiple documents more often than expected, we have a basis of possibly significant relationship.

Co-citation analyses are created by examining the number of times different publications are referenced together in the same documents. This count is aggregated across the entire document collection. The resulting table implies relationship between documents that multiple authors often reference together. We can look further at the documents that cite certain sets of references together. For a given pair of references, we can examine terms (e.g.,

TABLE 9-6. Relational Analyses Based on Self-Joins

To Investigate	Combine Information from
Co-Citations	Citations with Citations
Co-Word	Topics with Topics
Authorship Networks	Authors with Authors
Institutional Networks	Institutions with Institutions

topics, institutions) of the documents that cite them. This can provide insight into how that pair of papers (or patents) are related. Vice versa, we can consider those of our documents that cite this pair of references as having some common interests. Co-citation analyses have proven fruitful at identifying research fronts.

Co-word analyses examine shared similarity in patterns of words or keywords rather than in patterns of citation. Here again, co-word analyses come in two flavors: We can understand relationships between words by joining via the document key, or we can understand relationships among documents by joining via the content key. Much of Chapter 10 builds on co-word analyses to ascertain and map term relationships.

Authorship networks have received significant attention lately because of the "small world phenomenon." Even large networks of scientists and engineers are bridged with relatively few mutual acquaintances (see Sidebar on Erdos Numbers). Effective communication and collaboration in such networks can be credited to the existence of a few "hubs" that channel information and bridge acquaintances. Such "hubs" become valuable partners because of their experience and wealth of contacts—ensuring that they become even more central to the network of collaboration. Centrality measures of networks attempt to rate the importance of hubs in a network. Such indicators can be created with "self-joins" on the author table.

Another intriguing measure is that of "betweenness." An individual scientist or engineer may not have an exhaustive list of professional contacts—but he or she may have precisely the right contacts to bridge gaps. Bridges are thereby built between institutions and between scientific and technical disciplines. Betweenness indicators to measure such bridges begin with self-joins.

Sidebar: Erdos Numbers

A vivid example of this kind of authorship network analysis occurs in the graph theory community. Graph theory is a field of applied mathematics that studies networked relationships of all sorts in the physical, biological, and social worlds. Not too surprisingly, graph theorists apply their analytic tools to their own discipline. A long-time "hub" of the graph theory network was the scientist Paul Erdos. Those scientists who collaborated directly with Erdos have an "Erdos number" of 1. Those who collaborate with the collaborators have Erdos numbers of 2. And so on, with each scientist receiving a number reflecting the number of steps to link himself or herself to the central figure of Paul Erdos. (Erdos himself had the number 0.) Graph theorists calculate their Erdos numbers for fun, or for prestige. Tech mining allows the calculation of indirect links—across a variety of disciplines—for specific applied purposes.

Institutional networks are not unlike authorship networks. Institutional networks aggregate publication or patenting by their staff. This information can inform collaboration, competitive intelligence, and funding determinations. Collaborations among industry, academia, government, and nonprofit sectors are becoming increasingly important to the conduct of science and technology. International collaborations are also growing in importance. The number and strength of such links can be studied through "self-joins" on the institution table.

Bibliographic Data Distributions

Our data are not very well behaved. Because we deal mainly with text data, our data sets often require different statistical approaches. Properties of text data deviate from standard statistical test assumptions in some regards. Chapter 10 explores discrete vs. continuous distribution attributes. More critically, the distributions of most text data are far from normal. To illustrate, our 173 record set has some 311 authors; of those, only 7 have more than 3 publications. We need to be aware of these highly skewed distributions. In one regard, this is welcome news, in that focusing on the "top few" authors, keywords, etc., may tell us all we need to know.

Appendix D provides background on the properties of the most important bibliographic distributions. The main thing that tech miners need to know from these empirical distributions is that the leading sources of information—whether they be organizations, authors, keywords, or citations—are concentrated unevenly. For instance, an information professional may apply a Pareto principle to decide which sources to acquire. The hope is that the top 20% of bibliographic sources will provide 80% of the pertinent data needed for analysis. Or, a tech mining analyst can feel reasonably confident in cutting off the number of terms in indexing documents or selecting keywords. Furthermore, the generalization is that to obtain the next X% of the content will increase the workload disproportionately. To calculate specific distributional values, refer to the respective equations in Appendix D.

Answering Technology Management Questions Through Basic Analyses

We assert that tech mining must be guided by the specific user questions to be answered. There is a dangerous tendency as analysts to generate information just because we can. Glance back at Table 9-5 to realize that one could go wild generating lots of "interesting" stuff that would befuddle, more than inform, a busy decision-maker. That said, what to do? Chapter 13 arrays 13 technology management issues that give rise to 39 questions, for which we nominate a couple hundred candidate tech mining "innovation indicators" to help answer. Table 9-7 provides a "mini" version of this structuring to get us thinking about how these various tech mining analyses can be directed to useful

TABLE 9-7. Tech Mining Issues, Questions, and Data-Based Responses

Tech Mining Issue	Sample Question	Possible Responses
Project initiation	How do we stack up against others researching or developing this technology?	Profile the leading institutions in this field in terms of numbers of publications, patents, and researchers. For each, indicate the main topics it emphasizes.
Competitive technological intelligence	What is the development trajectory?	Plot trends in publishing and patenting activity for major topic clusters. Fit S-curve growth models (if appropriate) and project ahead.
Mergers & acquisitions	Which possess appealing intellectual property for this technology?	Tally the patent activity of leading small companies and key individuals. Characterize in terms of topical emphases.

ends. It suggests a way for you to think about your data in terms of what your user needs to know.

To get started, we advise performing *requirements analysis*. Interact with the target users to identify what they want to know. Chapter 14 offers suggestions on nurturing such interactions throughout the course of the tech mining activity.

We would not advocate diving in and generating every possible table and relationship. That said, we can point out certain analytical elements likely to have extensive uses:

- Activity—number of documents (publications, patents) by year
- Activity breakouts—number of documents on particular topics, or by particular institutions, by year
- Institutions—leading organizational contributors to the target R&D area, with breakouts for each ("profiles" on what each of them emphasizes, their leading researchers, how much recent activity, etc.)
- Institution of particular interest—breakouts by particular topics (spotlight the leading contributors on given topics)
- Researchers—identify the leading individual contributors to the target R&D area, with breakouts for those leading individuals (e.g., topics each emphasizes)
- Researchers—breakouts by particular topics to spotlight leaders in various subareas
- Topics—tabulate the leading topics (e.g., by keywords or other content indicators)

The next subsection illustrates a sample tabulation.

TABLE 9-8. Borlaug-Citing Sample Tabulation

Entity	Description	Distinct Instances	Most Frequent	Example
Documents	Publication abstracts	173	"2" (1 duplicate)	"Greening pesticides: A historical analysis of the social construction of farm chemical advertisements"
Year	Publication date	16 years	23	2001
Topics	Abstract phrases	4849	344	Crop, crops
Authors	Authors	311	10	W. H. Pfeiffer
Institutions	Organizations	108	25	CIMMYT
	Nations	25	87	United States
Sources	Journals	104	15	*Crop Science*
Citations	References	6741	17	Jensen, N. F. (1952), *Journal of Agronomy*, Vol. 44.
	Cited journals	3229	181	*Science*
	Cited authors	4642	280	N. E. Borlaug

Sample Tallies for the 173 Borlaug-Citing Publication Records

Returning to the DeBrand inquiry, Table 9-8 shows overall totals and example instances. These are the basic documents and terms available.

Table 9-8 shows that there were 173 publication abstract records in the sample, with 1 duplicate. These were published in 16 different years, with the most (23 articles) in 2001. Note the tremendous number of distinct phrases found in the 173 records. The articles average about two authors each. As noted earlier, Pfeiffer authored the most (10 papers). The most prolific organization is Pfeiffer's CIMMYT in Mexico. The authors were located in 25 different countries, with the greatest number of affiliations from the U.S. (87). The 173 papers appeared in 104 journals, led by *Crop Science* with 10. Our sample's papers cited 6741 other papers—an average of 39 apiece. Those cited papers had a total of 4642 authors and appeared in 3229 journals. The sample cites a wide variety of work.

Figure 9-4 presents Bradford distributions of the publishing journals (for the 173 articles) in the Borlaug sample and for the 3229 journals cited. Bradford distributions show the concentration of research information across journals (see Appendix D). The graph uses log scaling on both axes and "logarithmic binning" to handle ties in rank. The Bradford distribution provides a fair description of these data. The intercept of the cited journals (the c in Equation D.2) is 627. The intercept of the journals that published our articles is

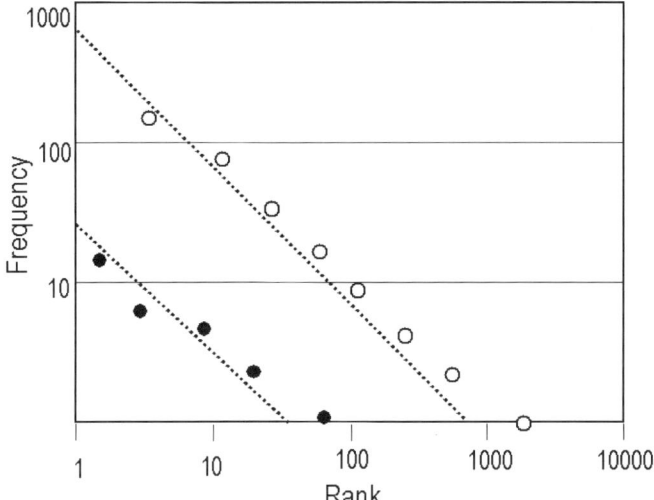

Figure 9-4. Bradford distribution of journals in agronomy sample

23. In other words, the cited journals are 27 times more frequent than the publishing journals. Table 9-9 shows the two sets of journals.

Suppose we seek to combine the two lists to make a single list of leading journals for DeBrand Foundation monitoring purposes. The cited journals include the prestigious general science publications, *Science* and *Nature*, in addition to subject-specific journals. The journals that published the sample match, but not perfectly, the citing journals. Using what we have learned about the respective distributions, we create a function to combine the information provided by the two lists. A weight of 27 for each published article vs. 1 for each citation gives roughly equal weight to the two sources of information. The journal *Crop Science* is at the top of the combined list, with significant contributions from both the cited and publishing list. *Euphytica* and *Phytopathology* are journals number two and three, respectively, on the list. The phenomenon of "scattering" embodied in the Bradford law lets us know that the relative productivity of sources will diminish rapidly. The DeBrand Foundation now has a list that they may use to monitor the literature on agronomy and human nutritional welfare.

Further Ideas

Ways to apply such basic analyses and extend them are myriad; we offer a few ideas. The line between basic and advanced analyses is not sharp. As set forth

TABLE 9-9. Top Journals in the Borlaug Sample of Agronomy

Rank	Journal Name	Weighted Score	Number of Citations	Number of Publications
1	*Crop Science*	560	155	15
2	*Euphytica*	425	74	13
3	*Phytopathology*	278	116	6
4	*Plant Physiology*	193	112	3
5	*Science*	181	181	0
6	*Plant Pathology*	165	30	5
7	*Plant Cell Tissue and Organ Culture*	135	0	5
8	*Theoretical and Applied Genetics*	134	134	0
9	*Journal of Agronomy and Crop Science*	119	65	2
12	*Plant and Soil*	108	0	4
12	*In Vitro Cellular & Developmental Biology— Plant*	108	0	4
12	*Field Crops Research*	108	0	4
12	*Current Science*	108	0	4
12	*Nature*	108	108	0

here, we call lists and matrices "basic," even though these may be multitiered and conceptually intricate.

You can look at the core set of research cited to determine how recent the referenced literature or patents is. You could probe how the researchers or institutions collaborate. You may also want to examine the leading journals and conferences to orient your monitoring activities (Table 9-9). Examination of topical terms (e.g., keywords, title phrases) could help you investigate how concentrated the research activity is, on few or many subtopics. In instigating a doctoral dissertation, for instance, this could help locate your activity with respect to what others are doing, see how much research has been done relating to candidate topics, and check that your topic isn't outdated.

As mentioned earlier, temporal relationships hold special interest. You might examine how publication or patent rates and content types are changing over time. Chapter 13 discusses advanced measures, innovation indicators—such as changes in keyword richness—that can indicate technological maturation.

The basic analyses contribute to the various types of technology analyses (Table 2-2). We mention a few particulars. Technology road maps involve designing feasible technology futures; this demands understanding of the developmental progress of a target technology along with its important constituent technologies and the systems into which it would be applied. Tech mining on these elements is vital. This entails tracking topical emphases over time by major contributors (especially one's own organization and its competitors).

CTI keys on "who's doing what." Identification of the key actors in the technology development system (see Chapter Resources) and determination of how they interact is essential knowledge. Analysis of R&D publication and patenting can be complemented by examination of accessible information on R&D inputs and by expert opinion.

Technology "knowledge management" involves evaluating technology competencies inside and outside an organization. It also concerns retaining and using this knowledge well. It may also involve locating compatible capabilities and IP. Collaboration, licensing, or other forms of closer integration then become possible.

Process management involves getting individuals together to discuss prospective solutions to problems. In national or regional technology foresight, tech mining can help by identifying active researchers. Internal to a large organization (e.g., a multinational corporation), self-profiling of related activity can help extend conventional wisdom on who to involve. Institutional and individual "knowledge networks" revealed through tech mining efforts can be further explored with techniques such as "actor analysis" (see Chapter 14).

Authors, institutions, and nations are the primary targets for creating S&T indicators. Aggregate publication, patenting, and citation measures at the suitable level contribute. Because indicators gain value as time series, it becomes important to generate repeatable measures.

Tech mining analyses can enable recognition of new topics through scanning of content together with time markers. The momentum behind new technology thrusts can be examined by monitoring the numbers of emerging publications in a given area and changes in content emphases.

CHAPTER 9 TAKE-HOME MESSAGES

We have come to the conclusion of this complicated chapter on basic analyses. The chapter is about cleaning the data and describing "what's happening" through various basic analyses. We have provided hard-earned advice as to "what works." But we've also offered an underlying conceptual framework (the Burt matrix, entity relationship diagrams and analyses, and prominent bibliographic distributions). From these, we've tried to stimulate your thinking about many possible analyses of S&T data (centered on Table 9-5).

We emphasize that querying a collection of articles or patents should help answer questions about S&T activity patterns, with implications for technological innovation prospects. Chapters 12 (on patents) and 13 (publications and patents) elaborate on how to bring tech mining analyses to bear as MOT decision support aids.

But, as we have seen in other chapters, analyses reach beyond merely querying data. Chapters 10 and 11 show how predictive models can help generalize tech mining results beyond a collection of articles to understand technology development prospects.

The primary messages of this chapter are:

- The first step of any analysis should be to understand user requirements.
- Next, "wallow in" your data to get a personal feel for the R&D activities involved.
- You can accomplish some basic bibliometrics without tech mining software.
- Clean your data. Tech mining software can help significantly in cleaning and filtering data to enable valid analyses. You can merge searches, remove duplicate records, and consolidate term entities (e.g., variations on the same author's name).
- Build thesauri to expedite consolidation of terms that vary and occur repeatedly. Keep augmenting these, and your tech mining operations will keep getting better and easier.
- Tech mining analyses are based on the Burt matrix. This arrays your search set on two dimensions:
  - documents (records)
  - terms (the data fields represented on each record)
- Relationships are discerned based on co-occurrences of terms across documents, or on document similarities on certain sets of terms (c.g., keywords).
- Begin with Level 1 analyses—lists of selected entities (terms) occurrences over the search set documents (e.g., "Top 10" keywords list).
- Move on to Level 2 analyses—crossing two lists together. The resulting tables and matrices break out who's doing what in many helpful ways (see Table 9-5 for ideas).
- Higher-level analyses can be made by combining three or more sets of terms (e.g., indicate which topics the leading authors emphasized over time).
- Most tech mining variables show distinctively skewed distributions; most activity concentrates heavily in the leaders (see Appendix D).
- Tech mining analyses variously serve product or process, and prediction or prescription, motives.
- Don't generate lots of tech mining analyses for their own sake; be sure to start with the MOT questions to be answered and generate only those analyses that help answer them.

Chapter **10**

Advanced Analyses

Chapter 9 discussed basic analyses, those deriving directly from the data. This chapter discusses advanced analyses, those that go beyond direct measurements. These address unobserved variables and model data behavior. They yield such useful results as depictions of relationships, causal models, and predictions. The chapter offers key concepts to guide tech mining practice.

10.1. WHY PERFORM ADVANCED ANALYSES?

Advanced analyses extend beyond basic, descriptive analyses to infer *relationships* in the data. We go further than we can strictly observe or count in the data (e.g., a list of the Top 10 inventors in a field), to patterns that are often hidden from the casual eye (e.g., the inference that company C is patenting to block others' entry into a field). Advanced analyses help check that the patterns and relationships generalize to other situations. A wise choice of model, with judicious empirical support, can lead to understanding relationships to draw inferences for technology management action.

At the heart of advanced analyses is a process of summarizing, or *modeling*, the data to find salient patterns. By definition, modeling deemphasizes noise or inconsequential detail, revealing more robust (repeatable) patterns in science and technology activity. For this reason modeling helps generalize. It helps determine patterns that are likely to extend into the future and across

Tech Mining: Exploiting New Technologies for Competitive Advantage, Edited by Alan L. Porter and Scott W. Cunningham.
ISBN 0-471-47567-X   Copyright © 2005 by John Wiley & Sons, Inc.

changes in particulars of the R&D endeavor. Modeling also leads us to examine "hidden" variables in the data, a topic that we pursue in the next subsection.

Modeling begins with an appreciation of pattern. What patterns do we anticipate finding in the data? What things are likely to be repeated across time, place, and content? Tech mining approaches can validate hypotheses—interpretations about technological innovation that we choose knowingly, weighing their strengths and weaknesses. We discuss the role of representation in the material that follows. Representation is, at its heart, a structure or pattern of data that we hope will be confirmed by analyses.

To set expectations, this chapter aims to provide a conceptual foundation for you to use in thinking about analyses that go beyond straightforward compilation and presentation of the data. We believe that this is the boundary between basic and advanced analyses, but we recognize that the distinction blurs in practice. This chapter is largely conceptual, although it concludes with an example. We view this chapter as a "basic treatment of advanced analyses." The present scope does not allow full development of the particulars of the many statistical and modeling approaches we note. For those just wishing to get to the main types of relationship-finding analyses, Section 10.3 is the place.

Hidden Variables or Constructs

Why should we believe in "hidden variables" in the data—things that we cannot count or measure directly? The truth of the matter is that we believe only with reluctance. Summary statistics (which are not based on unobserved constructs) do help us comprehend the underlying structure of noisy data. However, belief in a few well-chosen constructs (variables not directly measurable) can enhance our comprehension of underlying S&T behavior and interpretation of the observed data. This section points to six key constructs. These can help us understand R&D activity and help draw inferences on motives and future extensions.

What we would really like is strong predictive theory covering research, development, design, implementation, on through effective commercialization. We have considerable "technological innovation process" experiential reports and some experimentation on which to draw for our conceptualizations. And we do so—for instance in the "innovation indicators" mentioned throughout this book. Postulating "hidden variables" is a lot like having a theory about what is important in the data. But unfortunately, we don't have exact theory of how S&T knowledge reflects in measured outputs like scientific papers and patents. Given the lack of a strong guiding theory, the smart tech mining analyst remains somewhat skeptical about empirical inferences.

What do we mean by "hidden variables," more formally known as constructs? Recall that Chapter 9 drew distinctions among documents, terms, and years as the essential tech mining data. Others would modify these somewhat. For instance, NEC Research Center (producers of Citeseer, also known as

ResearchIndex) distinguishes documents, terms (content), and actors (authors, etc.). In the patent domain, one could refine the emphases somewhat differently (see Chapter 12). Anyway, we consider the primary distinction to lie between documents (records) and terms (the content of particular fields contained in the documents—keywords, authors, etc.—see Chapter 9). Hidden variables are conceptual variables that help explain these observed data, especially the terms and their relationships. From another domain, imagine that we are staring at mounds of data on worker tasks, behavior, and performance. Without psychological constructs such as motivation, incentives, and learning, we would be hard-pressed to make sense of the worker data. Analogously, we look to tech mining constructs to help account for patterns in our documents, terms, and years.

Table 10-1 lists six key constructs not directly measurable from the data. It notes observable variables (measures) that pertain and describes the logic of the underlying constructs. The tenor of the constructs shown leans toward "sociology and philosophy of science." We find that these notions help understand the data and related observations of tech mining. For instance, when we tally terms (e.g., keywords) for a body of work, we are really looking at clues to the generation and building of S&T knowledge. You can certainly devise additional constructs to be explored in tech mining.

Hypotheses About the Data

The six constructs invite consideration of underlying S&T mechanisms. The directions of such inquiry depend on the tech mining motivation. One can pursue many possibilities, from surmising national scientific program intents to competitive technological intelligence ("CTI") probes into why a competitor is patenting certain technologies. We offer a few illustrations to stimulate your thinking on how these constructs (Table 10-1) could serve your interests.

We might pursue *prestige* considerations to understand what makes a particular R&D institution so effective. *Life cycle* notions can help us interpret trends in publication and patent activity. *Invisible college* aspects are more powerful in explaining scientist than engineer behaviors. Identification of such research communities ("small worlds") can help us target monitoring efforts. We may also want to become members of these communities ourselves to ensure access and influence. *Learning* models can help get at what main themes and capabilities are being combined by a given R&D institution. This might suggest we attend to certain of these ourselves. Tracking work (publications and patents) over time could also help predict likely further (future) specializations of the institution under examination. *Structure of knowledge* information could point us toward seminal ideas that could indicate breakthrough technologies. *Knowledge production* concerns some of the institutional factors that lead research institutes to develop research programs, and to grow and maintain these programs over time. This may help us understand institutional thrusts and how they are nurtured.

TABLE 10-1. Key Tech Mining Constructs

Construct	Empirical Measures	Description
Prestige	Citations	Esteem shown by the pertinent professional community—e.g., journals cited heavily by other prestigious journals; authors heavily cited by their peers.
Life cycle of ideas	Publication or patenting trends; citation trends	Attention to particular issues rises and falls (life cycle). New tools or new discoveries may trigger an "avalanche" of R&D in certain fields or initiate new fields. We may even track the growth and spread of ideas with "epidemiology" models.
Invisible colleges	Topical content emphases; methods emphases; coauthorship & citation patterns	Schools of thought develop around differences of theory or method. Content analyses can discern use of key concepts. Shared ideas also reflect in human networks. Certain authors become central to the community of interest, as reflected in coauthorships and in citation to their work as a hub of knowledge.
Learning	Topical or methods emphases	Knowledge increments over time. Learning may involve incorporation of new theories or concepts, methods, and/or empirical results into a field of inquiry. Cumulation of new ideas may reflect through the content of publications or patents of an R&D domain.
Structure of knowledge	Topical or methods emphases (terms used); popularity of journals or conferences	Technical knowledge reflects structures (approaches, concepts, ideas, methods) associated with particular disciplines, fields, or schools of thought (see invisible colleges). The respective technical vocabularies often differ and are reflected in the terms used in publications. Prevalence of a field's papers in particular journals or conferences can indicate dominant knowledge structures.
Knowledge production	Publications & patents; author & inventor affiliations over time	Institutions maintain a "research memory." Continuity in a research organization's emphases over time derives from selective hiring, special training, and collaboration. Internal and external networking (funding continuity, enduring collaborations) reinforces ongoing research emphases and fosters productivity.

The "Monkeys" sidebar revisits a classic scientific debate. Imagine trying to ascertain key elements and implications in "real time." How can tech mining algorithms tabulating numbers and content of scholarly papers and patents hope to find out what is meaningful? The story (loosely) suggests the importance of constructs to comprehend the meaning in scientific discourses. We must be alert to invisible college, prestige, and structure of knowledge factors at work. In the evolution debates, interpretation in terms of learning (acceptance or rejection of key theory) and life cycles (uproars in this case) can help us understand what is happening. Not to liken researchers to monkeys, but we are individualistic, using terms somewhat differently. But if you consolidate over many documents, you can model meaningfully.

Sidebar: Monkeys, Typewriters and Shakespeare

In the middle of a college campus, during the height of the Victorian era, an agnostic and a bishop met. Monkeys were discussed, and God as well. There was even talk of typewriters—although it would be seven more years before typewriters became widely available to the public.

This was the debate before the "Great Debate." For sixty-five years later this event would be recreated as the Scopes "Monkey Trial." However in 1860 when this debate was held, the British nation was gripped by fever and rioting about a movement to revitalize the Church of England. Meanwhile, the works of Charles Darwin were growing in repute, leading many to question the traditional teachings of the Church.

The debate was held at the student union of Oxford University. The British Association for the Advancement of Science proposed a discussion of Charles Darwin's new book, "The Origin of Species." Bishop Samuel Wilberforce was a wily man, not to be bound by the ecclesiastic arguments of the time (many centered there in Oxford). His very slipperiness earned him the nickname "Soapy Sam." Darwin himself was chronically ill and could not attend. Darwin's champion was the agnostic and educator, Thomas Huxley. Huxley was a skeptic of all that could not be observed or otherwise empirically verified. Fervent in his belief, Huxley abandoned his own biological research to popularize the ideas of evolution throughout Britain. The actual circumstances of the debate are perhaps lost to history. The story has it that there was much shouting—and a woman who fainted at the rudeness of it all.

During the course of the debate Wilberforce was reputed to have asked Huxley "Which one was it—your grandfather or your grandmother—who descended from an ape?" Huxley was uncowed. He asked Wilberforce to give him "the service" of six monkeys and a typewriter. Given enough time, said Huxley, the monkeys would not only reproduce the works of Shakespeare, but all the books in the British Library as well.

10.2. DATA REPRESENTATION

Given a large collection of publication and/or patent abstracts, we can approach these deductively or inductively. *Deduction* begins with a theory or hypothesis that we pose a priori. We then examine the data to see whether they support or counter that conceptualization. We provide two quick illustrations—first, suppose that we theorize that firm F is trying to block entry into a particular technological arena. One hypothesis is that they will pursue an intellectual property ("IP") strategy of filing "blocking" patents to keep others out. We then examine their patenting portfolio to see whether this is consistent with a blocking strategy. (Chapter 12 elaborates on empirical measures and how these relate to alternative patenting strategies.) As a second illustration—suppose that we have a set of categories of S&T content that we can interpret with regard to stages of technological innovation (life cycle stages). We might use a thesaurus to then classify a set of data into those categories to assess the relative maturity of a target technology.

An *inductive* approach first turns to the data. We listen to the data, and then try to categorize and interpret. Taking the same two illustrations, here we would examine the patenting portfolio just mentioned first and then consider what the patterns noted might mean in terms of some underlying institutional strategy. For the second case, we could allow a software program to group the content terms (e.g., keywords) with no a priori restrictions (e.g., use PCA, principal components analysis, to generate keyword clusters). Then we would examine and interpret the resulting categorizations. If the inductive reasoning proves productive, we might label the resulting formulations a "model." In the future, we could then assess this model via other data sets—that is, on that next iteration we shift into the deductive approach and test conceptualizations on new data. (Actually, this reflects the classic scientific method in iterating between hypothesizing and empirical validation and hypothesis refinement, followed by further empirical investigation.)

We suggest that tech miners need to be overtly aware of which of these approaches they use. Both deductive testing and inductive formulating are valuable, but muddling them can twist us into circular reasoning. Put another way, we should not allow the data to generate hypotheses that we then claim those same data validate.

Model Selection and Assessment

Models come in myriad forms and sizes. A model is just a simplification of reality that helps us understand real processes in some way. There are *qualitative* and *quantitative* models. Our efforts to understand technological innovation aims constitute one form of qualitative modeling. Now, however, our attention is on quantitative modeling of tech mining text and numeric data.

Another distinction lies between *deterministic* and *stochastic* models. Deterministic models fit their parameters, so that "running the model" always generates the same results. Our distribution equations from Chapter 9 could be deemed deterministic models. Stochastic, or probabilistic, models incorporate uncertainties. So, running these models can give a range of possible outcomes. Most tech mining entails stochastic modeling.

A third distinction concerns whether the model's orientation is *static* or *dynamic*. Static models address a set of variables under given conditions (state). Dynamic models generate results that "unfold" through time. These may project future states from past and current information.

Tech mining can involve a wide range of modeling approaches, from quite casual to highly formalized. Our attention focuses mainly on *quantitative, stochastic* modeling. We seek to model large quantities of text data for the purposes of understanding how these came to be generated. We sometimes raise dynamic considerations—asking what we might predict for the future. As noted in Chapter 2, sometimes we go further, seeking to prescribe. In modeling terms, this suggests "what if" examinations, leading to recommendations that if we do X, desirable outcome Y is likely to result. For instance, if we add Company C's IP to our own, we will be positioned beautifully for application A, whereas if we acquire Company D's IP, we could tackle application B.

A "good" model does two things: It fits the data with accuracy, and it uses relatively few parameters. Two sample issues arise in fitting PCA models (see upcoming sidebar). One is how many keywords (or other terms) to include. Another is how many principal components (factors) to extract. For instance, we might compare an 8-factor model with a 12-factor model. We consider various goodness-of-fit statistical measures (discussed a bit later), as well as our subjective judgment as to which yield interpretable factors.

Many metrics have been created to determine how accurately a model fits. One metric is *likelihood* (see sidebar). It describes how likely it is that the model could generate the data. Building a tech mining model requires that we address uncertainty, error, and variation. Many, but not all, models, allow us to generate sample data. (We discuss more about "generating" data shortly.)

Modeling Using Probability and Likelihood

Probability and likelihood are two different quantities. Both quantities are numbers scaled between 0 and 1. Despite this they measure very different things. In this section we give a brief discussion of probability and likelihood, and we introduce a few concepts of probabilistic modeling.

Those readers desiring a more detailed (and more rigorous treatment) of the subject are recommended to consult the work of Edwards (1972), a clear if theoretical treatment of the concept of likelihood. Ross (2002) provides a good overview of probabilistic models, without an excess of theoretical detail. Baldi (Baldi et al., 2003) gives a timely overview of applying probabilistic models to data collected from the Internet.

Probability is a measure of data given a model, and likelihood is a measure of models given data. Probability and likelihood are therefore two sides of the

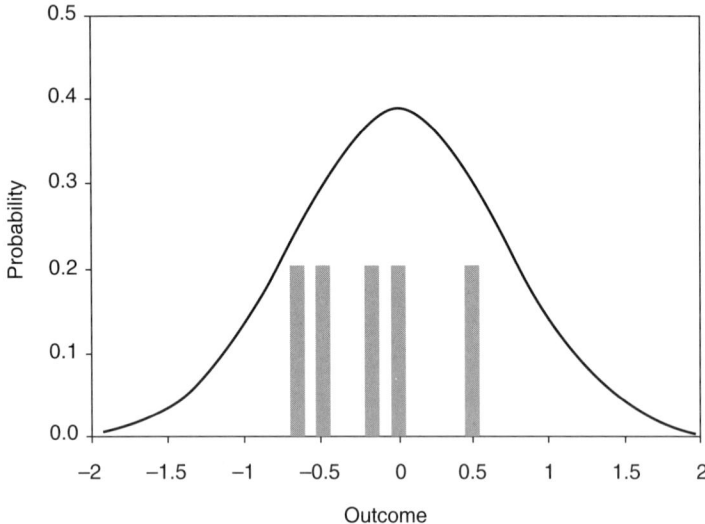

Figure 10-1. A probabilistic model of data

same coin. Let's examine this concept a little further with three examples. Figure 10-1 shows a set of values, plotted on the X-axis. The values are clustered near 0 but are both positive and negative. A few of the values are scattered rightward toward higher positive numbers. Although we have only six data points, we may form a hypothesis of the underlying process that generated the data. We ask—if we continued to observe the data, what more might we see? What would the distribution of observations then look like?

Figure 10-1 overlays the data with a simple model that might have generated the data—the unit normal distribution. This distribution is symmetric and centered on 0. Events far from 0 are comparatively rare in the distribution. Note that the distribution shows the relative frequency of all events—the probability that a particular value might be an observed outcome. We don't really know the relative frequency of our observations, whether they are typical events or otherwise unusual. Each event in the data is given an equal weight of one-sixth.

However, now that we have made this very simple model we can look through each of the observations and ask: What is the probability of this data point, given the assumptions of our model? Figure 10-2 examines the probability of the unit normal distribution generating a value of 0.5. As we have seen, values near 0 are more probable—values far from 0 are less probable.

We may also look at this from another perspective. Given all the possible models that might have generated this data, which is the most likely? Let us entertain for instance three different models—normal distributions centered on –1, 0, and 1. Each of these distributions has the same spread or variation.

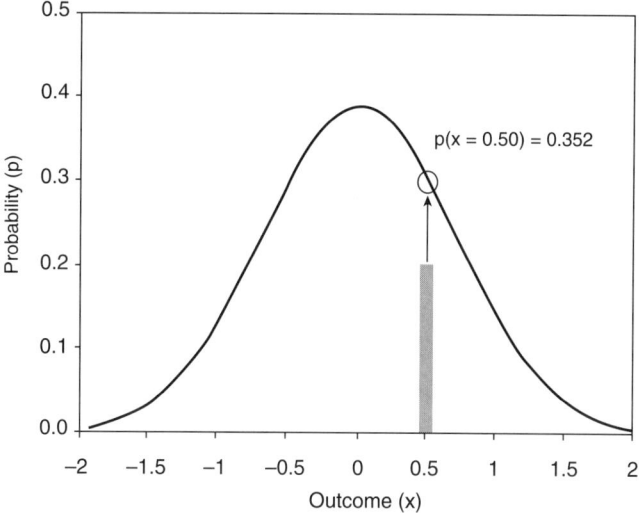

Figure 10-2. The probability of the data given the model

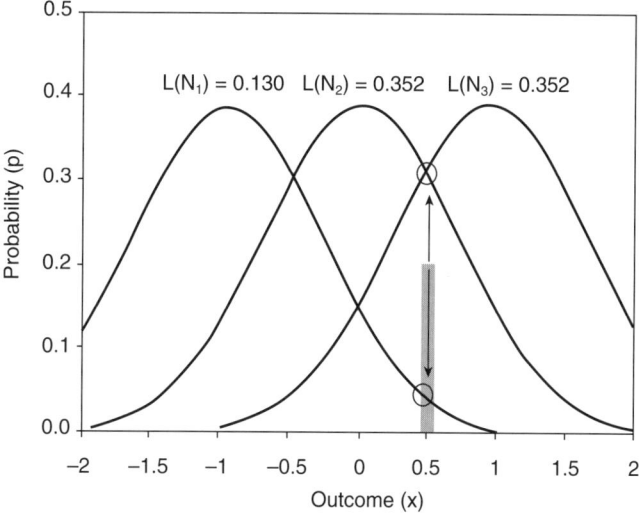

Figure 10-3. The likelihood of the model given the data

Which of these models is the most likely given the data? For this simple example, the data consist of but a single point at 0.5.

Figure 10-3 shows that the models centered on 0 and 1 are equally likely. (The likelihood values are 0.352.) In contrast, the model centered on –1 is comparatively unlikely—a value of 0.130.

If we had no other basis for choosing between the models we might choose models in proportion to their likelihood.

If we had more data points, we could use this information to further our calculation of likelihoods. Each "independent event" contributes more evidence toward our assessment of the model. Likelihoods are compounded through multiplication. A credible model is one in which all events are well described under the model. A single unlikely event discredits the whole of the model.

Equation 10.1. Compounding the likelihood across all events in the data

$$L_{\text{total}} = \prod_{i=1}^{N} L_i$$

Likelihood as a product of all points in the data is shown above as Equation 10.1. It is convenient for multiple reasons to calculate the log likelihood of the data. Taking the log of the likelihood enables us to sum across the data (instead of multiplying). The likelihood of many models is vanishingly small (for indeed, many models can produce an immense variety of possible outcomes); it is advantageous to examine the logarithm of the value rather than the absolute quantity. Such infinitesimal quantities are thereby easily comprehended. Finally, because likelihoods are bounded inclusively by 1 and exclusively by 0, the log likelihood is always a quantity between 0 and negative infinity. Because the logarithm is a monotonic transform of the original data, no data are lost by this alternative scaling.

Equation 10.2. Summation and log likelihood

$$LL_{\text{total}} = \sum_{i=1}^{N} \ln(L_i)$$

Figure 10-4 takes this reasoning a bit further. What if we extended our model search to all normally distributed models with unit variance? Which of the models would then be the most likely? Our model, the normal distribution, requires two parameters: the central tendency of the model (the mean) and the spread of the model (the variance, or standard deviation). We extend our search to find the most likely mean of the data, while keeping the variance fixed at 1.0.

Figure 10-4 illustrates that the single most likely model is that of the normal distribution centered on 0.5. This is the maximum likelihood estimate of the mean. Although this is the most likely value of the model, we might also entertain a variety of alternative models about the data of varying levels of credibility.

In this example we have gone from examining a single parameter of the model to now considering hyperparameters of the model: What range of values

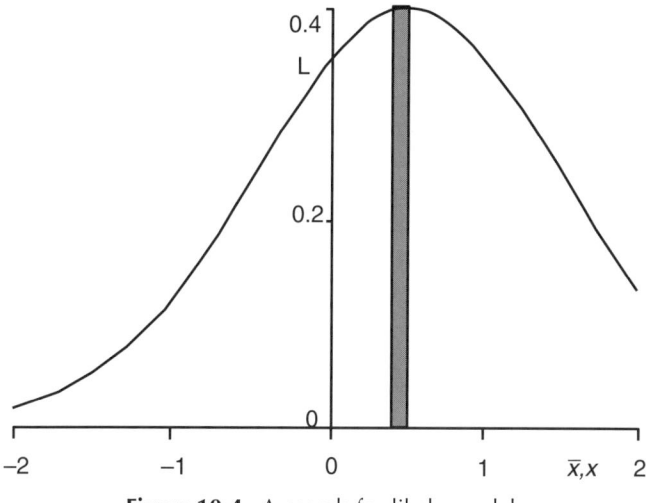

Figure 10-4. A search for likely models

of mean for our model is most likely? We have here, in essence, a model of possible models. We might consider Figure 10-4 to be a likelihood distribution—but instead by convention we turn back to probability. The hyperparameter is described by its own random distribution. The likelihood is a sound basis for searching for the best class of models to describe the data; and with more data we might begin to narrow our search, and to eliminate the uncertainty about the model under consideration.

When and where does the search for models stop? Ultimately, there are an infinite number of possible models that might describe any given set of data. For indeed, we might extend our search from parameters of the model to whole families of models through the use of hyperparameters. An exhaustive search through all classes of models is not feasible, nor would it necessarily produce valuable results. Modeling, then, is also a process of judgment, where explicit knowledge of the subject must be introduced.

Another way of restating likelihood is to consider the following question: Knowing only the model (with its parameters fit to a tech mining data set), how well can you recreate the data? Because model processes are noisy, uncertain, or incomplete, model outputs vary.

Therefore, likelihood is described by a probability. A probability of 0 means that the data could never have been generated by the model. (The probability of generating the data may be vanishingly small, but it never actually results in a likelihood of 0.) A probability of 1, on the other hand, says the model can always recreate the data—it is a perfect summary of the data. For example, recall the "documents by terms" data of Table 9-3 (the Burt matrix). If we

---

Sidebar: Likelihood and Tech Mining

We have previously alluded to a hypothetical case in which we suspect a competitor is pursuing a "blocking strategy" with their IP. Suppose we compile a data set of their patenting in a given technological domain. We might inquire as to the likelihood that the patenting pattern observed is consistent with an underlying blocking strategy. This implies that we have a stochastic (probabilistic) model of how a blocking strategy would reflect in patenting. If the observed patents are highly consistent with this predicted pattern, we assess the likelihood as high. Conversely, if they fit the prediction poorly, the likelihood approaches zero. Over time, as we refine our model and validate it, we could translate it into an algorithm that might be programmed so as to automatically flag patenting patterns observed by other companies with a high likelihood of being caused by efforts to block others. This could serve as an "alert" in our CTI to examine such situations further.

---

specified the correlations among all the terms, how well could we generate a "twin" to this matrix? In general, we do not aim for exact precision; rather, we want to get the important patterns right.

Higher-quality models have a higher likelihood. Often the logarithm of the likelihood—the log likelihood—is given instead of the actual likelihood. This log transformation provides advantages in that very small likelihoods are scaled to negative numbers. In addition, the likelihood sums across various components of the model; as a result, it is easy to determine the most and least satisfactory parts of the model.

An accurate model is not always a succinct model. In fact, any sufficiently detailed model may simply be restating exactly what is found in the data. This phenomenon—known as overfitting—means that the resultant models are cumbersome, but also not robust. Overfitted models may not be representative of the underlying processes; specific details have been reflected, not principles. Understanding the succinctness of the model requires that the analyst examine the parameters used in fitting it. The more parameters (variables) used, the less succinct the resulting model. This is essentially just Occam's Razor, a cornerstone of scientific reasoning.

EXHIBIT 10-1   *Occam's Razor—The Principle of Parsimony*

"One should not increase, beyond what is necessary, the number of entities required to explain anything."

Sidebar: Understanding Model Succinctness:
Principal Components Analysis (PCA)

Consider a study in which the tech mining analyst collects data about keyword usage across articles. One hundred keywords, indexed across 5000 articles, results in 500,000 (100×5000) data points as the basis for the analysis. The analyst applies PCA to discover common patterns.

PCA uses a reduced set of dimensions ("components," or factors—it is a basic form of factor analysis) to represent the data. Based on the interplay of whatever "terms" are studied (here, keywords), it fits a first linear combination of the 100 keywords that explains the most variance possible. Then it creates a second component that best explains the remaining variance, and so forth. The resulting handful of components take the place of the 100 keywords so that we can better understand data patterning. Results can be depicted as maps (see Chapter 16). Each keyword "loads" on one or more components. Each article gets a "component score" showing the components to which it most closely relates. In addition, PCA attempts to center and rescale the data using the mean for each term and the standard deviation for each term. The following model parameters are involved:

- Model treatment: Means (one for each keyword) (K)
- Model treatment: Standard deviations (one for each keyword) (K)
- Document loadings: Number of documents (D) multiplied by the number of components (C)
- Keyword loadings: Number of keywords (K) multiplied by the number of components (C)
- Total number of parameters: $C*(D + K) + 2K$

In our example, the number of keywords is fixed (100) and the number of documents is fixed (5000). How many components could reproduce the data without error? It turns out that 98 components would generate 500,000 points of data—as much as the original data source! If the analyst were to use 98 components (and perhaps even far fewer components) the resultant model would be overfit.

As the preceding sidebar suggests, accuracy and succinctness are competing criteria for model building. The most accurate models use more parameters—and the most succinct models use fewer. Fortunately, there are techniques to help balance accuracy and succinctness to choose the most informative models. One such criterion is Akaike's information criterion (AIC; Eq. 10.3). Every increase in model parameters must have a good "payoff" in terms of model accuracy. Otherwise, the model is less informative than it ought to be. Analysts may choose a model so that a minimum score of AIC is achieved. (AIC is always a positive number.)

Equation 10.3. Akaike's information criterion (AIC)

$$AIC = -2*likelihood + 2*parameters$$

There are other measures of model accuracy and succinctness. PCA models can be compared in terms of their groupings' cohesiveness and entropy, for instance (Watts and Porter, 2003). Regardless of the specific measures of model adequacy chosen, the same essential trade-offs should be considered.

Modeling Data Relationships

In inductive analyses, we are interested in modeling two kinds of relationships in the data: "parent-child" and "sibling" relationships (Cunningham, 1996; Kongthon, 2003). When we do not have a priori categories, we often would like to understand hierarchical relations. Certain forms of clustering strive to do this. "Parent-child" relationships offer one way to consider hierarchy. For instance, to note that "high-energy physics" (the child) is a form of "physics" (the parent), we are really saying that physics is composed of a number of sub-disciplines and high-energy physics is but one of these fields.

Sibling relationships concern association. For instance, to note that TU Delft and Georgia Tech are siblings is to say that they play somewhat similar R&D roles—but TU Delft does not work exclusively with Georgia Tech, nor vice versa.

Relationships may be characterized as parent-child, sibling, or a mix. All that is needed for a pure parent-child relationship is probably a good "data dictionary"—for instance, to relate all research institutions to their respective nations. These are well suited for "tree-based" analyses (Section 10.3).

Relationships are rarely exclusively sibling, if only because any two sets usually vary in magnitude. Nonetheless, advanced analyses are useful to characterize, structure, and predict what elements of the data set are most likely to be close siblings. Many of the term relationships at any one level can be considered siblings—for example, studying the shared R&D interests of multiple organizations.

Figure 10-5 demonstrates these relationships with Venn diagrams. Each circle represents a set of articles. Sibling sets, as noted, have some overlap. In tech mining, this translates to co-occurrence in the data (e.g., articles by these

SIBLING   PARENT-CHILD   MIXED

Figure 10-5. Relationship types

two organizations tend to share keywords). Parent-child relationships find one set completely contained within the other. Mixed relationships combine attributes of both. Tech mining data rarely (if ever) fall completely into either parent-child or sibling.

Drawing Valid Inferences from the Data

We have already discussed the role of advanced analyses as summarization of the data. A generative model is a recipe that describes how to recreate the data. Not all analytic techniques involve a generative model, but there are advantages in so doing (Exhibit 10-2). First, generative models make explicit hypotheses about the data. The resultant analysis may then be confirmed or denied as a reasonable description of the data. Analyses without a generative model risk being "unfalsifiable"—the analyst may produce interesting results but have no way to assess their merits. Second, generative models require the analysts to be explicit about what they do or do not know. Knowledge, or the lack thereof, can be expressed in terms of probabilities. Third, this points toward measures of model quality—the likelihood of the observed data, given the model. Fourth, generative models can create mock outputs or "parodies" of real scientific data. This can assist knowledge management purposes through possible simulations. Fifth, generative models can highlight which part of the data is most unusual or surprising. Because the chief tech mining problem is "information overload," models that highlight what warrants our attention are especially helpful. The "Threats to Validity" sidebar adds another perspective on checking model validity (see also Section 15.3). The "Parodies" sidebar (Section 10.3) illustrates data generation issues.

EXHIBIT 10-2 *Advantages of Generative Models*

(1) Explicit Hypotheses
(2) Codifying Knowledge and Level of Uncertainty
(3) Explicit Measures of Model Quality
(4) Parodies Real Data
(5) Highlights Surprising Results

> #### Sidebar: Threats to Validity
>
> Don Campbell and colleagues (Campbell and Stanley, 1963; Cook and Campbell, 1979) devised an incredibly helpful set of considerations to check the validity of inferences drawn from empirical data. They distinguish four types of threats to validity. Here we synopsize those and add questions for the wary tech mining analyst to ask:
>
> *Continued*

- Internal Validity—whether the observations (the data) are potentially biased
  - Is our data set possibly distorted in ways that could affect important conclusions (e.g., we are missing "gray" or "black" literature)?
  - Might certain events have altered publication and patent behavior in ways we should take into account (e.g., 2 companies merged)?
  - Can we think of ways that publication or patenting behavior might be a distorted reflection of actual R&D activity (e.g., a company discourages publication)?
- Statistical Conclusion Validity—whether sample sizes are sufficient to be confident that results are not due to chance
  - Is our data set large enough to draw robust conclusions?
  - Have we sampled all the important R&D outputs? (e.g., in our fuel cells tech mining, should we have searched Chem Abstracts in addition to engineering-oriented EI Compendex?)
- Construct Validity—whether the underlying constructs are properly reflected by the observable measures
  - Do our constructs really relate to the observed measures as we posit? (e.g., have we gathered data to correctly represent the R&D domain we think we are addressing?)
- External Validity—whether the findings for the given data set generalize to other situations
  - Are the data suitable to draw conclusions for our tech mining target interests? (e.g., if the data are historical, do they pertain to current conditions? if we study R&D publication data because we have them, and these heavily reflect university activity, can we say anything meaningful about companies' R&D?)

Variables and Distributions

Important features of models include random variables, data transformations, and "plates"—variables at a comparable level of aggregation. Let's start by considering the idea of a random variable. Random variables are the outcome of a random process; these processes are described by distributions. Random distributions describe the range of possible outcomes, and the likelihood of any given outcome actually occurring.

Technological innovation is far from random. However, parts of the system are not fully understood and many aspects have a stochastic (noisy) element. Observers might not be able to identify the key terms, important relationships among terms, and the relative placement of one publication among others of its kind. We can, however, create random variables to represent this imprecise knowledge and use suitable models to estimate this information.

S&T publication and patent measures include some discrete variables and some continuous ones (Exhibit 10-3). Discrete variables take on only certain values (e.g., number of authors of a paper); continuous variables take on any

EXHIBIT 10-3   *Examples of Continuous and Discrete Variables in Tech Mining*

Variable Type	Example	Given
Discrete	Probability of Citation	Given a particular author, journal, date or publication
	Probability of Keyword	Given a particular abstract, journal or collection of records
	Probability of Collaboration	Given a particular author, institution, or keyword
	Probability of Numbers of Authors	Given a particular paper
Continuous	Probability of Numbers of Publication	Given a particular keyword, date, or collection of articles
	Probability of Numbers of Keywords	Given a particular paper, date or collection of articles
	Probability of Numbers of Journals	Given a particular paper, date or collection of articles
	Probability of Numbers of Authors	Given a particular paper, date or collection of articles

TABLE 10-2. Important Discrete Distributions in Tech Mining

Distribution	Discussion
Binomial distribution	Models the probability of a single event with a fixed outcome. Multiple Bernoulli trials with binomial (0/1) outcomes. The distribution is discrete because there is a fixed outcome (true or false; yes or no).
Multinomial distribution	Models the probability of choosing from multiple outcomes. Multiple trials of multinomials are known as the categorical distribution. The distribution is discrete because there is a finite set of outcomes to choose from.
Poisson distribution	Models the probability of discrete events occurring with a specific rate. The distribution is discrete because the outcomes are limited to occurring in unit quantities. (For instance, authors either publish a complete article or no article at all.)

values along a continuum. In tech mining, we most often address situations in which sufficient discrete values occur that we can approximate by a continuous distribution (e.g., number of papers of a large research community). Examples of discrete random distributions include the binomial, multinomial, and Poisson distributions, introduced in Table 10-2.

Many data attributes of tech mining interest are appropriately modeled as continuous variables. Some of the more frequently used continuous distributions include the normal (or Gaussian), gamma, Dirichlet, lognormal, and the power law (Table 10-3 discusses three of these). The choice to treat discrete quantities (such as papers, citations, collaborations, institutions, and keywords) as continuous variables is somewhat arbitrary. With sufficient numbers of observable values, continuous and discrete modeling should yield similar results. Beware using continuous variables whose distribution encompasses negative and positive values (e.g., mean of 0) for entities that do not actually take on negative values. Also, strongly consider using "thick-tailed" distributions (such as the lognormal or power law) instead of the more popular normal distribution to represent bibliometric data because, as already noted, these tend to be highly skewed.

Probability distributions are related to outcomes, and measurable tech mining data, through "link functions." Often these links are simple multiplicative relationships. They help "multiply" the findings—showing, for instance, how a range of outcomes can be explained by a small subset of the variables. Links can also help scale the variable. (The random variable itself typically has scaling parameters, so it need not be transformed.) Scaling transformations, where a quantity is scaled and an offset is added, are known as "affine" transformations. These are linear transformations.

There is some concern that the phenomena in tech mining are very complex and therefore quite nonlinear. Although this may be true, there is considerable merit in starting with simple assumptions and testing to see the merits of the resulting model. Sometimes we transform the data (e.g., taking the logarithm) to better match model to data (e.g., to cope with long-tail distributions). One important role involves transforming continuous quantities into discrete outcomes. Discrete outcomes have distinct floors and ceilings (often the func-

TABLE 10-3. Important Continuous Distributions for Tech Mining

Distribution	Discussion
Normal (Gaussian) distribution	A two-parameter distribution with a symmetric bell shape. Occurs when many additive causes contribute to creating an outcome. Related in fundamental ways to the binomial and Poisson distributions.
Lognormal distribution	A distribution with a leftward-skewed shape, which is bounded at 0. Occurs when many causes multiply to create a single outcome. Related in fundamental ways to the normal distribution.
Power law distribution	Distribution with "thick tails"—high-magnitude events are relatively frequent. Occurs whenever the "winners" are rewarded, or are given cumulative advantage. Related in fundamental ways to the lognormal distribution.

tion is limited by 1 at the maximum and 0 at the minimum). A transformation may ensure that the continuous function does not result in unrealistic outcomes. One of the most useful transformations for this purpose is the logistic equation.

Equation 10.4 transforms a variable with an open-ended range of outcomes to a variable with a bounded set of outcomes. The equation showing x as a function of y is known as the "link" function. The equation showing y as a function of x is known as the transformation. (The probit function may sometimes be handy, too—it relates the continuous range of the normal distribution to a 0-1 discrete probability function.)

Equation 10.4. Logistic link and logistic transformation

$$x = \frac{1}{1 + Ce^{-ry}}$$

$$y = \ln((x)/(1-x))$$

Tech mining models may address different levels, on a hierarchy, of variables. For instance, we might be analyzing both journals (higher-level aggregation) and papers (lower-level aggregation). Or, we have previously noted the example of the specialty of high-energy physics falling under the higher aggregate, discipline, of physics. This tiered approach allows you to address differences among entities, by level. For instance, we could note that high-power transmission falls under the discipline of electrical engineering (EE) and that publishing practices differ between physics and EE. Graphical depictions conveniently distinguish entities by level of aggregation—so, physics and EE would appear on a higher "plate" than the specialties, high-energy physics and high-power transmission.

We have now discussed the fundamental components of generative models—probability distributions, transformations and links, and plates or aggregates of variables. In Section 10.3 we examine how these components are combined to create standard models. In subsequent sections, we discuss how custom analytical models may be built for special purposes by selectively combining these components.

10.3. ANALYTICAL FAMILIES

We have argued that advanced analyses serve to summarize the data and reach conclusions that may generalize beyond the tech mining data sample. Our choice of representation for the data is one form of hypothesis. We hypothesize a structure in the data, and then we use analytical techniques to discover and fit the data to the structure. As we will see there are several kinds of "families" (alternative modeling approaches) widely used in tech mining. These are inductive approaches—seeking to infer structure from observed regularities in the data. When we begin the analysis with an already-accepted data structure (deductive approach), we don't fit a model to the data; instead, we fit the data into our a priori model. For instance, one of the innovation indicators discussed in this book asserts that extensive corporate R&D publication is an indicator of commercialization potential for a technology. To score this indicator, we apply thesauri to tally the extent of publication by industry, academia, and government/nonprofits/others. In contrast, this section focuses on inductive approaches.

We address, in turn, dimensional (spatial) models, usually presented as maps, clustering, trees, and (briefly) causal models. Chapter 11 discusses trend analyses, another data fitting approach. We seek to distinguish these analytical approaches and recommend when to use each kind. But note that terminology and usage blur. For instance, the approach most heavily discussed in this book, PCA, fits into the dimensional models family, but we routinely treat its outputs as "clusters." Don't be put off by this, but recognize that the underlying statistical approaches have much in common and that outputs can be represented in multiple ways. The message of this section is to think through your tech mining objectives and then choose from among the analytical approaches one that best serves those objectives. Put another way, don't be captive to a single approach.

Term Mapping (Dimensional Analysis Techniques)

Tech mining data are usually of very high "dimensionality," meaning there may well be thousands of words, documents, journals, or other items in need of analysis. Dimensional techniques (Exhibit 10-4) share a common approach in handling data. The techniques all represent the data with relatively few, underlying dimensions. These dimensions can be graphically depicted as "semantic spaces" or "science maps." Dimensional analyses help the analyst find and communicate major patterns in the data by dramatically simplifying focus. Dimensional techniques can be recognized through their emphasis on creating "spaces," factors, or other reduced dimensional forms of representation.

Concerns arise from poor data compression metrics. These may obfuscate underlying relationships or even introduce artifacts. Metrics that define distance or similarity among entities make a difference in model outputs. For instance, in using PCA on keywords, one could use keyword correlations or a

---

EXHIBIT 10-4 *Dimensional Analysis Techniques*

Correspondence analysis	Latent semantic indexing
Factor analysis	Multidimensional scaling
Independent components analysis	Principal components analysis
Kohonen maps	

cosine function (or other similarity measures), resulting in somewhat different representations of the data.

Spatial techniques are quite effective for understanding networked S&T structures. Such structures involve sibling relationships, so are most applicable when comparing things of similar scale. That is, map "plates" of items together; if you want to break out sublevels (lower aggregates), be sure to distinguish these clearly in the visualization. Another set of issues arise in how one depicts distances in a 2-D or 3-D mapping. Some depictions convey what's happening more effectively than others. In most situations there is not a simple "right" way, but rather several reasonable representations. Picking one or a few highly informative ways to map a data set combines science (statistics) and art (a sense of what best informs the target users). But if several topics are being compared, consistent representation is vital. In many semantic spaces, the axes themselves are arbitrary—be sure your users understand this. Examples of spatial techniques applied in tech mining appear at the end of this chapter and in Chapters 12 and 16. The sidebar notes data generation issues.

Sidebar: Simulating the Data

Suppose we have performed a dimensional analysis, say, using PCA. We stare at one of the dimensions—call it "Factor A." Suppose Factor A is composed of five high-loading keywords. We now work our way back from those loadings to create "mock" documents. We flip through a set of these simulations of the real data to see whether anything appears ludicrous. If we spot keywords appearing together that make no sense, this is a "red flag" to reassess our model.

Another illustration: A police artist listens to descriptions of criminal suspects by witnesses and tries to render a realistic model. As the artist compiles responses to a whole set of questions—e.g., size of forehead, ear location, cheekbones—the artist periodically shows the model to the witness to check whether it is coming out "right." Be sure to test your models' depictions on knowledgeable folks.

Clustering Techniques

Clustering techniques (Exhibit 10-5) share a similar goal—the discovery of "natural types" in the data. These can be identified by an emphasis on discovering natural clusters, groups, or prototypes. Effective clusters group the

---

EXHIBIT 10-5 *Clustering Techniques*

Agglomerative algorithms	Mixture models
Hierarchical clustering	Probabilistic clustering
K-means	Single-link clustering

most similar items and exclude the most dissimilar ones. The result of a clustering analysis is the discovery of a set of distinct groups.

The most serious challenge these techniques face is defining the nature of similarity. A wide variety of potential similarity measures have resulted. Perhaps the most basic is co-occurrence—items in common. For instance, two articles that share more keywords in common are more similar.

An advantage of clustering techniques lies in their very simple representational structure. The idea of groups, or types, is very easy to communicate. For instance, we have mentioned "typing" research organizations as academic, industry, or government. The constitution of the groups is also easily communicated—very simply stated, it is a list of all, or the leading, members belonging to that group. A drawback of the technique is that it may be too simple to effectively represent many phenomena.

Clusters come in various types (Chapter Resources provides a few pointers). Some approaches generate "exemplars"—prototypes of the items in a class; others represent aggregates—a sum total, or collection, of all the members in its class. Some clustering techniques allow for "soft assignment"; others don't. With soft assignment an item may belong to multiple groups. Soft assignment with exemplar-type clusters can effectively represent sibling relationships, much like spatial techniques could. Clustering techniques are quite versatile and may also be used for representing "parent-child" relationships. Clusters can indicate relationship between the collection and its members, allowing the analyst to understand hierarchical characteristics in the data. Such analyses may be quite similar in character to trees, to which we turn in the next section.

Techniques for grouping data can be used in cleaning the data. For instance, you might suspect that the query has yielded some irrelevant documents. Clustering can help you identify by grouping (and eliminating) the irrelevant articles. Examples of clustering applied to tech mining appear at the end of this chapter and in Chapter 16. The sidebar introduces geographically oriented clustering.

Sidebar: Regional Cluster Analyses

Our Georgia Tech colleague, Phil Shapira, has applied tech mining techniques to help study "innovation clusters." In this case one associates R&D outputs (patents and/or publications) with particular geographic regions (e.g., countries or, particularly, metropolitan areas). A thesaurus helps link the particular organizations generating those patents (or publications) to particular regions. This tends to be relatively straightforward for major universities, research institutes, and large companies, but more challenging for the residuals (e.g., small companies and other organizations). Of course, one must resolve uncertainties such as how to credit a patent by a multinational that involves invention carried out in multiple labs, with assignment in another place and initial patenting elsewhere. But, for the purpose of identifying concentrations of types of technological innovation activity (e.g., Herfindahl index) to ascertain the best opportunities for certain cities, complete precision is not needed. Shapira and colleagues have used tech mining to identify the best technological development opportunities for Georgia cities.

A "natural" for such regional analyses is to map results as suitable overlays on geographic models (geographic information systems—GIS depictions). Showing concentrations of particular R&D activity can have effective technology policy implications, too. We have explored such possibilities with the U.S. National Library of Medicine. In that case, the interest was in showing whether regions confronted with emissions of particular hazardous substances were or were not the locales wherein research on dealing with those hazardous materials was taking place.

Tree-Based Techniques

Tree-based techniques (Exhibit 10-6) involve successively dividing data into groups or classes. The tree is "complete" when the links among the various classes are fully explored. Much like a natural tree, tree-based models reach from aggregates of data to specific instances or classes. Trees may be created "trunk out" from the most aggregate level to the most specific (the "leaves"— small group of documents or terms). Alternatively, trees may be created by starting with individual articles and aggregating up until the entire collection has been subsumed. Tree-based techniques can be identified through their emphasis on denoting family or "compositional" relationships.

The most serious challenges for tree-based techniques are assumptions of relatedness. These approaches seek to create a parsimonious description of structure. Armed with a good theory about how the data were generated, and a succinct and well-formed description, the analyst may then assume that items that are close together on the tree are more closely related. Unfortunately, there may be multiple competing trees, all offering succinct descriptions of the data. Worse yet, when new data are introduced, the shape of the tree may change radically. "Pruning" the tree—that is, including neither too much nor too little structure—is an important consideration for building robust models.

EXHIBIT 10-6   *Tree-based Techniques*

Association rules	Hyperbolic trees
Classification and regression trees	KD-trees
Decision trees	Minimum spanning trees
Discriminant analysis	Nearest neighbors
Hierarchical models	Probabilistic trees

TABLE 10-4. Three Analytical Families

Family	Strengths
Dimensional analysis	Generate a few variables to capture most of the key information; Facilitate graphical depiction to help understand relationships
Clustering	Conceptually straightforward—group like terms; Can generate exclusive groups or allow multiple membership groups
Trees	Provide useful hierarchy information (parent-child); Can be represented multiple ways

Trees have interesting representational characteristics. They can be efficiently represented through a set of rules or as a standard textual outline structure. This can communicate well to those who might not like spatially oriented techniques (visualizations, maps). However, trees can be represented graphically, too. Graphical plots often have some danger of misleading, as they suggest proximity among items that may be neither more nor less related than other items. In addition, tree-based techniques may produce such richly branched or cross-linked structures that they are difficult to present graphically.

Trees involve successive divisions of the data, and these are an integral part of the final representation. In this way, tree-based techniques convert simple partitioning (like cluster analysis) into more complex structures. Cluster analysis addresses the basis for similarity among items. Tree-based approaches are often more concerned with describing key differences among the branches in the data.

Trees share some similarities with dimensional approaches as well. Both sort the data along dimensions of similarity or relatedness. Ironically, the basis for comparison in tree-based techniques may be richer and more multi-dimensional than that of a strictly dimensional approach. For this reason, trees offer a more succinct way of representing data than dimensional approaches—only the key discriminating factors in the data need be highlighted in the rules and branches of the tree.

Trees work well to represent parent-child relationships. Items of very different scales can be shown on the same map—for better or worse. For example, some use hyperbolic trees to represent the branching citation structure of the World Wide Web. A personal home page can appear alongside such luminaries in the Internet world as "Yahoo," "Google," or "CNN.com"—pages read and cross-linked by millions. Examples of tree-based techniques used for tech mining appear at the end of this chapter and in Chapter 16.

Table 10-4 highlights key advantages of each analytical family.

Additional Analytical Approaches

Each of the techniques listed in Exhibits 10-4, 10-5, and 10-6 could be explored in detail, but that is beyond our intent and scope. Chapter 11 is devoted to trend analyses. We have noted that this section's approaches are inductive; the contrasting approach is deductive, trying to fit the observed data into preset categories.

We call attention to *causal modeling*. Unlike the other approaches addressed in this section, the driving motive is not classification but understanding of cause and effect. Such models may build upon classification and trend analyses. Multiple regression is one popular form of generating causal models. For instance, imagine that we want to know the relationship between U.S. Department of Energy (DOE) funding and publication activity. We could gather data on funding (amounts, thrust areas, by year) and model against publishing activity (by field, for DOE grant recipients and others, lagging the funding by different years). Certainly, correlation (or regression) does not imply causation. Generating effective causal models for tech mining data requires compelling constructs and system depiction. One needs to explain how observed data are more consistent with a causal explanation than with alternative possible explanations. This sort of thinking, carefully vetted, can be superb input into management of technology ("MOT") decision processes.

We have introduced a wide variety of analytical models that can be combined in many ways to achieve particular analytical ends. You can string together models of different types much like beads on a necklace. The combined models are often more effective than their separate parts. The ability to jointly estimate two separate models creates a more effective, more succinct result. A strength of the probabilistic approach is the ability to mix and match modular components. As an example, you could create a "mixtures of factor analysis" model (also known as a "mixtures of experts" model). Such a model includes a cluster model at the aggregate level and a factor analysis model at the lower, more specific level. Tech mining software could separate records for each cluster and then perform factor analysis on just those records to see what topics (keywords) that cluster uses most heavily. This might help recognize distinct kinds of content across a broad or diffuse field. Or, as another example, such a model might be used to recognize and detail differences among research institution emphases.

Various types of information can be combined with *temporal markers* to identify change. The number of approaches is extensive, starting with just plotting an activity measure versus time (see, e.g., Chapter 11; also Fig. 16-3). "Time slicing" into interpretable periods (e.g., before and after a policy shift) can show up differences. For other purposes, smoothing trends over time statistically may elucidate what is changing (e.g., using exponential smoothing to reduce the noisy splatter over monthly patenting data).

We have explored ways to compare PCA maps over time. Particular maps vary greatly because of the nature of multidimensional scaling (MDS), so direct comparison over time slices is not straightforward. Human-interpreted and -drawn maps of the key factors dominant in successive time periods can be very effective. We used this approach in studies for the Army Environmental Policy Institute to track changes in military noise issues with developments in technologies to reduce noise (e.g., the emergence of active noise suppression methods). We prepared a series of 5-year maps showing noise factors prominent for each period. We then extrapolated into the future on the basis of apparent trends and expert views on what forces were driving change in this area. Thus tech mining term mapping helped forecast technological and issue changes.

Record topographies are wonderful to visualize change over time. For example, we could mark one organization's patents on a landscape by using different colors for different time periods. Showing this interactively where first the earliest period's patenting is lit and then successive periods are added can convey the diffusion of that organization's interests compellingly.

Our colleague Bob Watts is researching advanced model quality metrics. These include (1) percentage of keywords included in PCA or PCD (principal components decomposition) "factors," (2) entropy, and (3) the richness of linkages among keywords. These metrics could, in turn, be used as indicators of technological maturation (Watts and Porter, 2003). The PCD algorithm applies metrics that provide a standard solution to a min-max problem (Watts, 2001). In other words, it uses an optimization routine to determine how many factors (principal components) to extract. The PCD process includes as many of the analyzed records (i.e., abstracts) in the derived "factor groups" as possible. In addition, the PCD algorithm strives to maximize both the number of factor groups and the group-defining descriptors (i.e., the high-loading "keywords" for each PCA factor group). The algorithm also attempts to minimize the number of documents associated with more than a single factor. This solution approach conceptually equates to minimizing the entropy and maximizing the cohesiveness of the factors formed (Borner et al., 2003; Steinbach et al., 2000). One advantage lies in reproducibility—given identical documents, the automated PCD analysis will repeatedly derive the same set of factors (groups of keywords).

The sidebar considers how one might want to represent (model) S&T, generally speaking. Section 10.4 offers empirical illustrations for a number of the approaches introduced.

Sidebar: Ideal Representations of Science and Technology

Science is often discussed in apparently contradictory terms. Commentators note the increasing specialization—as John Ziman once put it, scientists "know everything about nothing." Scientists appear to work on disconnected topics, making little reference to the works of one another, particularly beyond their disciplinary bounds. In contrast, others highlight the densely connected research networking—emphasizing the people, institutions, and papers that create connected hubs for the preservation and exchange of ideas. Such networks, in this perspective, are core to S&T advance.

The ideal S&T representation might balance these two depictions. On one side, it would seek to identify individual scientific specialists as the keys to advance. On the other, it would seek to model networking as key to the spread and diffusion of scientific knowledge. Can we combine both in one ideal modeling and information representation? We offer this as a challenge to the tech mining community.

10.4. DEBRAND TRUST ADVANCED ANALYSIS EXAMPLE

We return to our DeBrand Foundation case data of 173 articles citing the agronomist Norman Borlaug (see Chapter 9). The main tech mining objective here is to create a monitoring system to discover significant new research, beyond this small sample, in the area of human welfare and agronomy. A second objective, therefore, is to expand the original search to discover additional relevant articles.

Given these objectives and the small cross-disciplinary data set, we favor an inductive approach to discover patterns in these data. *Clustering* techniques seem most in order. By identifying various types of articles, the DeBrand Foundation hopes to deepen its knowledge of relevant research.

The analysts know that the breadth of the topic is not well-suited to being captured in keywords. Borlaug is cited across disciplines, by authors of very different interests and orientation. Instead, the analysts chose a mixture of cited authors and cited journals to get at a suitable range of scientific and policy knowledge elements. (This is a mix of two kinds of basic analyses; see Chapter 9). Table 10-5 shows the 20 most frequently cited authors and most frequently cited journals in this literature.

Table 10-5. Top Twenty Cited Journals and Authors for the Borlaug Collection

Rank	Journal	Author
1	*Science*	Borlaug NE
2	*Crop Sci*	Wolfe MS
3	*Theor Appl Genet*	Christou P
4	*Phytopathology*	FAO (United Nations Food and Agriculture Organization)
5	*Plant Physiol*	Zadoks JC
6	*Nature*	Mundt CC
7	*Plant Soil*	Jensen NF
8	*Proc Natl Acad Sci USA*	Worland AJ
9	*Bio-Technol*	Browning JA
10	*Plant Cell Rep*	Rajaram S
11	*Euphytica*	Malhi SS
12	*Soil Sci Soc Am J*	Vanderplank JE
13	*J Food Sci*	Law CN
14	*Agron J*	Finckh MR
15	*Plant Mol Biol*	Baligar VC
16	*Annu Rev Phytopathol*	Wang HL
17	*Adv Agron*	Khush GS
18	*Plant J*	Simmonds NW
19	*Nat Biotechnol*	Marschner H
20	*Genetics*	Leonard KJ (tie) Sanford JC*

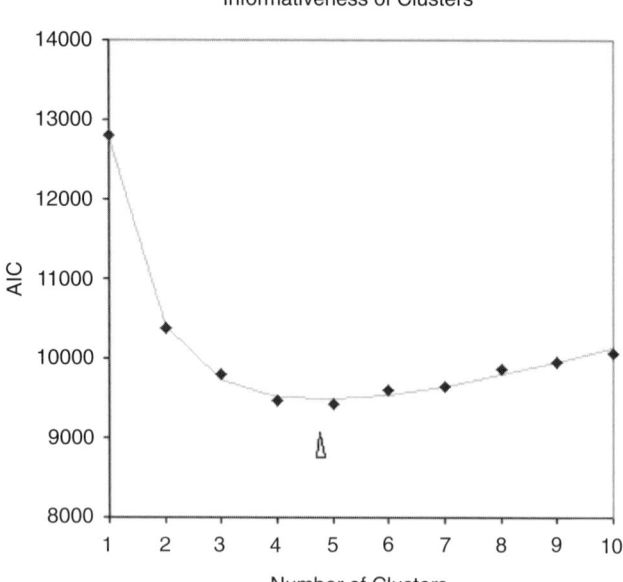

Figure 10-6. AIC for the Borlaug clustering model

Two general- interest scientific journals, *Science* and *Nature*, rank highly—affirming the broad scope of this inquiry. Not surprisingly, Borlaug leads the list of most frequently cited authors—the collection is based on citation of his work. More interesting are the other highly cited authors—Wolfe, for instance, and the United Nations Food and Agriculture Organization. These have been "learned" from the data as possibly significant sources of agronomy literature.

In preparation for clustering, each of the 173 articles is indexed to identify whether it cites each of these Top 20 journals and authors. Because authors and journals may be cited repeatedly, the index for each article is discrete numbers greater than 0. This indexing leaves only 5 of the 173 articles without any indices at all. (These are popular news articles that don't formally cite scientific content.) The indexing strategy is therefore successful in revealing a dense network of cited literatures that spans the article collection.

The analysts choose Poisson clustering to group the data. This probabilistic technique assumes that the data have been created from a mixture of "*N*" distinct types of literature, each citing the authors and journals with their own characteristic rates. (McLachlan and Peel, 2002, discuss probabilistic modeling in detail.) The tech mining analysts run a set of analyses, varying the number of clusters extracted. They compare results using AIC as the criterion (recall Equation 10.3—the lower AIC, the more informative the model). Figure 10-6 suggests that the best model contains 5 clusters (clusters are discrete—we're not going to have a model with 4.7 clusters!). Models with fewer clusters do not adequately describe the data. In contrast, the models with more clusters make greater assumptions about the data and require more parameters to fit the data—these models are too elaborate given the limited nature of the sample.

A "responsibility matrix" shows on an article-by-article basis the assignment of documents to underlying clusters. The clusters are "responsible" for describing the characteristics of their respective documents. The probabilistic clustering technique allows for "soft clustering"—the possibility that a given document might be a mixture of one or more clusters. But, as can be seen in the partial matrix (Table 10-6), most documents appear to be "pure types," best explained by one cluster.

The key findings from the cluster analysis are shown in Table 10-7.

The table summarizes the five clusters in terms of the raw numbers, the key authors, and the key journals associated with each cluster. First of all, it is interesting to note that the generalist journals *Nature* and *Science* are read across all the clusters identified—citations to these sources can be eliminated as a source of discriminating power. Likewise, by definition, Borlaug is widely

Table 10-6. Sample of Responsibility Matrix

	Cluster 1	Cluster 2	Cluster 3	Cluster 4	Cluster 5
Article 1	0.00	0.00	1.00	0.00	0.00
Article 2	1.00	0.00	0.00	0.00	0.00
Article 3	1.00	0.00	0.00	0.00	0.00
Article 4	0.00	0.00	1.00	0.00	0.00

TABLE 10-7. Key Findings from Cluster Analysis

Cluster Number	Tentative Label	Number/ Percentage of Articles	Key Authors	Key Journals
Cluster 1	Genetics	18/10%	Christou P Leonard KJ Sanford JC	*Genetics* *Natl Biotechnol* *Bio-Technol* *Proc Natl Acad Sci*
Cluster 2	Plant science	15/9%	Marschner H Khush GS Baligar VC	*Plant Sci* *Soil Sci Soc Am J*
Cluster 3	General interest	102/59%		
Cluster 4	Phytopathology	39/20%	Wolfe ME Zadoks JC Mundt CC Browning JA Finckh MR Leonard KJ	*Phytopathology*
Cluster 5	Food science and agronomy	4/2%	Wang HL Malhi SS Worland AJ	*Agron J* *J Food Sci*

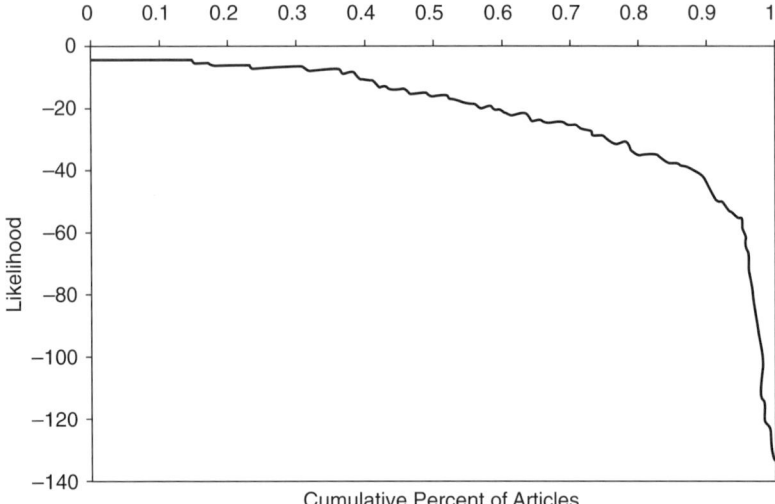

Figure 10-7. Articles and likelihood

sampled across the collection and is therefore not a significant discriminator among groups.

Secondly, it is interesting to note that a very substantial cluster (Cluster 3) represents "general-interest" articles of a topical character. These articles contain relatively few citations to particular established authors or journals. Isolating these articles is helpful as it clarifies the characteristics of the more specialist clusters that remain. It also suggests that our recommended monitoring strategy reach beyond specialist literature.

Cluster 1 represents biotechnology articles. Three authors and four journals stand out as particularly representative of the group. Cluster 2 consists of a group of plant science and food science articles. Plant horticulture and nutrition are at the forefront in these articles. Cluster 4, a substantial group, talks about phytopathology. There are a number of significant articles, but at least in this sample, only a single journal stands out as a source of literature. The fifth group is relatively poorly sampled—four articles or roughly 2% of the total. This group emphasizes research on food science and human nutrition. Journals include the *Agronomy Journal*, and the *Journal of Food Science*. The leading author (in this very limited sample of articles) is SS Malhi.

As discussed, the likelihood can also serve to highlight surprising, or poorly explained, aspects of the data. With the probabilistic clustering model, each record of the data receives a "log likelihood" score—the more negative the number, the less likely the record could have been generated by this model. Figure 10-7 shows the distribution of likelihood across all articles. (The quantity shown is actually the log likelihood.)

As can be seen there exists a small, but substantial "tail" of poorly explained articles. Examining these articles on a case-by-case basis (data not shown)

TABLE 10-8. Reweighting the Journal Index

Added to the Journal Index	Dropped from the Journal Index
J Agr Food Chem	*Plant J*
J Sci Food Agr	*Nat Biotechnol*
Can J Soil Sci	*Genetics*

often gives revealing results, and offers opportunities to improve the model. First, we note that two of the six least likely articles come from the "food science" cluster—a cluster we have acknowledged needs additional exploration and data collection. Second, the single most unlikely article comes from a particular set of authors (name withheld) who cite themselves far in excess of the rates at which other authors in their discipline cite their work. We can compare the generative model (likelihood estimates) with the actual to fine-tune; for instance, in recommending a monitoring strategy we might shift the journal set slightly (Table 10-8).

In sum, this case analysis has proven productive. It identifies research domains we might monitor based on the clusters that inductively arise from the 173-document set. Could we have used dimensional analyses (term mapping) or tree-based analyses? Yes, but we do not feel these would have worked as well. Continuous dimensions to research emphases would be awkward in terms of specifying which areas to monitor. Trees provide more linkage (hierarchy) detail than we want here. Space does not really allow us to demonstrate all these analytical options. The intent of this section is to provide one illustration of advanced analysis—probing more deeply into a data set to get at underlying and interpretable patterns.

CHAPTER 10 TAKE-HOME MESSAGES

- Advanced analyses summarize the data to help discover patterns and relationships.
- Consideration of constructs (hidden variables) can help to understand underlying mechanisms at work.
- We offer six key constructs to help stretch tech mining thinking beyond observables.
- Deductive approaches fit the data into preset models to ascertain how well they fit.
- Inductive approaches seek natural categories emergent from the data themselves.
- We contrast model types: qualitative and quantitative, deterministic and stochastic, and static and dynamic.
- Weigh two evaluative criteria for your models: accuracy and parsimony.

- "Likelihood" offers a useful concept in assessing model effectiveness in representing data.
- The AIC helps in assessing model quality.
- Recognize two basic data relationships: parent-child (hierarchy) and sibling (same level).
- Generative models that recreate (simulate) data sets can serve to check model reasonableness and to generate mock data for decision explorations.
- One ought to consider whether discrete or continuous variables can best capture the essential data attributes.
- Three analytical families merit strong consideration in striving to understand your data: dimensional analyses (term mapping), clustering, and tree-based techniques.
- Combinations of analytical approaches add further power.

CHAPTER 10 RESOURCES

Others (cf. Teichert and Mittermayer, 2002) and ourselves (cf. http://tpac.gatech.edu) compare alternative clustering and related approaches in the tech mining context. As we write, Alisa Kongthon and Cherie Courseault are completing dissertations on these topics at Georgia Tech. Zhu et al. (1999) and Watts and Porter (2003) discuss facets of these issues more deeply as well.

Chapter 11

Trend Analyses

Time is a variable of special importance to tech mining. This chapter presents a range of approaches to trend analyses, in order to:

- Describe technological changes occurring in the past
- Predict future technological changes, over time

11.1. PERSPECTIVE

This is the third of our analytical chapters. Chapter 9 presented basic analyses (relatively straightforward reflections of publication and patent abstract data sets). Chapter 10 introduced advanced analyses (considerations in uncovering underlying relationships among the data). Here we address trend analyses (ways to understand and represent changes over time). The intent of this chapter is to foster insight and present options, more than detailing "how to" particulars. We do introduce a new example of nanotechnology publication data, through which to illustrate various time series analysis approaches.

Chapter 9 distinguished documents (records), terms (content of the information fields), and time (usually years). This chapter focuses on analyses pertaining to *terms changing over years*. One can also map documents over years. That is another way to get at technological activity changing over time, for instance, in tracking when a company began to patent in each of several specific domains.

Tech Mining: Exploiting New Technologies for Competitive Advantage, Edited by Alan L. Porter and Scott W. Cunningham.
ISBN 0-471-47567-X   Copyright © 2005 by John Wiley & Sons, Inc.

EXHIBIT 11-1   *Steps in Trend Analysis*

1. Spell out your objectives and perspective; decide if you want to model your data.
2. If so, identify candidate models.
3. Collect and treat (possibly transform) the data.
4. Fit the models to the data.
5. Project forward with the suitable model or models (if you are interested in forecasting).
6. Check model fit and perform sensitivity analyses.
7. Interpret.

This chapter notes the straightforward trend analysis possibilities but does not expend much energy on these (see Chapter Resources). It instead introduces a number of ways to measure change over time. We suggest ways to deal with data uncertainties prevalent in tech mining. Much of the discussion is couched in terms of regression, a basic statistical technique. (Any lack of familiarity with statistics should not seriously hinder your understanding of the ideas presented.)

Time series data are prevalent in tech mining, and a variety of questions can be answered in many ways, so we don't present a lockstep procedure. That said, Exhibit 11-1 offers a set of steps worth addressing in some order. However, the chapter is not organized neatly by steps, so you will find discussions and illustrations interspersed as we raise various tech mining trend considerations. Previous chapters addressed many data issues.

If you are doing trend analyses, adapt these steps to your situation. For instance, you may instigate trend analyses on inspection of your tech mining data to find potentially important shifts over time. That may lead you to obtaining more historical data to deepen the time series. If you are the consumer of trend analyses, just check that the analysts have performed some approximation of these. For instance, you probably don't want to be swamped with lots of alternative trends and projections, but you deserve assurance that the analysts have considered alternative formulations.

Description Versus Prediction

This chapter discusses several techniques for analyzing changes in publication and patent data over time. Two reasons for performing trend analyses contrast sharply—description and prediction. *Description* means succinctly summarizing the patterns and trends found in the existing data. *Prediction* seeks to extend those trends into the future to help anticipate likely science and technology advances.

Trend description helps address many technology management questions. Technology monitoring requires ongoing efforts to summarize incoming data.

Describing trends and changes in them can help to understand what's changing in S&T. This can help decision-makers get abreast of shifting emphases in a research area, or of changing emphases by a rival organization. At the least, trends help gain an overview of R&D activity in ways not otherwise viable, say by reading thousands of patent abstracts. Describing trends in the data is a vital part of technology monitoring.

Prediction aims to understand where S&T activities are heading. This is a daunting task. Publication and patent data provide useful information, but we recognize that these are "outputs" of an *S&T system*. As such, they don't tell us directly about changes in the underlying mechanisms. Consequently, we need to be wary in offering predictions. We recommend combining publication and patent trend analyses with other empirical analyses (e.g., R&D funding patterns) and with expert judgment about the research domains under study to generate strong predictions.

Trends of What?

As just noted, we focus on trends in observed publication and patent activity (but most of the principles apply to other such data). We have time series of data on these outputs, not on the "real" processes generating them. *So beware extrapolating observed trends in situations where the driving forces are apt to change.* For one thing, publication or patent patterns for emerging technologies are likely to hold only for relatively short time periods (i.e., a few years). For another, we need to reflect on the likelihood of new techniques, conceptual breakthroughs, empirical discoveries, or market shifts altering the historical patterns we observe. We need to remind ourselves that R&D does not plow forward in a simple linear fashion to commercialization. In addition to science impacting technology, new technology (e.g., tools, methods) often opens up new scientific opportunities. Furthermore, R&D outputs generate new ideas and opportunities to pursue on side routes. Disappointing application results may feed back to develop a capability further, leading to different applications. Technological innovation results from highly interactive, multivariable systems. Some observers would distinguish three forms of technological change: incremental, transitional, and transformational (Kash and Rycroft, 2000).

We address trend analyses separately in this chapter, and not in Chapter 9, because we want to press beyond "data description." At the simplest level, we just plot activity versus time to see the pattern. At a second level, we break out subtopics from the overall activity of an R&D domain to track shifts in topical (or other) emphases. For instance, at a first level we plot general nanotechnology research publication by year (see Fig. 11-1 below). At a second level, we might plot occurrences of selected subtopics (e.g., from Table 11-3 later in the chapter, we might compare research intensity on nanofilms with that on nanotubes—not illustrated). Tech mining software expedites such sub-data set trend analyses by making it easy to group and separate records.

At a third level, we consider ways to analyze changing publication and patent patterns over time to help elucidate underlying mechanisms. Toward these ends, we will introduce approaches to estimate rates of change, ways to combine multiple entities in trend analyses, and means to examine the fit of particular growth models to the time series data.

A potential weakness of the time series approach is the fairly simplistic depiction of scientific growth that results. We might determine, for instance, that a given area of technology is growing at 35% yearly. Unfortunately, we do not learn the causes of such rapid growth, nor do we know whether the growth can be sustained in the future. Nonetheless the relative success of these simple models of S&T adoption suggests that many fields operate under conditions of relative stability. The collective action of multiple scientists, regularity in funding patterns, and publication continuity (active publishers tend to keep publishing) helps ensure relatively steady growth for many S&T fields. When the driving forces remain essentially the same—a situation known as "stationarity"—trend projection is valid. Determining stationarity calls for insight into what is driving R&D in the domain and any impending limits (physical, funding, or market). The greater the understanding of the technological systems in question, the more assured you can be of getting this right. This cries out for involvement of substantive experts in making projections for an emerging technology.

The prototypical tech mining time series data concern some "content" over "time." Content—that is, terms—means patent classes, title phrases, keywords, inventors, organizations, and so forth. But any of the content of our document sets can be analyzed over time. For instance, we might track changing coauthorship patterns or national interest in a subtopic. "Over time" usually means exploring how these measures change year over year. Rarely do the data warrant using finer periods (e.g., months) because of anomalies at that level (e.g., a substantial portion of a year's research publication might derive from a conference that happens to fall in a given month). On the other hand, we sometimes find value in consolidating multiple years as "time slices" to see important changes better. Unfortunately, for "emerging technologies," we rarely have the luxury of extended time series for such purposes.

So approach simple trend descriptions and predictions with caution (for a classic treatment of the pitfalls see Ascher, 1978). Think through what factors and forces could alter historical patterns. Trend forecasts do best in S&T domains showing a cumulation of incremental changes. The prominent example is Moore's law—describing sustained exponential growth trajectories in many facets of silicon-based semiconductor advances from the 1960s over the following 50 years (more on this in Section 11.2). As we write, however, prognosticators speculate on how much longer these trends will hold, mainly because of impending physical limitations as component dimensions shrink toward the nanoscale.

TABLE 11-1. Nanotechnology Publication

Years	Observed (& Estimated) Annual Publications	Cumulative Publications	Log of Annual Publications
1988	3	3	1.099
1989	0 (5)*	8	1.597
1990	3	11	1.099
1991	20	31	2.996
1992	9	40	2.197
1993	18	58	2.890
1994	34	92	3.526
1995	30	122	3.401
1996	38	160	3.638
1997	42	202	3.738
1998	75	277	4.317
1999	88	365	4.477
2000	135	500	4.905
2001	175	675	5.165
2002	293	968	5.680
2003	304 (405)*	1373	6.004

*Values have been estimated.

11.2. AN EXAMPLE TIME SERIES DESCRIPTION AND FORECAST

In this section we illustrate trend analyses on a sample research publications data set generated by searching on "nanotechnology" in the title, keyword, or abstract fields in the *Science Citation Index (SCI)*.* We start with simple analyses, then introduce regression analysis, and proceed through a range of more advanced analyses.

Basic Nanotechnology Trends

Table 11-1 arrays the number of nanotechnology articles by year. The table clearly indicates dramatic growth. (We discuss the two estimated values later and address the other columns shortly.) Figure 11-1 plots the same data. Some users prefer the tabular presentation to get the precise numerical values; others perceive the pattern more effectively from the visualization. Many alternative visualizations are possible—consider line charts, histograms (bar or column charts), and three-dimensional column charts to compare trends for several entities (Chapter 16 illustrates such options.) Find out what information representations work best for your target users.

*For a more considered tech mining treatment of nanotechnology R&D activity patterns, see Hullman and Meyer (2003). This exercise only illustrates trend analyses; it taps one database with a simplistic search. Much molecular level S&T research does not use the term "nanotechnology."

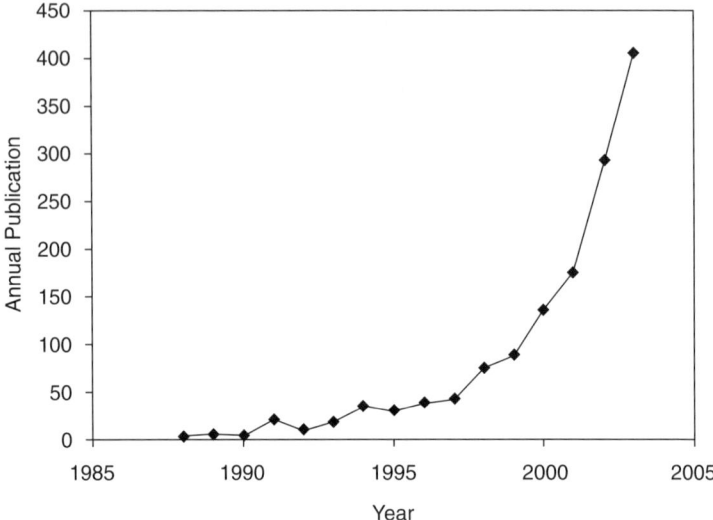

Figure 11-1. Annual nanotechnology publication

Consider "second-level" trends we might calculate. Instead of charting the overall activity, we could break it out for a partial set of certain terms. For instance, we might identify the "Top 6" universities (or whatever) for this data set and then chart the publication activity for each of them separately. Better yet, we could compare the trends for those Top 6 in one figure, possibly looking for those with the longest duration of, or the most recent, activity.

Also decide how you wish to treat "time." For some purposes, you may want finer than yearly data, possibly to pick up seasonality. This is most likely for economic aspects of technological innovation, such as sales of a technology-enhanced product versus the base product. As already mentioned, sometimes we find it helpful to bundle years together as "time slices." These can be chosen to reflect significant contextual changes. In examining Iraqi engineering R&D (Porter, 2003), we used natural break points (e.g., occurrence of the Gulf War) to make up periods of roughly comparable duration (or total activity) to spotlight marked shifts in R&D publication.

Figure 11-2 plots the *cumulative* data. (This just sums all the activity year by year, beginning in 1988, up through a given year.) This "transformation" is useful in visualizing the "S-shaped" growth curves, to be discussed further.

Basic Regression Notions

Here is a quick primer on linear regression ("regression" for short)—a basic statistical technique that proves quite helpful in trend analyses. Equations 11.1 develop regression notions. Part (a) shows an ultrasimple relationship. Suppose Y is the English system unit of length, yards, and X is feet. This equa-

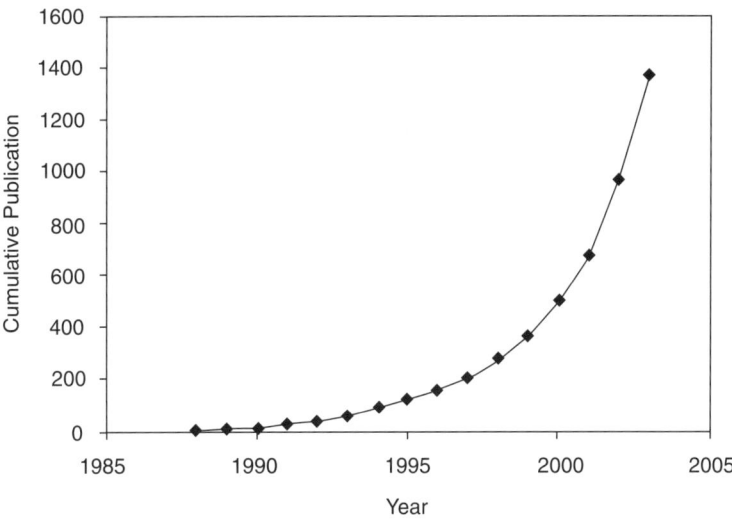

Figure 11-2. Cumulative nanotechnology publication

tion can calculate the number of yards in any given number of feet (e.g., 15 feet equals 5 yards)—Day 1 of high school algebra! Linear regression is just an extension of this.

Equations 11.1. Linear regression

(a) $Y = (1/3)X$

(b) $Y = bX$

(c) $Y = a + bX$

(d) $Y = a + bX + e$

(e) $Y = a + b_1X_1 + b_2X_2 + e$

(f) $Y = a + b_1X_1 + b_2X_2 + \ldots b_NX_N + e$

Part (b) generalizes the relationship between Y (the dependent variable) and X (the independent variable). The constant "b" could take on any value. When X is multiplied by b, one gets Y. Returning to high school algebra, we sometimes had to solve one or more equations like this to find "b."

Part (c) adds a constant. This allows one to displace the relationship by the amount "a." Again testing your recall (if you're American), part (c) could be used to relate degrees Fahrenheit to degrees Celsius. In this case, $a = 32$ and $b = 9/5$. To express 40 degrees C in degrees F, we solve part (c) to get 104 degrees F.

Part (d) brings us into real statistics! Here we add an error term, e, reflecting that our equation may entail imprecision. Unlike the relationship between degrees Fahrenheit and degrees Celsius, other relationships are often noisy. For instance, consider American men's weight as a function of their height. Weight tends to be greater for taller people, but there is a lot of individual variation. The greater the scatter about the line [part (c) is an equation of a line], the larger "e" is. If all the points fall exactly on this regression line, "e" is 0. If the points are a "shotgun blast" of randomness, with no relation to the line, $b = 0$.

Part (e) is getting slightly sophisticated. This is multiple regression. Here we use two independent variables to help predict Y. For instance, we might refine our "weight as a function of height" equation by adding gender as a second "X" term.

Part (f) just expresses this in a more general way. Here one could have as many independent variables as desired. That's regression. Solving these equations to fit actual data yields specific values for the "a" and "b" coefficients that constitute many of our trend models. This involves minimizing the sum of squared deviations of the actual data about the calculated regression line. MS Excel or any statistics package does the calculations.

Linear regression attempts to model the relationship between two variables by fitting a linear equation to observed data. Before attempting this statistical fitting, a modeler should first consider whether or not there might be a meaningful relationship between the variables of interest. This does not necessarily imply that one variable causes the other (for example, higher SAT scores do not cause higher college grades), but that there is some significant association between them. The commonly used method of least squares calculates the best-fitting line for the observed data by minimizing the sum of the squares of the vertical deviations from each data point to the line. If a point lies on the fitted line exactly, then its vertical deviation is 0. Because the deviations are first squared and then summed, there are no cancellations between positive and negative values.

Many tech mining variables have integer values. But regression can generate noninteger values. Reality frowns on partial patents or publications. Usually you can just round to the nearest integer. Better yet, deemphasize "point" forecasts in reporting tech mining results in favor of likely ranges.

Growth Models

In technology forecasting, we have found four growth models particularly useful (Exhibit 11-2). This Exhibit shows how Equation 11.1 (d) would be adjusted to represent each type of growth. The linear model requires no transformation, just use Equation 11.1(d). The other three are inherently nonlinear, but we can transform them into a linear form to use linear regression to fit their model parameters. We then can transform them back if desired (e.g., take the antilog in these cases). Exponential growth is properly fit by a straight line if one takes the logarithm of Y (either natural log or log to the base 10). The two S-shaped growth curves are fit by the somewhat more intricate transforms shown. All these models pertain well to cumulative growth data.

Annual growth data are obtained by taking the difference of subsequent cumulative growth data. An annual growth data model results from taking the

---

EXHIBIT 11-2 *Four Growth Models*

Model	Linear Transformation
Linear	Y
Exponential	ln (Y)
Gompertz	ln (L/Y) where L = the upper growth limit
Fisher-Pry (Pearl)	ln [(L − Y)/Y]

---

derivative of a cumulative growth data model, so relationships differ accordingly. Figure 11-3 (a) plots hypothetical annual data that grow exponentially to a peak, then decline to zero. Plot (b) shows the corresponding cumulative data. This "S-shape" reflects slow initial growth, followed by rapid ascendancy, then slowing growth that asymptotically approaches the limit (L). Variations of this pattern are most common for growth in technological capabilities and also for market diffusion processes. They are not universal, of course.

Many other equations also describe growth relationships. A particularly popular S-shaped curve for representing market growth is the Bass model. Systems of equations can be formed to describe more complex, interactive relationships. For instance, the Lotka-Volterra equations represent Y as a function of X, but also X as a function of Y (Porter et al., 1991). The next section investigates exponential growth modeling.

Which model, using which form of data, best suits publications and patents? Unfortunately, the answer is not clear. Growth in the "hard sciences" seems better fit to annual publication rates, whereas the "soft sciences" seem better aligned to models based on cumulative growth. Why? Perhaps in faster-changing domains, one finds many active researchers feeding off more recent literature for ideas, new tools, etc. Scientists read (and cite) less older literature than do social scientists. We suggest you inspect both the annual and cumulative publication or patent trends. Consider which seems most suitable to the case at hand.

Do S-shaped curves model publication or patent growth well? On the one hand, publication in a fast-growing domain like nanotechnology cannot grow at exponential rates forever. However, to fit the S-curves, one needs to specify a limit—that is, an upper bound to growth. In many technology domains one can identify real limits to growth (e.g., physical limits on microelectronic effects). In most market domains we also have reasonable grounds to set limits (e.g., 100% as maximum market penetration). Publishing or patenting have no actual limits, but we can generate conditional limits for modeling purposes. For instance, based on our sense of the number of active researchers and the extent of funding we might pose 1000 articles per year as a limit for a horizon of interest, say 5 years. This may help us generate a reasonable fit and projection over the next couple of years. A good idea if we take this route is to try out several reasonable limits and observe how model outcomes vary as a result (e.g., set L = 800, 1000, and 1200 in the equations of Exhibit 11-2).

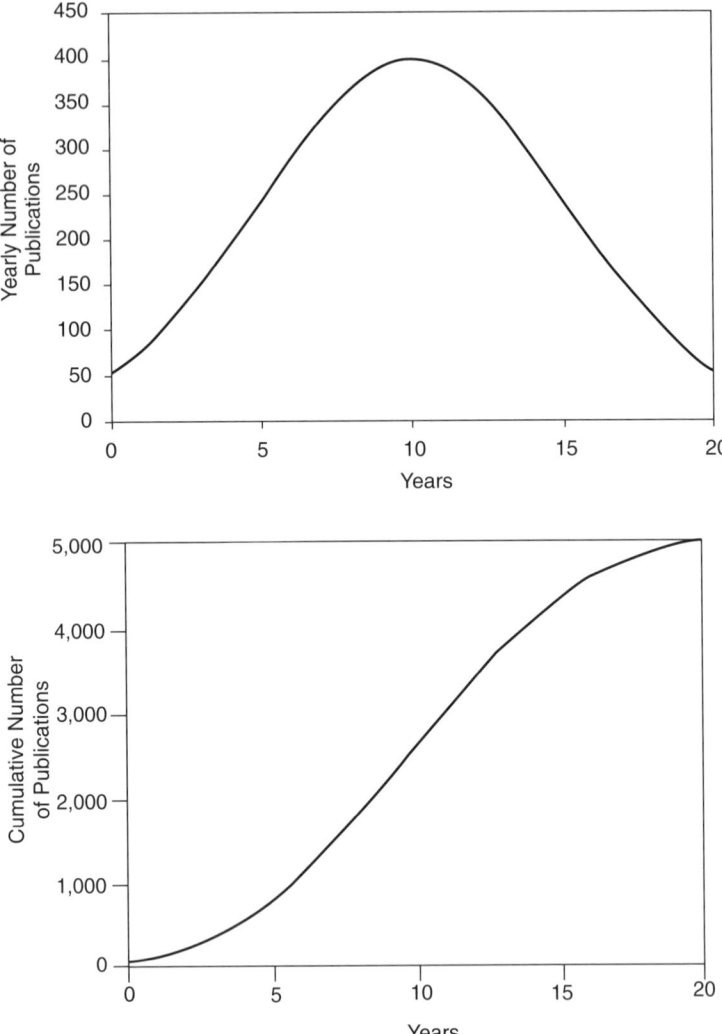

Figure 11-3. Annual vs. cumulative data

Another good idea is to fit several models (e.g., both S-curves) as a form of sensitivity analysis—observe results and interpret by offering a reasonable range of projections.

If an S-curve appropriately describes growth of the research domain being studied, we then fit one or more such curves. These can provide highly valuable insight into impending maturation of the domain. Even as the number of patents or publications continues to increase, but at a decreasing rate, we can signal to management that this domain may be losing momentum. This is vital information for R&D program management to begin to redirect efforts early on.

What happens to publication in a hot emerging technology domain like nanotechnology? A paradox to ponder—if you scan across research domains at one point in time and compare their growth rates, smaller domains tend to be growing faster. But if you track one domain over time, adjusting your search algorithm to capture relevant research emphases, growth often continues unabated as the domain grows. Why? Many factors interact. Subtopics tend to grow and wane—you get annual rates that rise and fall, often roughly like a bell-shaped curve (normal distribution). Cumulation of these subtopics over time may generate an "envelope curve"—a succession of S-curves building one upon another. Another tendency is toward specialization. As interests evolve and particularize, terminology changes. So, a search on "nanotechnology" 5 years from now is apt to capture considerably less of the related activity than it does today. In other words, we add indexing, or searching, complexities on top of the increasing specialization of the research.

As the number of publications or patents under study becomes large, differences in modeling approaches become less influential. For nanotechnology, we have more confidence in the recent trend than in the early years of the time series. Note in Figures 11-2 and 11-3 the relative smoothing of the curves after the early period.

Exponential Growth

Although the S-shaped growth curves just discussed are more common in technological innovation, another simple and effective time series model is to assume that the growth rate is constant over time. The most famous exponential growth case is known as Moore's law. Gordon Moore, founder of Intel, predicted in 1965 that a number of key semiconductor performance metrics would keep improving exponentially—and they have for five decades! The constants differ somewhat and, in fact, show some changes over time, but Moore's law has profoundly captured the exponential growth powering "IT" and our "Information Economy." For example, the density of components on computer chips doubles roughly each 2.5 years.

Exponential growth assumptions are expressed mathematically by Equation 11.1 (d) with $\log(Y)$ replacing Y, and in alternative forms by Equations 11.2. Yearly publication is Y; the change of publication over time is the derivative of Y; and the rate of publication change is b. Note that the independent variable (the general "X" in Equations 11-1) is specified as time "t." Equations 11-2 nicely represent the case in which "publication increases by a constant multiplier each year." The resulting differential equation represents exponential growth. (Or, when b is negative, it represents exponential decay.) The variable C represents a constant of integration—in this case, the level of publication at year zero. (We may arbitrarily set "year zero" to whichever year we wish. For this example, we set it to 1987.)

Equations 11.2. Constant publication growth

$$\frac{d^2Y}{dt^2} = b$$

$$\frac{dY}{dt} = bY$$

$$Y = C\exp(bt)$$

$$\ln(Y) = \ln(C) + bt$$

We have already seen that our nanotechnology publication activity shows strong growth. Is an exponential growth "model" suitable? Visually, both annual growth (Fig. 11-1) and cumulative growth (Fig. 11-2) appear exponential. Figure 11-4 shows the linearized version of the annual growth, reflecting the log (no. of publications). Note that after the earliest years, the data fall quite nicely linear. (Not shown, the logarithmic version of the cumulative data also are nicely linear after the first few years.)

We want to estimate the rate of change of nanotechnology publication. To use linear regression, we can take the last form of Equations 11.2. We have 16 data points (Table 11-1): t values are the years; $\ln(Y)$ values are the log (no. of publications)—Column 3 of Table 11-1. The regression shows a good fit, as reflected in the R^2 (amount of variance in the values about the line accounted for) of 0.95. The growth rate (coefficient "b" in the last of Eqs. 11.2) is 31%. (See sidebar on "compounding growth.")

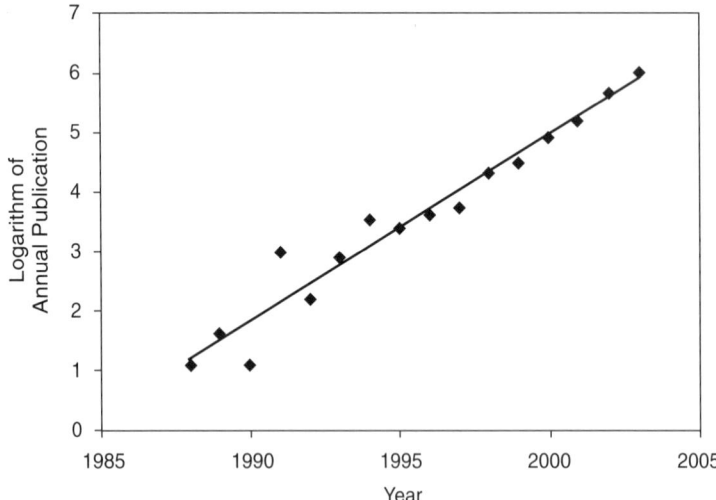

Figure 11-4. Logarithm of annual nanotechnology publication

> ### *Sidebar: Compounding Growth Rates*
>
> Note two possible ways to express the growth rate. One is to take the coefficient from fitting the growth equation. This is like receiving interest on an annual basis. If you get 31% interest on $100, you have $131 at the end of the year. But we may prefer to think of continually compounded interest. That is, if the bank pays you interest daily, instead of yearly, at the same rate, you end up with a bit more than $136. Whichever rate you use, the nanotechnology literature shows a startlingly rapid rise in publication.

This same regression line can be extended into the future to predict publication rates. Doing so anticipates 493 publications in 2004 and 674 in 2005. Obviously, one should not believe in any such precision. Indeed, 50% confidence intervals for 2005 stretch from 510 to 891 publications. Confidence intervals widen as one extrapolates farther into the future (these are discussed further in a later section). (Regression routines to do these calculations are available in MS Excel or in statistical packages.)

Exponential models are one of our four attractive growth models. Exponential projections would not be expected to hold over an extended horizon. Scientific interests specialize and shift, and the number of scientists is limited. However, for the short run, they make sense. They offer an appealing simplicity in that they can be given as a compounding growth rate. Nearly all of us understand such notions from experience with checking account or credit card interest rates.

Which growth model should we use—exponential, linear, or S-shaped? (See previous section.) The tech mining analyst must consider:

- What fits our concept of what is going on?
- Which of the four (or more) models fits the data best? (judge this based on appearance and statistical goodness of fit)

Conceptually, what makes sense? Tech mining analysts should engage persons who understand the S&T in question to check. To gauge fit, one should consider data issues too—the next section addresses these.

Data Considerations

Publication and patent data sets that span many years give us a rich source of tech mining information because they show how emphases change over time. However, they also pose a host of issues in dealing with time series data. This section identifies a number of issues and suggests how to deal with them.

Chapter 9, in particular, treats issues in cleaning data. Were this not done, you can well imagine problems being caused for trend analysis—for example, inflated values due to duplicate records. Consider also general data considerations, including:

- Data quality—What basis do you have that reliability is satisfactory?
- Data relevance—Do the data measure what you want to analyze?

Missing Values

We sometimes have gaps in the data. Some statistical approaches can just ignore these and go about fitting a suitable trend. If not, you need to estimate the missing values. Such procedures can become quite complicated. Consider how you are using the results to determine how much care is warranted in coming up with a superior estimate. For instance, in our nanotechnology data (Table 11-1), we deal with a related data issue, the zero value for 1989. Given that we are much more interested in recent activity developments, we shouldn't expend much effort on this "ancient" value. Put another way, the pertinent conclusions and tech mining recommendations are not very sensitive to how we treat this value.

What we have done to generate an estimate for the 1989 nanotechnology publication is to fit the regression and use its calculated value here. One approach is to iterate this estimation. For example, we could first insert the average of the adjacent series values (3 for both 1988 and 1990, so 3). Then fit the regression, obtaining an estimated 1989 value of 4. Repeating the process, converges to a value of about 4.5, which we round up to 5.

Outliers

Another issue concerns how to deal with unusual series values. Note that the 1991 nanotechnology publication value is considerably greater than the adjacent values. Our first question is whether we feel our tech mining conclusions would be highly sensitive to this. In this case, the answer is "no," so no further refinement is needed. Were we concerned, we might next examine the actual data to see whether we can tell what is happening. For instance, publication data often show bursts of activity as special conferences convene on topics not routinely addressed. Were that the case, we accept the data and turn to how best to analyze them.

One powerful approach to deal with "bursty" series is smoothing. There are many variations, such as moving average (e.g., replace each year's value with an average of it and the previous 2 years' values) or exponential smoothing (places more weight on more recent data).

Another approach can combine statistical treatment and judgment. Note, for instance, that Poisson regression is an explicit technique for dealing with bursty data. Visual inspection can flag an outlier data point that appears erroneous. We can check this by statistical measures (e.g., comparison with a regression estimate for that year). We can also ask knowledgeable persons whether they can explain the apparent outlier. Explanations could provide insight into mechanisms, unusual actions, or research breakthroughs.

When we have an outlier that seriously affects the tech mining, we have several choices. We can discard the outlier and treat it as a missing value. Or we can adjust it as illustrated for the next topic.

Partial-Year Effect

Typically, our most recent data point(s) are incomplete. This causes an apparent decline in publication rate and may lower estimates of growth rate. Most critically, it may lead to erroneous conclusions that the R&D domain is in decline.

There are multiple causes including *incomplete indexing* by the database. In the past, lags of a couple years were not uncommon between publication or patenting and inclusion in the database. This could be compounded by version lags—for instance, if your organization acquires the database on CD quarterly or annually. We also see differential lags for various patent offices. Furthermore, a differential occurs between publication outlets that provide electronic version access to particular databases and other conferences or journals that are indexed only after actual paper publication and distribution. Although major S&T databases have reduced the indexing lag, it is not unusual for publication tallies to continue to increase slightly for several years.

Tech mining almost always grapples with whether to include the most recent *part-year* data. On the one hand, getting the most up-to-date data is vital. Tech mining data already suffer from lags between actual research and publication or patenting. On the other hand, the most recent yearly tallies are artificially low. Again, choices include leaving out the most recent, incomplete year to combining this last part year (especially if it is a small portion of the year) with the previous one to help compensate for the indexing effect. In general, a better approach is to adjust the data according to the fraction of the year to date that is included. In the "nanotechnology" example, the 2003 data were collected through August 2003. We called this about 75% of the year, and so corrected the observed value by multiplying by 4/3 (Table 11-1).

Zeroes

A special estimation issue arises if we are fitting an exponential model—the logarithm of 0 is undefined. Our judgment on how critical the estimates are influences how simple or complicated we make the resolution. We have lots of choices:

- Omit that value (best if relatively few zeroes)
- Assign a very low value and proceed (sacrifices some accuracy)

- Model the data in their cumulative form
- Estimate a nonzero value; then fit the exponential growth model; next replace the initial estimate with the new one provided by the model; iterate until suitable convergence is obtained (as presented in addressing missing values)
- Estimate the model in its exponential (nonlinear) form. (This involves maximizing the log likelihood of the normally distributed error about a function estimating the function—e.g., publication activity.)
- Use Poisson regression. This stretches beyond our scope, but is worth noting in dealing with series with more than a few low numbers (i.e., near 0). Poisson fits and forecasts never admit negative outcomes, whereas regression based on normal distributions can. This is especially likely in presenting forecast ranges (confidence intervals). Managers could be put off by "negative publications or patents," so be sure to present lower limits as 0. We illustrate the use of Poisson regression in the next subsection. Agresti (2002) offers a complete introduction to Poisson regression, as well as other techniques for handling categorical data.

An "Emerging Technology" Nanoexample

One of the early proponents of nanotechnology, Eric Drexler, anticipated a "wet" route to nanotechnology. He anticipated that nanotechnology might progress by emulating organic life. Let's examine a sample data set that loosely pursues this interest by using the query: both "nanotechnology" and "genetics" as keywords. This yields a small subset of 30 of our nanotechnology documents. Table 11-2 shows the publications by year for this query.

As can be seen, in the early years of the time series, publications are very sparse. The many zeroes, low numbers, and dramatic growth suggest that Poisson regression would be a good approach to fit an exponential model. We choose, for convenience, the year 1990 as "year zero." Exhibit 11-2 shows the estimated parameters for the last of Equations 11.2 for our exponential growth model using Poisson assumptions of noise. The forecast estimates an astounding 61% growth rate for publications in this area. The equivalent normal (Gaussian distribution) regression (not shown) estimates 45% growth and, more significantly, yields a much less likely explanation of the data (based on log likelihoods).

EXHIBIT 11-2   *Model Parameters for "Nanotechnology and Genetics"*

Intercept	Parameter	
[year 0 = 1990]	−3.770	
Slope	0.470	Annualized growth 61.1%
Log likelihood	−19.466	

TABLE 11-2. Publications per Year for the Query "Nanotechnology and Genetics"

Year	Year Index	Publications
1985	−5	0
1986	−4	0
1987	−3	0
1988	−2	0
1989	−1	0
1990	0	0
1991	1	0
1992	2	0
1993	3	1
1994	4	0
1995	5	0
1996	6	0
1997	7	0
1998	8	1
1999	9	1
2000	10	2
2001	11	7
2002	12	10
2003	13	8

For amusement, we project these two growth rates out to 2020. From the same time series, the Poisson projects 38,000 vs. the normal distribution, exponential model regression's 5800 publications per year. For obvious reasons, we would not expect these long-term projections to pan out. We extrapolate just to indicate that these two ways to fit an exponential growth model to a single data set yield different results. Figure 11-5 shows only the Poisson-based exponential model projection, with confidence intervals.

In tech mining, dealing with "emerging technologies" often confronts us with data series like this—very short series (if we ignore the single 1993 publication) with low numbers. Be alert to the distributional assumptions in considering how best to model such data. We also emphasize that any resulting model should be treated as crude. The next year's data point is likely to alter the model and its projections significantly. So, interpret findings with due caution. On the other hand, tech mining provides great value by early identification of such new research fronts. The 30 publications in 10 years is a "ho-hum" research specialization, but interpreted via the trend model, one sees a potential blockbuster brewing in nano/genetics.

Confidence Intervals

Confidence intervals are an important part of time series projections. As noted above, publication or patent counts are variable, as they reflect the interplay of many forces and factors. The forecasts based on these data cannot be taken

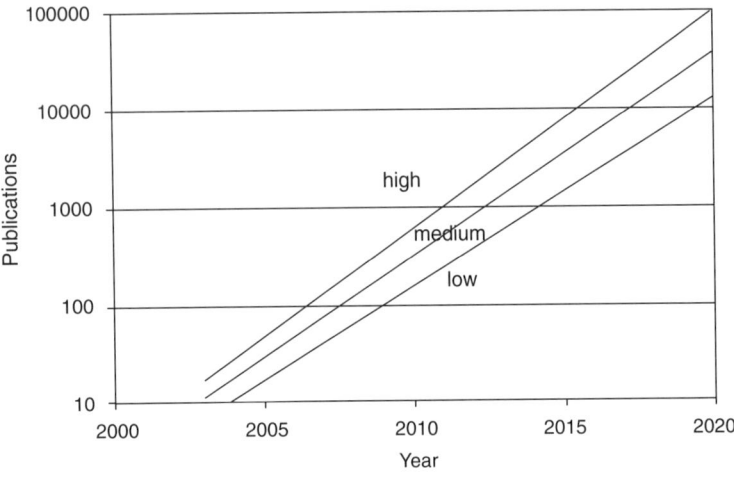

Figure 11-5. Forecast bounds for nanotechnology and genetics

as absolute. Providing confidence intervals communicates this uncertainty and natural variability to decision-makers who may use the forecasts. Confidence intervals (CIs), or prediction intervals, are a function of the assumed data distribution. They include three terms—a term that increases the further out one projects the trend, another that decreases as the data set size increases, and an increasing function of the scatter of the actual data about the regression line. A 90% CI implies only a 10% chance that the actual value would fall outside these bounds (based only on statistical considerations). A 50% CI has a 50% chance of proving wrong, so this CI fits more snugly to the regression line (cf. Porter et al., 1991).

Figure 11-5 shows 90% confidence intervals for our forecast of "nanotechnology and genetics" publications. The middle forecast is much more plausible than either the high or low forecast. Note that the CI bounds are not symmetric about the regression line because the model is nonlinear (it's exponential). Also, the bounds widen as the extrapolation lengthens—the longer the forecast, the higher the uncertainty.

Recall that the 2020 "forecast" using normal instead of Poisson regression (last section; not shown in Fig. 11-5) differed greatly. These "naïve" forecasts do not address the causal forces driving publication rates. The tech mining analyst would want to interact with subject experts to consider what the drivers are and how they are likely to change. Uncertainties in research output measurement and complexities in the R&D system combine to endorse active, ongoing monitoring. These considerations also demand careful interpretation of emerging technologies—active research does not guarantee successful innovation. Put another way, were a forecast of "nanotechnology and genetics" made today, it should be frequently revisited and updated.

11.3. MULTIPLE FORECASTS

In the previous sections we developed trend models and projections for a chosen technology. We presented logistic (S-curves) and exponential growth models and compared models with different distributional assumptions. Although the example applied the techniques to nanotechnology, any variable (or for tech mining, any term of interest) could be modeled with similar techniques. In fact, we fit exponential models for both overall nanotechnology publication activity and also the subset of "nanotechnology and genetics."

We could also have modeled these research outputs with a linear model (but that would obviously have been stupid—see Figs. 11-1 and 11-2). More appealing, we could have tried S-curve models and compared their fit and projections to the exponential models. Indeed, various Gompertz and, especially, Fisher-Pry models fit the nanotechnology data quite well (about as well as our exponential models). We strongly encourage comparing such multiple models as a form of sensitivity analyses.

We now suggest "multiple" in another sense—to compare the trends of multiple entities. Variations on this approach can be used to compare rates of progress across technologies. Here we compare multiple values from the nanotechnology publication content. We can apply this to answer questions such as:

• Which subtopics of nanotechnology are growing the fastest?
• Which nations are pursuing nanotechnology the most vigorously?

This multiple-indicators approach, which we will illustrate in this section, aids in scanning the technological environment to identify high-growth research domains or rapidly shifting organizational emphases. It can help benchmark one's activity against leaders in the field. This points toward another tech mining principle—relative (comparative) indicators are generally more effective than absolute (stand alone) indicators.

Cross-Sectional Plots

Often we are most interested in recent change in a target R&D domain. Also, sometimes we lack much early data. A handy alternative to trend plots is to produce a cross section of publication outputs for each of two years. Figure 11-6 shows one such cross section. Plotted on the figure are one hundred of the top keywords occurring in the collection of nanotechnology articles. Plotted on the graph are the numbers of publications containing each of those keywords in 2001 versus 2002.

The graph is scaled logarithmically on both axes. This presents more than three orders of magnitude between the most frequent and the least frequent

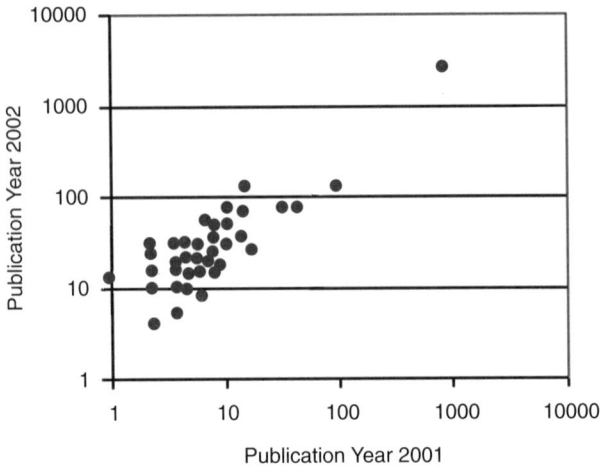

Figure 11-6. Nanotechnology publication 2001 vs. 2002

keyword in the selection. This cross-sectional plot can help identify high-growth subareas. To do so, we would add labels for the more interesting key-words, particularly those showing much higher values in 2002. Terms falling along the diagonal occur as frequently in 2002 as in 2001. Terms above the diagonal increased in popularity, suggesting heightened interest (see Fig. 11-8 below).

Figure 11-7 shows three idealized patterns of growth encountered in cross-sectional plots. The patterns are shown as clean lines; real plots are noisy (Fig. 11-6). The idealized patterns help distinguish types of S&T growth. The first pattern is when nearly all subtopics, or whatever variables (terms) you are examining, increase over time—this is labeled "increasing growth" in Figure 11-7. This suggests a uniformly expanding research domain. The second, "no growth" pattern evidences similar output year over year for all the subtopics. The "declining growth" pattern finds all the subtopics shrinking in the later year.

Now imagine Figure 11-7 with the lines angling. Let's focus on the growth side (above the diagonal). If the "increasing growth" line sloped more steeply than the diagonal, that would mean the large subtopics were gaining momentum relative to the smaller ones. This would fit a situation where R&D were concentrating in certain prime subtopics at the expense of fringe interests. This "big get bigger" phenomenon is so well known in bibliometrics that it has its own name, the "Matthew effect," from the biblical proverb. Mechanisms such as focused funding and peer reinforcement can fuel this.

Looking at the nanotechnology cross-sectional plot (Fig. 11-6), we see the expected noise. Not every keyword grew over time, nor did the keywords grow in equal proportions. Fitting a regression line shows a pattern of the

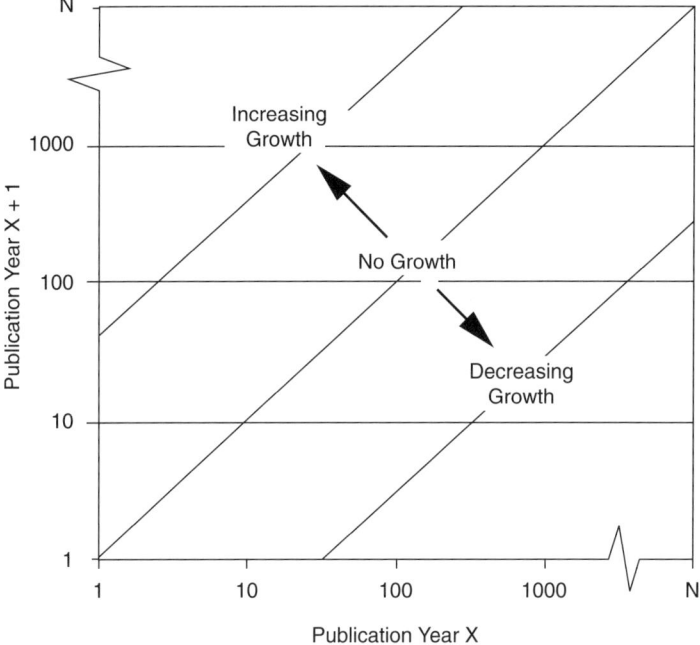

Figure 11-7. Three patterns of growth

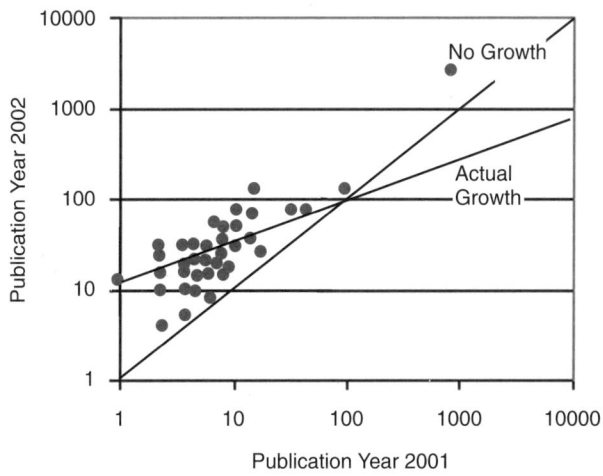

Figure 11-8. Patterns of growth in the nanotechnology sample

"small get bigger" phenomenon (Fig. 11-8). This may reflect, in part, a size phenomenon—it is easier for low-frequency terms to show a high percentage shift increase. Put another way, we are more likely to find low-frequency keywords well above the diagonal; Figure 11-8 shows just that pattern. Interestingly, the two most frequent keywords in the sample—"nanotechnology" and "microscope"—contradict this. They are above the extrapolated line as well as the no-growth diagonal line. This means that the terms are growing relatively more vigorously.

Note that you could represent these data other ways, too. As the nanotechnology publications are increasing so rapidly, from 175 to 293 papers from 2001 to 2002, you might recode values as a fraction of the year's total. Such a relative scaling could help focus on fast-growing topics. We have noted the tendency of low-frequency terms to have a propensity to be especially fast growing. So, you might use a threshold to exclude very low-frequency terms (e.g., 1s and 2s). Our message is to explore alternatives to help you discern and then communicate phenomena pertinent to the technology management questions you are pursuing.

Multiple Regression Techniques

Cross-sectional, biannual scatterplots visualize the data simply. However, they only utilize two years (or two time slices) of the data. Plots of publication or patent activity for each of several leading terms over time provide a rich alternative (cf. Fig. 11-4) but can get visually complex if we include many terms. If one wants to assess growth rates for many terms over the entire time period, multiple regression works well.

Multiple linear regression employs multiple independent or dependent variables as per Equation 11.1(f). In the following example we have multiple independent variables predicted by a single dependent variable. We seek to predict the pattern of keyword occurrences using time as the key variable. In this example we look most closely at predicted time coefficient—the growth rate.

Figure 11-9 gives an example of growth rates across a selection of leading nanotechnology keywords. This particular display accentuates distributional characteristics. For other purposes, one would want to provide the term identities, particularly for the fast-growing keywords. The figure illustrates that there is a range of growth within the single topic of nanotechnology. Many keywords within nanotechnology are growing at rates of 18–27%. Relatively few keywords are growing much slower or faster. This gives a good sense of the rate of advance of the core fronts of the field.

Recall earlier that the average growth rate for all of nanotechnology was 31%—yet the average keyword within nanotechnology grows more slowly (22%). Why is this? In this particular example, a few high-magnitude, high-growth terms dominate the sample (Table 11-3). This pushes the overall growth rate higher. Nanotechnology, like most fields of science and technology, is a very heterogeneous topic. Results such as Figure 11-9 suggest that the growth rates of keywords within nanotechnology are themselves random variables. Point estimates of growth do not appear to capture the concept that some areas of nanotechnology grow much faster than others.

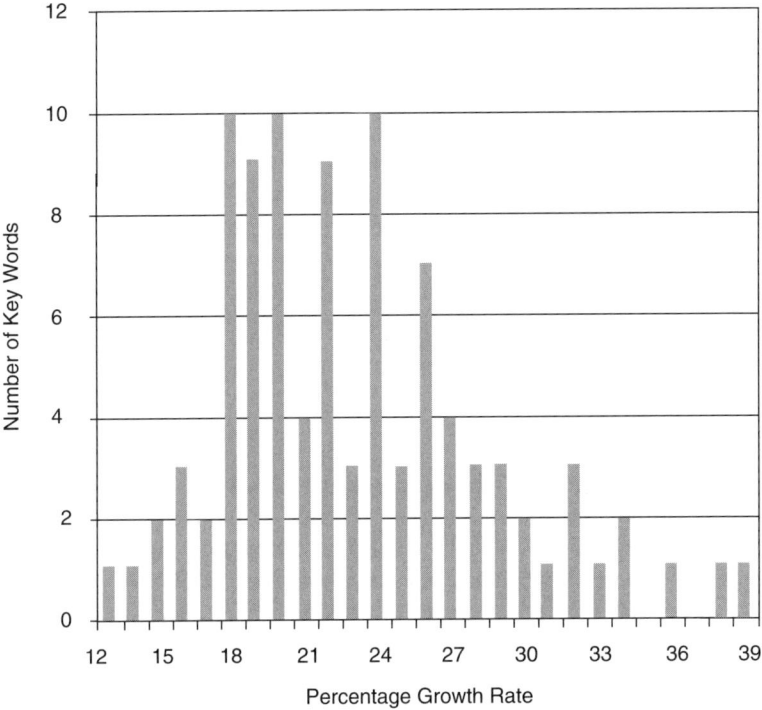

Figure 11-9. Sample growth rates of nanotechnology keywords

TABLE 11-3. Highest-Growth Terms

Keywords	Percentage Growth
Molecule, molecules, molecular	38%
Nanotechnology	37%
Film, films	35%
Nanoparticle, nanoparticles	33%
Surface, surfaces	33%
Nanotube, nanotubes	32%
DNA	31%
Microscope, microscopes, microscopy, scanning tunneling microscopy, atomic force microscopy	31%
Polymer, polymers	31%
Self-assembly, self-assembled	30%

The advantage of multiple linear regression is that it allows a rapid scanning of many keywords to identify the highest and the lowest growth in the sample. In this nanotechnology literature, the highest-growth keywords are shown in Table 11-3.

Topping the list are keywords involving variants of the word "molecule." "Nanotechnology" itself displays vigorous growth. Several other terms involve molecular structures—films, tubes, and surfaces—these may be the focus of up-and-coming nanotechnology research. A list such as this may suggest new areas of investigation, or ways to expand or improve data queries. Exploration of these results with knowledgeable researchers could prompt deeper exploration of "why" these patterns present themselves.

11.4 RESEARCH FRONTS

We gain a different perspective by "clumping" keywords by their dominance of particular time periods. These "research fronts" evolve over time and may reflect major research thrusts. This form of temporal analysis thus clusters terms (e.g., keywords) and clusters years (into time slices). (Note that Henry Small and his ISI colleagues have long identified research fronts based on co-citation, rather than our use of co-keyword co-occurrence.)

Figure 11-10 illustrates what such an analysis might provide. We see three groups of research activity—the "fronts." The first is centered about the year 1989, and the second and third about 1994 and 1999, respectively. Each successive front increases in level of publication. In addition, each front can be identified by a distinct set of topics, evolving over time as the field adapts its agenda to embrace new discoveries, concepts, or tools.

An important aspect is to associate keyword identifiers to research fronts. For instance, in *VantagePoint*, we might proceed as follows:

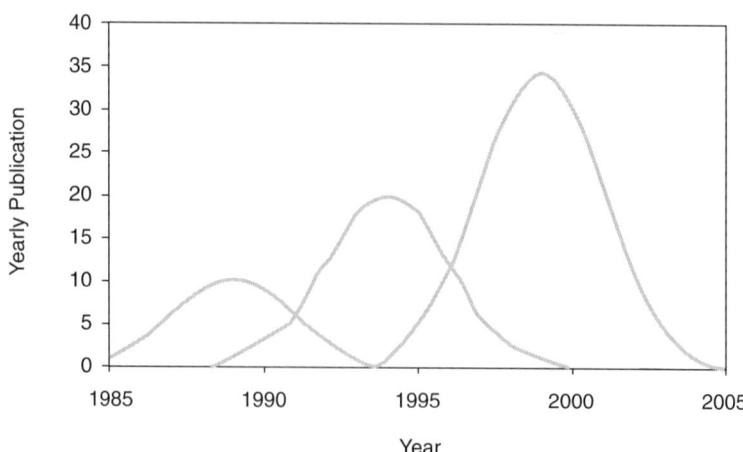

Figure 11-10. Research fronts

1. Create a matrix of the leading 100 or so keywords vs. year.
2. Visually discern concentrations of keywords by time periods.
3. Time slice the data into those time periods.
4. For each time period, perform principal components analysis (PCA) to identify the leading activity cluster.
5. For each such cluster, compile the occurrence of its high-loading keywords over the entire time period.
6. Plot the equivalent of Figure 11-10.

Other approaches can achieve similar ends. Note the iterative nature of grouping terms, and grouping years, and revisiting each. A table of keywords showing their association with the research fronts nicely complements Figure 11-10. Such analyses need not be limited to examining keywords (or other units of content). We could apply this, for instance, to examine research institutions prominent in each front.

Examining how a given research front develops over time affords a different perspective based on the same underlying data. Figure 11-11 illustrates a potential "takeover curve." To generate this, we could identify the keywords that define a given research front (e.g., the most recent one in Fig. 11-10), then select all the nanotechnology articles that use any of these keywords. We then examine this subset of records. We plot time against some activity measure:

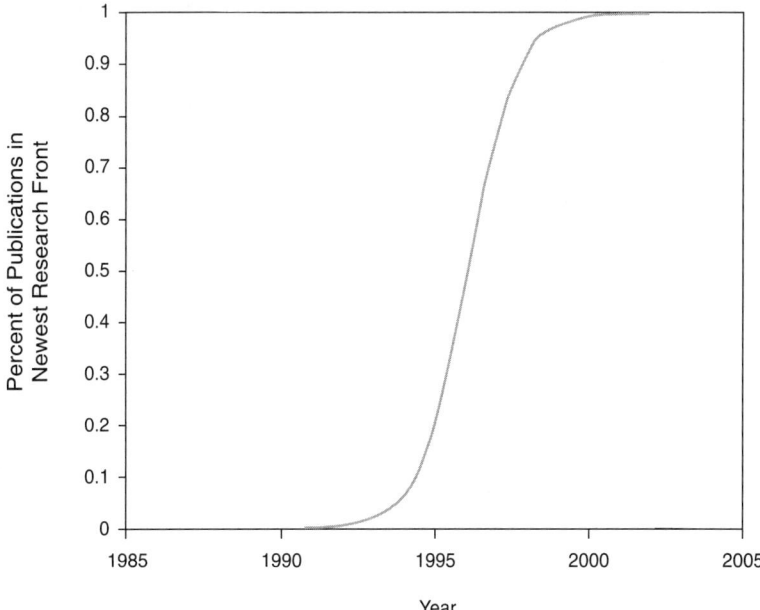

Figure 11-11. Takeover analysis of research fronts

"relative activity in research front" (Fig. 11-11). The pattern of growth is typically S-shaped, showing a leading edge of researchers adopting new research ideas. The growth picks up rapidly in this example throughout the 1990s. Tech mining may pick up early signals of the generation of a critical mass of attention to an emerging technology. Some call this a "small world" and deem it an essential precursor to successful technological innovation.

These adoption curves are familiar from the literature of technology forecasting. A variety of such curves pertain depending on the nature of the technological adoption processes. Some authors have suggested that these curves have much in common with epidemics. Barriers to the adoption of new technological ideas (e.g., software) may be substantially lower than the barriers to adopting new physical technologies.

11.5. NOVELTY

Novelty is an interesting parameter to study with time series methods. First, it gives an indication of how new topics emerge over time. Second, it provides a valuable early warning bell that our queries may be growing dated in the face of new and emerging science.

Figure 11-12 shows the cumulative total of new keywords introduced in our nanotechnology sample. Note that we have counted only the top 100 keywords. (As per Zipf's law, additional keywords much beyond 100 tend to occur in very few records.) Elsewhere in the book we discuss "innovation indicators," one of which is "keyword richness." Briefly, the notion is that the specialization of terminology in a technological domain may signal significant maturation. From this point of view, we see a dramatic surge from 1992 to 1996. We could investigate further by identifying the nature of the new terminology (e.g., whether it pertains to research topics, new tools, materials, or applications).

Rather shockingly most of the keywords were first introduced over a decade ago! Our study may be dated before it has even begun! Figure 11-13 provides another perspective on the cumulative number of new keywords introduced over time. The vertical axis plots this variable as a fraction of the total number of articles published. As can be seen in the nanotechnology sample, the introduction rate of new keywords is declining dramatically over time. The function fits exponential decline.

Is the novelty of new nanotechnology research actually declining? In all likelihood, the answer is no. Patterns like Figure 11-13 could, in part, reflect database indexing. The parameters of our search query and the resulting keyword selection may be growing more and more inadequate over time. This chart serves as a helpful warning signal to begin considering ways to expand our query and to increase the recall of the original selection. As the science of nanotechnology grows, we must be careful to expand and improve our terminology or risk not keeping up—and worse, sampling unrepresentative publication data.

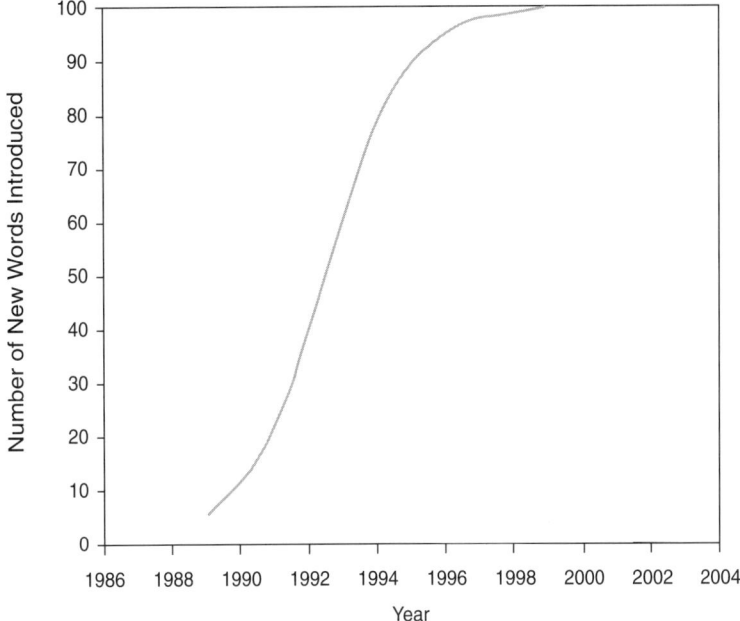

Figure 11-12. Nanotechnology, total keywords introduced

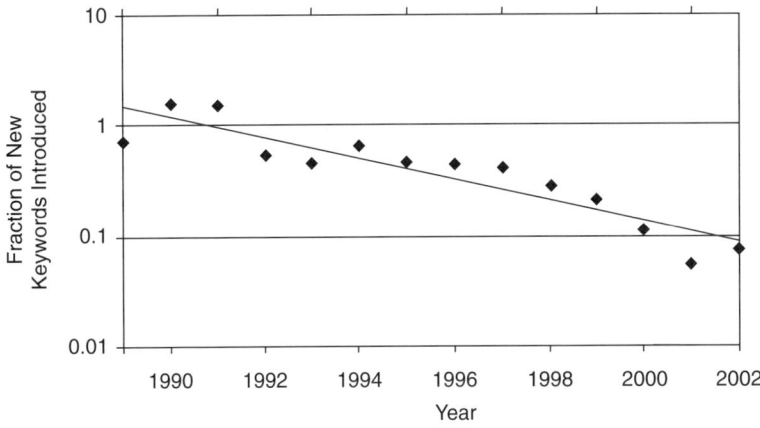

Figure 11-13. Nanotechnology keywords introduced as a fraction of total articles

CHAPTER 11 TAKE-HOME MESSAGES

This chapter has tried to provide ideas on how trend analysis can enrich tech mining. We invite you to address the temporal dimension to your time series data sets:

- Consider whether description and/or prediction contributes to your technology management aims.
- Consider the driving forces that could alter the trend projections in determining how far to extend these and how to interpret implications.
- Determine which form of the time series data are most suitable—annual (intuitive and directly informative) or cumulative (often best suited for growth models).
- Determine whether to examine overall activity trends and/or differential trends broken out for particular research topics (or by other variables, such as research organization).
- Choose how to treat time to best convey important shifts in activity; time slices provide a good alternative to yearly units.
- Decide whether to present simple plots, tabulations, and/or estimated equations (especially giving growth rates).
- Information visualization options include line plots, histograms, and 3-D multiple-term profiles.
- If your data extend over an order of magnitude, consider plotting the log (data).
- If you want to focus on recent change, consider cross-sectional scatterplots comparing two time slices.
- Remember that assumptions about underlying distributions and variability may strongly impact your forecasts.
- Recognize the importance of confidence intervals in producing robust forecasts; avoid thinking of singular valued forecasts as "the future."
- For sparse data series, consider modeling based on the Poisson distribution.
- Term frequency, over time, can be represented several ways. Analysis of changes can help detect novelty (new topics emerging).
- Tracking research fronts over time consolidates much detail into actionable form.
- Favor relative over absolute tech mining indicators—comparisons are extremely illuminating.

CHAPTER 11 RESOURCES

Technology forecasting books (e.g., Martino, 1993; Porter et al., 1991) and journals (e.g., *Technological Forecasting and Social Change*) treat technology trend analysis methods in some depth. Nearly all statistics texts present techniques such as correlation and regression. Advanced statistical treatments of forecasting and trend analyses differ in orientation, emphasizing lengthy time series of measures with analytical interest usually in predicting very few time periods ahead (cf. http://hops.wharton.upenn.edu/forecast/).

Chapter **12**

Patent Analyses

Patent analysis provides key information concerning the technology environ-ment. This chapter introduces a rich set of analytical possibilities. We begin by discussing the basics of patents and what patents offer for analysis. We then discuss steps in patent analysis, building upon the basic tech mining process. We provide a detailed case study of patent analysis in the automotive indus-tries, focusing on "fuel cell" technologies. Finally, we examine two advanced examples of patent analysis: patent citation analysis and the TRIZ technique.

12.1. BASICS

Patents protect one's intellectual property ("IP"). Other IP protection comes in the form of trademarks, copyrights, and trade secrets. Trade secrets may involve a process or device, or even a compilation of information used in one's business for competitive advantage—such as Coca Cola's "formula." Each brings to bear varying degrees of legal punch. This chapter focuses on mining patents, but the concepts extend in varying degrees to generating intelligence from trademarks and copyrights as well.

As a rough rule of thumb, a patent these days reflects on the order of $1 million in R&D investment. So, if you see 50 distinct patents in a particular domain, this suggests some $50 million being invested. This frame of reference can especially help in gauging a small company's activities.

Tech Mining: Exploiting New Technologies for Competitive Advantage, Edited by Alan L. Porter and Scott W. Cunningham.
ISBN 0-471-47567-X   Copyright © 2005 by John Wiley & Sons, Inc.

Four main fields traditionally rely most heavily on patents for intellectual protection: mechanical, electrical, chemical, and thermodynamic. Biological patenting is becoming very important, too. And we find other coverage developing, including software and business processes. As a technology matures, the type of patenting tends to shift. Initial patents cover a basic technology or product. Then particular application patents may follow. Spreading even further in time may be process patents to cover better ways to produce the product.

The patenting process works roughly as follows:

1. An individual or an organization determines that commercial interests warrant seeking protection of an invention. Working with a patent attorney, they file the patent application with one or more patent offices.

2. The patent application asserts certain claims reflecting the utility (scope) of the invention; a patent examiner reviews prior art to determine whether the invention is novel (at least in that patent office's jurisdiction) and whether the claims are valid given the prior art.

3. If approved, a patent is issued. In exchange for revealing the idea and how to implement it so others can build on this knowledge, legal protection is granted for (generally) the longer horizon of 17 years from patent issue or 20 years from patent application.

The data used in tech mining derive from this process. *Patent applications* are the earliest source, coming available sooner than granted patents, but not always published (patent office regulations vary). Granted patents themselves contain several important parts. The "front page" summarizes the patent and provides the content for *patent abstract records* (see sidebar). Patent *claims* contain essential information on the intended purposes and can be located in patent claims databases. Full patents, including text and diagrams, provide the greatest detail. *Patent citations* are references to other patents. The section on Patent Data considers the uses of these different forms of information.

The sidebar shows one of the fuel cell patent abstracts included in the sample analysis (Chapter 16 and this chapter). Many other fields are available in Derwent World Patent Index ("Derwent") abstracts, including other classification codes.* Full patents can be extensive, ranging at the extreme to thousands of pages. For tech mining purposes, we analyze the abstract record text fields, but with the general caution that patents and abstracts are written to convey as little useful information as possible, while asserting effective claims.

---

*http://thomsonderwent.com

Derwent rewrites the abstracts to enhance understanding (see "Sample Patent Abstract Record" sidebar).

Sidebar: Sample Patent Abstract Record

Derwent Accession Number	1984-090845
Family Member Years (most recent)	1984
Priority Years (earliest)	1981
Title	Molten salt chemical current source-operable at relatively low temps.
Basic Patent Year	1984
Inventors (Cleaned)	GAGYI P, NAGY A, PREPOSTFFY E, VINCZE P
Derwent Classifications (Cleaned)	L03-Electro-(in)organic - chemical features of con ductors, resistors, magnets, capacitors and switches, electric discharge lamps, semicon-ductor and other materials, batteries, accumu-lators and thermo-electric devices, including fuel cells, magnetic recording media, radiation emission devices, liquid crystals and basic elec-tric elements, X16-Electrochemical Storage
File Segment	CPI, EPI
Abstract	Molten salt ($NaCl/KCl/AlCl_3$) current source has a heated, scaled housing sepd. into two chambers by an ion-selective membrane. Specifically, the cathode is made of non-alloyed tol metal and the anode consists of a grid made of a conductive metal resistant to the melt and coated with a pyrite-contg. solid mixt. The appts. is relatively corrosion-resis-tant and is operable at relatively low temps.

We usually simplify patent dates to years. "Patent family" refers to a set of closely related patents, many of which may be essentially the same patent filed in different patent offices. So, "family years" range from the earliest to the latest (most recent). "Basic patent" here means the one that Derwent obtained first, and its year often indicates when it was filed as a patent application. "Priority patent" is the first application filed—this is often, but not always, the basic patent. Also keep in mind that the average lag from filing to issuance of a patent in the United States is about three years, with some applications pending for a decade or more. This lag is important for tech mining. The "Submarining Patents" sidebar notes that things aren't always quite what they seem.

> ### Sidebar: Submarining Patents
>
> Some companies play games with patenting to inhibit competitors. One trick is to "submarine" their own patent application by submitting a stream of modifications over several years. The United States only began publishing patent applications in 2001, and then only with approval of the applicant. Thus, one could keep the U.S. filing secret while still establishing defensible IP should a competitor arise. (Note: this doesn't work if one files internationally.)

12.2. WHY PATENT ANALYSES?

Effective technology management increasingly requires patent intelligence. Patent filings are on the rise and becoming more international. Licensing of one's IP has become a major revenue source. Informed IP management impacts legal, marketing, and technological interests. Patents and IP have gone from the periphery of business to the core as we collectively move into a knowledge-based economy. It is important to know whether your business actually depends on someone else's patent (see "Commodore's Demise" sidebar). Conversely, you can make a lot of money by licensing your IP to others.

> ### Sidebar: Commodore's Demise
>
> Rivette and Kline (2000) relate the tale of Cadtrak Corporation and Commodore Computers. In the early 1980s Cadtrak invented improved graphics processing but failed in marketing computer-aided design products. Then they discovered that a patent of theirs was vital to the graphics processing that became EGA and VGA displays. Many companies, including IBM, licensed the technology from Cadtrak. Commodore refused. Cadtrak took them to court and won an infringement ruling, along with a permanent injunction barring sales of Commodore computers in the United States. That contributed heavily to bankrupting one of the early leaders in personal computers.

Both internal and external technology management demand understanding of the patent environment. Internally, it can help formulate a smart R&D program. This can be followed by profitable exploitation of the resulting knowledge and attendant IP. From a technological intelligence perspective, patents tell about what other organizations consider worth protecting. Carefully interpreted, this can yield insight on those organizations' developmental trajectories. Mining patent information can also help forecast others' upcoming technology-based products and services.

Trippe (2003a) distinguishes two types of patent analysis: micro—detailed examination of a small number of patents—and macro—study of a large number of documents. He describes the latter colorfully:

> Instead of finding a needle in a haystack, today's searchers are becoming analysts and being asked to identify haystacks from space and then forecast whether the haystack is the beginning of a new field or the remainder from last year's harvest.*

We emphasize the macro level. One adjusts the scope of a patent search depending on its intended use (Akers and Khorsandian, 2003):

- State of the art: Find background art on a technology (e.g., in initiating a research program)—not as comprehensive as a search to be the basis for a legal decision.
- Patentability: Check whether background art (across all dates) anticipates a particular invention (needed in applying for a patent).
- Freedom to operate ("FTO"): interest in the claims of those patents still in force, in each country of interest, to determine how this impedes your own technology development.
- Validity: in-depth, focused search to uncover references that could invalidate other companies' patent claims.

The differences are major. For some uses, we want to probe a few directly relevant patents in extreme detail. For others, we want to map broadly to detect related interests. For tech mining purposes, we undertake patent analyses to facilitate technology management in many ways, including:

- R&D management: One might determine the "appropriability" (protectability) of potential project results (technologies) as a consideration in project selection.
- Technological intelligence: Identify new technological capabilities and key IP and assess possible product pathways to commercialize the technology.
- Identification of desirable IP: This can help target licensing or joint development opportunities, or even "M&A" possibilities.
- Mergers and acquisitions (M&A): Patent profiling using indicators introduced here can help evaluate the IP potential of one company versus others. In-depth probing, such as examination of inventor team longevity,

---

*http://www.infotoday.com/searcher/oct02/trippe.htm

external ties, and current employment of key inventors, can help assess the opportunity.
- Competitor intelligence: Profile one or more companies' IP by patent classes to benchmark relative strengths.
- International market analyses: Note concentrations of competitor patent families by countries.
- Human resources management: Strive to understand the roles and relative productivity of inventors both within your organization and in others.

Contrast patent analysis with our other emphasis—analysis of S&T publications. In a nutshell, these reflect mining at two stages along the technology development stream. R&D publications are usually "upstream" of patenting. Publications provide earlier-stage information on prospective emerging technologies (recall Chapter 6). Patents, in contrast, tap a later developmental stage activity. In particular, they reflect an organization's assessment of potential commercial value for the technology in question. The interplay of patents and business publications is also rich. The former speaks to technical aspects, whereas the latter informs of business intents concerning technological innovation. Publication and patent analyses complement each other in the tech mining repertoire.

Certain analytical aspects differ between publication and patent analysis. At some point you may want to go through the fuel cell analysis illustrations in this chapter (patents) together with those in Chapter 16 (mainly publications, but some patent analyses). Some differences to keep in mind, particularly in combining publication and patent analyses in a comprehensive tech mining:

- Publications are more likely to reflect academic activity, whereas patents tap industrial R&D efforts.
- Companies' policies on publishing and patenting vary greatly.
- The content of publication and patent abstract records are comparable, in part, but not identical (e.g., inventors vs. authors, patent assignees vs. author affiliations, publication vs. patent class codes).
- Publication abstracts usually try to explain what the researchers did and why. Patent abstracts often try not to convey corporate intents; this makes patent searching and analysis more challenging.

12.3. GETTING STARTED

This chapter takes the nine-step tech mining approach presented earlier and again in Chapter 16 as its base. It distinguishes aspects of patent analysis that demand special consideration over and above the general tech mining approach.

Patent Data

Chapters 6 through 8 address general issues in getting data. Here we discuss special patent data issues. Patent data come in several forms from many sources. Most basic is the patent itself, often available first as an application and later as an issued patent (if that indeed occurs). Government patent offices establish the patent information, increasingly available via websites. Our interests as tech miners key on abstracted information, including "front page" (abstract information), "claims," and "citations" (explored later in this chapter). Databases compile such information for one or more patent offices, often refining the data and providing access in bulk. (Retrieval of large numbers of records is cumbersome from most patent authority websites.)

We also note the existence of auxiliary patent documents. Some are publicly available, such as state-of-the-art reports, opposition claims, and appeals. Both patent applicants and opponents may also compile proprietary documents. Information on licensing and maintenance of patent rights (in what patent jurisdictions and over what time periods) rounds out the picture (Granstrand, 1999).

For certain purposes, you may want information on patenting in particular nations. For instance, in assessing Malaysia's knowledge economy, we found it fruitful to distinguish patenting by Malaysians from that by foreigners in Malaysia. The latter indicates appreciation of Malaysia as a market, whereas the former gets at indigenous technology development capabilities. Or one might track a competitor's country patenting patterns to gauge their global marketing intents. We note in passing the Patent Cooperation Treaty (PCT) that aims to facilitate international patent protection (See http://www.wipo.org; see also "national profile" discussion in Chapter 13's "Patent Indicators" section.) For tech mining, we usually concentrate on one or more of three key patent offices—European (EPO), Japanese (JPO), United States (USPTO)—or composite collections. We note patent database producers and distributors (some play both roles) in Appendix A.

As noted, patent data come in many forms. Good news—these are public information and therefore free. Bad news—you get what you pay for. Individual patent offices provide front page information, claims, and full patents, via websites and other means (e.g., on CDs). When you need to study a particular patent, you can go directly to the appropriate patent office website. But tech mining demands access to quantities of patents, not just a select few. The databases provide the key to convenient access because it would be onerous to retrieve large numbers of patent abstracts from the patent office websites. Furthermore, the databases add value to varying degrees by providing consistent formatting (e.g., field structured abstract records—see the sample record sidebar), additional coding, and search and retrieval capabilities.

Not surprisingly, price varies accordingly. For instance, U.S. patent abstracts can be compiled as a database and are commercially available quite economically. Delphion and MicroPatent go further to compile patent abstracts from the major patent offices in a single format, striving to provide consistent renderings of names and indicating related patents ("families")—but at a somewhat higher cost. MicroPatent and IFI also provide claims information. (Chapter 6 also discusses patent sources.)

Because we are using Derwent patent abstracts in our fuel cell example, we focus on them. Derwent abstracts are more costly because they rewrite the originals. In principle, patents protect IP in exchange for public disclosure so that others can learn from the invention. In practice, patent attorneys word applications with great care, making it quite difficult to discern the real capabilities and intents. Derwent also adds additional classification coding (more on this later).

Patent data pose serious challenges. Searches to meet legal needs must be just right, and we don't venture to address those special issues here. Finding chemical and pharmaceutical information involves special knowledge, too. For instance, one might make use of chemical structures, names, CAS Registry Numbers (American Chemical Society), and/or scientific terms. For our vitamin B12 illustration ahead (Section 12.4), Xu's search algorithm in MicroPatent incorporated 92 terms! MicroPatent offers abstracts and claims— one could search either or both, or pursue full patent records. The following section addresses another valuable resource—patent citations. The sidebar notes another increasingly important type of technological information resource.

Sidebar: Sequence Data and Biotech

Biotechnology with genetic modification efforts generates a whole new domain of IP considerations—sequence data. Sequence data present issues somewhat analogous to general patenting issues. The data behave like text in many regards, but not regular "English" writing. DNA or RNA sequences, for instance, can be interminable strings representing the nucleotides—A (adenine), C (cytosine), G (guanine), T (thymine), and U (uracil). Searchers may be looking for gene similarities, nucleotide strings near gene activators (codons), and so forth.

Tech mining enters in with similarity searching of protein sequences (Chemical Abstracts Service—CAS—provides over 1.8 million as of early 2003) and nucleic acid (RNA, DNA) sequences (over 22 million as of early 2003). Key sources include CAS Registry, BLAST (available through STN), GenBank, DGENE, and PCTGEN.

For tech mining, where should you search for patents? It depends on your objectives and resources. Inventors with sufficient resources generally patent in the United States because it is such an important market. Coverage is likewise often sought in Japan and Europe, with additional national patents added to the "family" to cover perceived competitive threats and key market targets. We recommend mining the accessible compilations of worldwide patents if you can afford the information. However, U.S. patents alone give an approximation (see Table 12-2 later in this chapter). Looking to the future, the Organisation for Economic Cooperation and Development (OECD) and others are investigating ways to analyze trioffice patenting (EPO, JPO, USPTO) to generate more consistent and inclusive metrics.

Searching

Tech mining searching should use patent indexes (classifications) as the primary search mechanism. This contrasts sharply with searching R&D publication sources, relying mainly on Boolean searching for specific terms. Term searching is very difficult in patent sources that have not been rewritten to improve clarity. Even in the Derwent records that have been rewritten, searching can be problematic. To illustrate—our "fuel cells" search incorporated two parts:

1. International Patent Classification (IPC) code H1M-8 (fuel cells; Derwent adds leading zeroes to standardize the format as H01M-008)

OR

2. The phrase "fuel cell(s)" in patent family title or abstract

This yielded 23,836 records (before we eliminated Japan-only patent families). Had we just searched on the H1M-8 classification code we would have missed 5245 of these (22%). Had we just searched using the terms, we would have missed 7920 (33%).

Patent classification codes were developed to aid patent examiners in finding prior art, but they also prove vital to electronic searching by users. Classifications are not globally standardized. Efforts are underway to harmonize JPO's FI/F terms, EPO's ECLA, and USPTO classifications, so one should monitor developments. Note:

- USPTO distinguishes about 150,000 sub-classes.*
- EPO is moving toward 20,000 top-level and 137,000 specific ECLA classes.
- JPO tracks some 180,000 FI terms and 320,000 F terms.
- WIPO's International Patent Classification (IPC) includes about 70,000 classes.

*For more information about USPTO classifications see http://www.uspto.gov/go/classification/help.htm#2b.

Concordances among these are not simple, and national patent offices are not fully consistent in their assignment of IPC codes. In general, examiner assignment of subclasses is not very reliable, so searchers may want to check related subclasses, too. Codes are updated frequently with new classifications added and old ones dropped.

Databases may add additional coding. The IFI CLAIMS and Derwent databases assign classifications based on what a knowledgeable reader sees as important. This provides better prospects for capturing most or all pertinent records (good recall) than searching raw records provided by the patent offices.

Gateways (e.g., STN, Dialog) offer the convenience of a "one-stop shop" providing a standard interface and search commands to access multiple databases. Dialog offers special operators that can facilitate patent searching, for instance:

(T)—to find multiple terms in a chemical name—e.g., amino (T) benzene, to locate those that incorporate "amino" and "benzene" in some combination within the chemical name

(L)—links two terms in the descriptor field, such as a particular CAS Registry number with some property of interest

Our fuel cell patent abstracts from Derwent actually offer more codes than the Derwent Classifications shown in the sample record. In addition, we can access IPC codes and Derwent Manual Codes—covering chemicals (CPI), electronics (EPI), and polymers.

A well-rounded search might begin by using an established protocol, augmented by review of key related patent classes. This could extend to tracking both exact and related terms identified in claims. Patent analysts should consider including citations (see "Citations" section). Obviously, your search strategy depends on the intended uses and available resources. You can spend a lot of money on your search, or you can obtain cheaper searches and spend a lot of time (and, hence, money) cleaning the data or floundering through the noise.

Until one becomes familiar with patent searching, guidance of an experienced patent searcher is invaluable. It is notable that the USPTO in its proposed 21st Century Strategic Plan would certify searchers.

Cleaning

Various tools can help perform patent analyses. Data cleaning tasks constitute a significant part of tech mining patent analyses. We illustrate with *VantagePoint.**

---

*Trippe outlines some of the common choices for patent analysis at his "Patinformatics" site: http://www.infotoday.com/searcher/oct02/trippe.htm. *VantagePoint* is also available in a version specialized for use with Derwent, Delphion, and Web of Knowledge (e.g., Science Citation Index) as *Derwent Analytics*.

Patent data providers take pains, to varying degrees, to clean the data. For instance, IFI maintains a standard registry of assignees and it indexes chemical documents with special attention to titles and text notations. It tries to update USPTO classifications to reflect subsequent classification changes and maintain a concordance with IPCs. CHI's assignee thesaurus contains about 45,000 name variants for the top 1800 patenting organizations! It is available through Delphion. Whichever patent sources you use, be cognizant of such aspects and assess the quality of the data.

Who Does Patent Analysis?

The "who" question parallels that of tech mining in general. To what degree should this be done by information specialists? patent analysts? researchers? technology managers? The answer blurs as access to patent information becomes easier for those who are not search specialists. Reducing the length of the chain of tech mining intermediaries to the end users is desirable. The danger lies in nonspecialists doing poor searches or failing to understand nuances in patenting—leading to significant misinterpretation. (For instance, U.S. examiners tend to cite more prior art than do EPO examiners, so direct comparisons are distorted.) The "Worry Factors" sidebar shares a few of Granstrand's 16 factors to be wary of in patent analyses (1999).

Sidebar: Worry Factors

(1) Company consolidations and deliberate use of multiple assignee designations, making it hard to correctly associate patents with companies

(2) Variable patent quality—may need to assess types of patents, blocking power, decoy patents, etc.

(3) Multiple patent classification systems that change frequently, making time series precarious

(4) Patent authority practices vary considerably; legal systems differ; and examination is uneven.

(5) Whether patent "counts" reflect individual patents or families, applications or patents granted, patents in force

(6) Uneven time lags from R&D to patenting by patent authority and company

---

To decide who should perform patent analyses, consider this mix of requisite knowledge and skills—(1) technical subject familiarity, (2) patent search and interpretation skills, (3) tech mining analytical skills, (4) familiarity with users and their needs, and (5) representation and communication skills. If these are concentrated in a single person, all is well! If not, suitable teaming should be arranged. Such teaming should entail cross-training so that the availability of the component skills diffuses within your organization.

The greater the number of persons familiar with many of the five skills just mentioned, the better the prospects for organizational tech mining learning—that is, the development of processes to use technology analyses effectively. Such development of shared knowledge is mandatory for robust tech mining. Don't underestimate the need for quality control. End users cannot easily distinguish rubbish, due to poorly cleaned or misinterpreted data, from valid tech mining inferences. Getting burned, even once, can cause managers to lose faith in all tech mining.

12.4. THE "WHAT" AND "WHY" OF PATENT ANALYSIS

Tasks and Resources

Xu (2003) gives a 3×3 framework: regarding IP management. Table 12-1 adds a strategic management task to his set to give 4 tasks $\times 3$ information resource types. Each cell gives one example to illustrate the intersection—for example, the upper right cell indicates that existing patents constrain a firm's options to develop technology. Ignorance of such constraints could lead a firm to expend precious resources pursuing down a blind alley.

Technology development tasks (Table 12-1, Row 1) include "white space" determination—an intellectually intriguing effort to "find something that's not there"—that is, to determine open opportunities not precluded by existing claims. Xu illustrates with a hypothetical case exploring Vitamin B12 possibilities. A white space emerges to develop a B12 derivative tolerant of high temperature because this destroys B12 effectiveness. (A contrarian perspec-

TABLE 12-1. Intellectual Property Tasks by Information Types, with Examples

Task	Information Resource Types to Mine: Technological	Business & Marketing	Legal
Technology, product, process, or service development	Identify white space	Assess market size, location, & positioning	Ascertain freedom to innovate
IP protection	Anticipate potential for novelty	Determine competition	Gauge scope of claims to assert
Commercialization of IP-based opportunities	Discover end use applications	Probe profit potential	Weigh patent status & term
Strategic management of technology	Assess technological environment	Assess competitive environment	Consider patents, standards, regulations

tive asserts that "white space" is voodoo—an artifact of where mapping happens to situate activity clusters. Most visualizations are somewhat arbitrary in that there is no one right way to depict a multidimensional reality in two or three dimensions. So, be sure to challenge interpretations, testing that they are grounded in technological and market facts.)

Another illustration from Xu aims at IP protection (Row 2). This begins with our firm's researchers coming up with a new retinoblastoma (cancer of the retina often seen in young children) gene with 17 phosphorylation sites. We review publications and patents to identify similar nucleic acid sequences, previously identified phosphorylation sites, and possible uses. Together, this information helps us decide whether to seek a patent or maintain a trade secret, what claims to assert if we do patent, and what markets to target if we pursue. We could go further to decide in which countries to seek patent protection, based on competitors' patenting and marketing locales.

For each of the four types of tasks, Table 12-1 reminds us to consider multiple types of information (see Columns). Strategic management (Row 4) weighs technological capabilities vis-à-vis the technological environment, but also market size and competitor positioning. Awareness of pertinent standards and regulations needs to enter the decision equation, too.

A common thread running through all four task types is *risk management.* Patent intelligence can be thought of from this perspective as helping managers estimate the relative risks and payoffs of alternative courses of action.

When we have IP, we have to decide how to exploit it. Options include manufacturing a product, licensing our IP to others, or trading for other patents more aligned with our core competencies. Tech mining can help by identifying upstream and downstream producers with potential intersecting interests.

Competitive Benchmarking

Elsewhere in this book (Chapter 13) we present ideas on how to translate particular tech mining measures into "innovation indicators." These measures track maturation of a technology and gauge its prospects for successful commercialization. Patent analyses contribute richly to such indicators. For instance, in Chapter 13, we discuss Ernst's (2003) patent-oriented indicators, including technological emphasis, technology status (e.g., share of activity in a domain but, more generally, any suitable indicator), and rate of technological growth.

Suitable combinations of indicators and more straightforward measures can generate informative breakouts. For example, plot technology status on the X-axis versus rate of change on the Y-axis. Depending on how many organizations we want to compare and their concentration, we might do a four-quadrant chart. Figure 12-1 illustrates.

We have plotted three hypothetical companies based on their standing in a domain of interest (say, the retinoblastoma gene engineering mentioned above). The technology status positioning reflects whatever indicator you

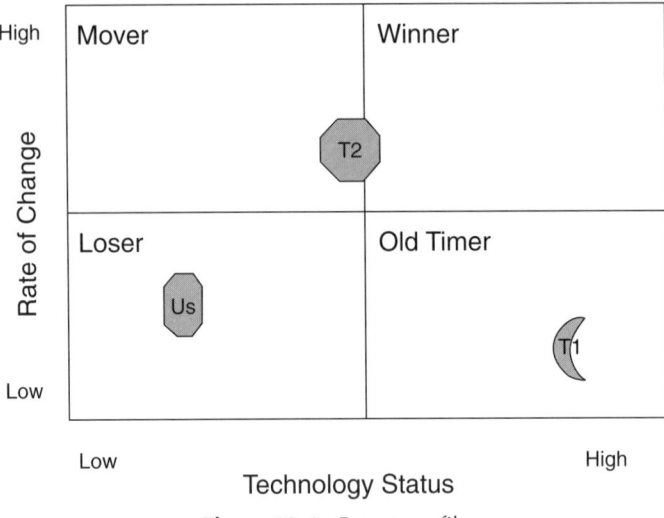

Figure 12-1. Patent profile

choose (e.g., drawing on technological share and patent quality). We then look to a suitable rate of change indicator (e.g., showing recent gain). The symbols reflect a third dimension also, technological emphasis, with the full octagon of T2 suggesting that this gene engineering is their main thrust, whereas this domain is only a small part of T1's developmental efforts, with Us in between.

What action does Figure 12-1 suggest? Scanning the X-axis, T1 ("Them") is the leader, followed by T2, with our company lagging. On rate of change, we find T2 leading with Us moving faster than T1. Implications? This paints a challenging picture. If we only focus on the leader, T1, we are gaining, but from a weak position. T2 adds an intimidating specter—they are ahead of us and moving forward more aggressively, too. So we might want to retarget our efforts to a nearby domain to sidestep T1 and T2, or look for a partner to enhance our joint capabilities, or get out of this domain altogether.

Pilkington (2003) offers an appealing variation of Figure 12-1 on which he plots "Quality" (citations/patent) vertically and "Productivity" (patents/organization or individual) horizontally, both divided at the mean for the technology set under scrutiny. He calls the High/High quadrant "Key," the High Quality/Low Productivity "Talents," and the Low/High group "Industrious."

We advocate adapting these notions to your organization's pressing technological issues and familiar communication forms. For instance, another graphic might compare our patent portfolios in all of our main areas of interest. We could indicate relative technology status and rate of change measures to identify more and less promising fields. Ernst (2003) also models a combination graphic that plots our relative technological strength side by side with our market strength to help assess prospects for a given technology-based product (or service or process) arena.

12.5. TECH MINING PATENT ANALYSIS CASE ILLUSTRATION: FUEL CELLS

This chapter exists because, although tech mining addresses similar questions whatever the data type, there are significant differences in analyzing patent data. One paramount difference is that *it is harder to analyze a technology than an organization from patent information.* As discussed under "Searching," topical patent searching is tough because of the deliberate obfuscation in writing patents. Contrast this chapter's examples with the fuel cell case analysis of Chapter 16 that mines publication and patent data oriented primarily toward *technology analyses.* Our fuel cell search works to identify the major organizational players, and we will pursue an *organization-focused example analysis* here. Depending on our tech mining sensitivities, one might want to probe further to look for organizations working on a related technology not identified as "fuel cell." Also beware subsidiary or special venture organizations not unified with the major companies identified.

You can certainly pursue technology-focused analyses with patents, but it is harder. To pursue questions like "What fuel cell technologies are likely to emerge as big winners in the coming years?" one might embellish the present fuel cells search along these lines:

1. Examine the leading 50 or so patenting organizations here; compare tallies with those organizations' overall patenting activity by searching in the source database (no need to download records, just count activity).
2. Retrieve the records of the leaders who focus heavily on this technology (e.g., we see that Ballard Power Systems concentrates on fuel cells whereas Siemens' interests spread widely).
3. Scan these recent patents to identify associated class subcodes that are not "fuel cells," but do relate—e.g., enabling technologies.
4. Now search on these subcodes for work by anyone (e.g., include Siemens here) to analyze what emerging technology pertains to fuel cells.

But now let's step through a sample analysis focused on organizations. We will try to answer "w4—who, where, when, and finally what?" To set the stage, Chapter 16 touches on European patenting with automotive interests (cf. Fig. 16-7). Suppose that our tech mining interest lies in understanding certain capabilities of three companies: Daimler-Chrysler ("Daimler" for short), Honda, and Ballard Power Systems ("Ballard" for short)—over the 10 years from 1993 to 2002. Actually, we have chosen these companies in part because they are three of the Top 5 overall fuel cell patenters for this period (Siemens is the overall leader with 312 patent families). That these three leaders share automotive application interests implies that this could be the leading fuel cell application arena. We next present selective results to show possibilities, not an exhaustive analysis.

Who? Where? When?

Imagine this section as "playing detective" with the fuel cell patent data to uncover possible technological thrusts by Daimler, Honda, and Ballard. Our attitude is somewhat casual as we don't have explicit driving questions to answer but are using this case to illustrate possibilities. Figure 16-7 triggers our inquiry. This figure suggests that Ballard is closely associated with DBB and XCELLSiS, based on copatenting. This leads us into a series of follow-on "who" and "when" questions.

In *VantagePoint*, we make an "Assignees by Assignees" matrix and sort on Ballard. Selecting the intersection cell between Ballard and DBB, we note in a Detail Window that the 17 joint assignee patent families' priority years peak in 1998 (8 patents) and end in 1999 (1). That prompts us to look at all 84 DBB patents—Figure 12-2 shows these rising and falling abruptly.

We probe further to find that the EXCELLIS profile is peculiarly complementary to DBB, showing a dramatic upsurge to 68 patent filings in 2000. A quick Google web search sheds light as we learn that Daimler and Ballard had a 4-year collaboration from 1993 to 1997. (Our data set shows 10 joint patent assignments peaking in 1996–97.) In 1997, reports indicate that they greatly

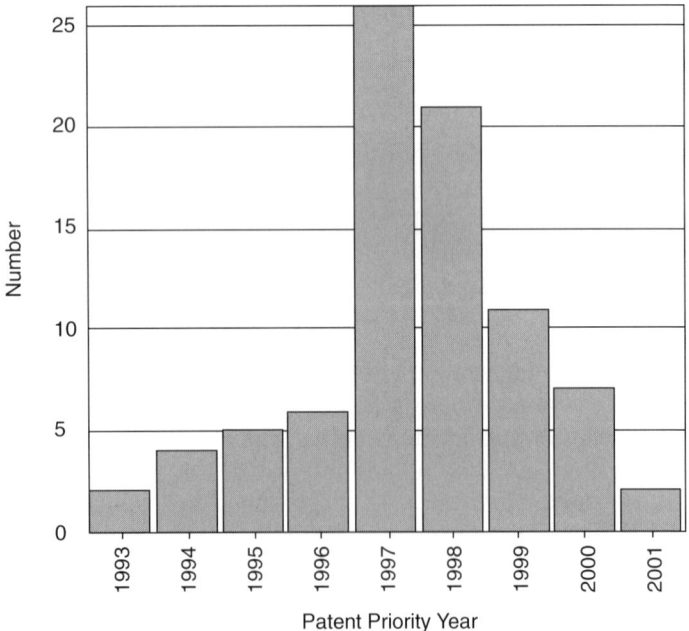

Figure 12-2. Temporal distribution of DBB patent priority years

expanded this collaboration with two jointly owned companies, XCELLSiS GmbH (formerly DBB Fuel Cell Engines GmbH!) and Ballard Automotive.

Furthermore, in 1997, Ballard and Daimler expanded their alliance to include Ford Motor Company investing in XCELLSiS and Ballard Automotive. The three together started a third venture, Ecostar, responsible for developing electric drives for electric vehicles (1 patent, priority year of 2002).

This vignette also hints at the difficulties in tracking company interests. Daimler's active patenting (149 families) appears to drop sharply from a peak in 1999 (priority year). However, we discover these three joint ventures, suggesting Daimler is really more, not less, committed to developing this technology. For Ford, we spot only 19 patent families (as Ford Global Technologies), with only two joint assignments—one with Ecostar and one with Ballard. Yet the sense here is that they too have very active efforts under way to advance fuel cells for automotive use.

Let's look at Honda, with 186 patent families. Can we ascertain a "knowledge network" of its fuel cell inventors? We map the inventors with five or more inventions (Fig. 12-3; compare to the map of Fig. 16-5 showing collaboration among another company's inventors). Figure 12-3 maps many small, relatively independent teams (not many cross-links). Stronger links (more co-inventing) are indicated by heavier lines. This pattern suggests a relatively new R&D operation—let's check this out.

Sticking with Honda, we note that their patent priority years are very recent. Of Honda's 186 patent families from 1993 to 2002, 91% of the patents are in the most recent three years. Only 4 of their top 49 inventors (those with 5 or more fuel cell patents) had a patent issued before 1999. "Fuel cells" is a very young Honda initiative! In contrast, Daimler also exhibits heavy recent activity (92 of its 149 families show basic patents in 2000–2003), but it shows notable activity dating from 1993 onward.

To check the longevity of the three companies' interests in fuel cells, we go back to the full fuel cells file. For any assignees, we find a trickle of patents issued (fewer than 10 annually) from 1967, with an explosion of interest starting in 1974 (129 basic patents that year). In comparison, Daimler's first patent appears in 1977, with another in 1989, and continuous activity from 1992 onward. Honda's first is in 1993, but we don't see multiple patents until 1997. Ballard first shows in 1991, with multiple patents from that year forward. So, Daimler and Ballard have been players longer than Honda.

Shifting gears, where do these three companies patent? Table 12-2 shows a very uneven distribution. Each company has its strongest presence in its home region. However, Honda also emphasizes American patent coverage, with some European spread. Daimler-Chrysler shows much less interest in North America (surprisingly?) and Japan. Ballard makes far greater use of the Patent Cooperation Treaty through WIPO than either of the other two.

We can probe "where?" on a finer scale. For instance, there are five joint Honda-Stanford patent families. All filed initially (the priority patents) in the United States, with the basic patent (first entered in the Derwent database) in

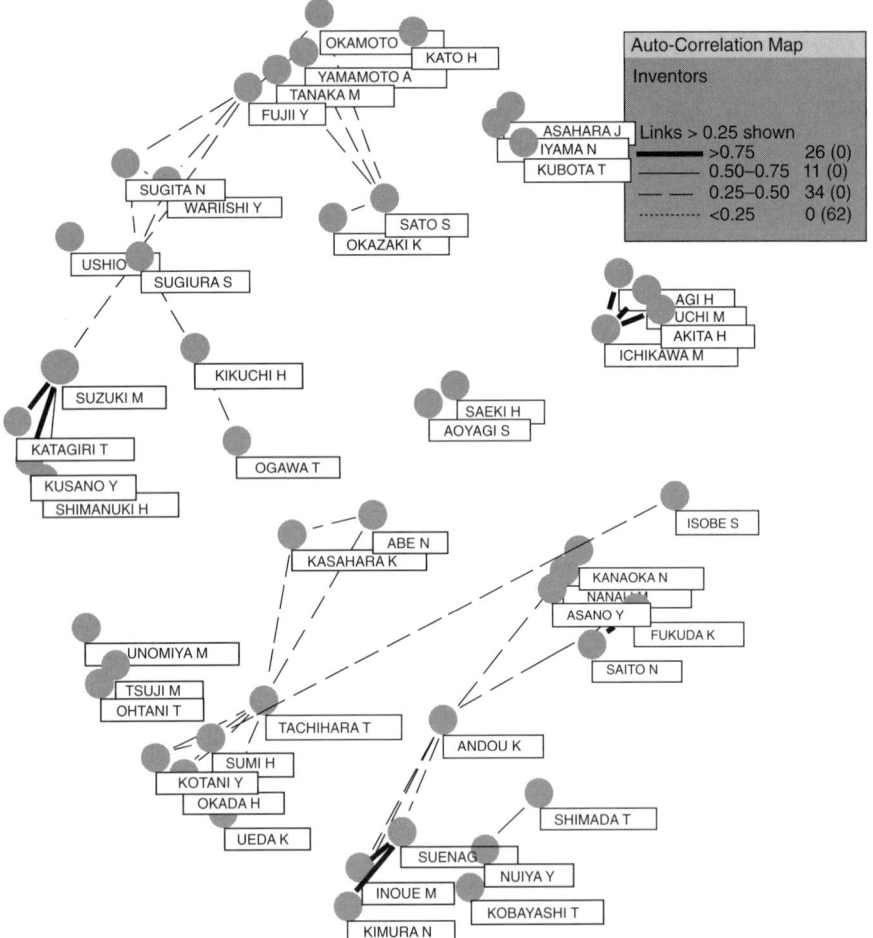

Figure 12-3. Honda's fuel cell knowledge network

WIPO. Intriguing is that all five show up for Australia—that's quite unusual for Honda (Table 12-2). Further analysis would be needed to confirm this because Australia automatically publishes a document if a WIPO application lists Australia as a "designated state," but the PCT applicant may not actually pursue an Australian patent.

We combine "where and when" questions to check Ballard's patenting vs. time. Figure 12-4 shows the seven patent offices where Ballard has heavy patenting (40 or more); all show more activity during the past couple of years than in Japan. In particular, note the rise in recent North American (U.S. and Canada) patents. This suggests that Ballard has given up on Japan, but such a conclusion requires confirmatory evidence (i.e., discussion with experts).

TABLE 12-2. Patent Family Distribution

	Honda	Ballard	Daimler
Japan	169	40	22
United States of America	160	126	58
European Patent Office	43	94	71
Germany	40	77	147
Canada	40	45	5
WIPO (PCT)	22	103	30
Australia	8	85	8
China	4	0	0
United Kingdom	2	14	4
France	1	0	5
Italy	0	0	2
Czech Republic	0	0	1
Hungary	0	0	1

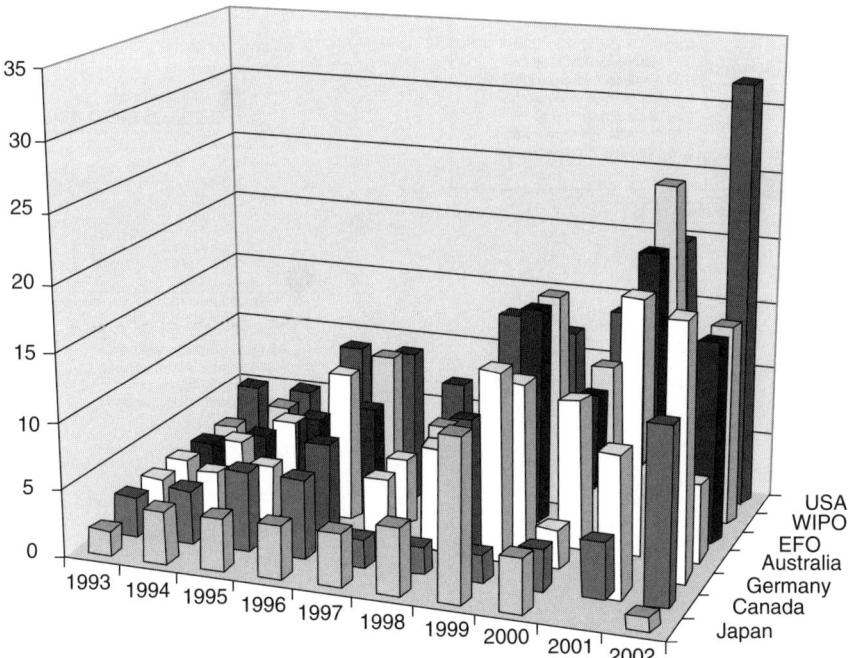

Figure 12-4. Ballard Power Systems patent family distribution over time

What?

We now consider the "what" question—what technologies are being patented? Scanning the Derwent Manual Codes and the abstract noun phrases for the inventor teams within Honda or Daimler, we have difficulty discerning

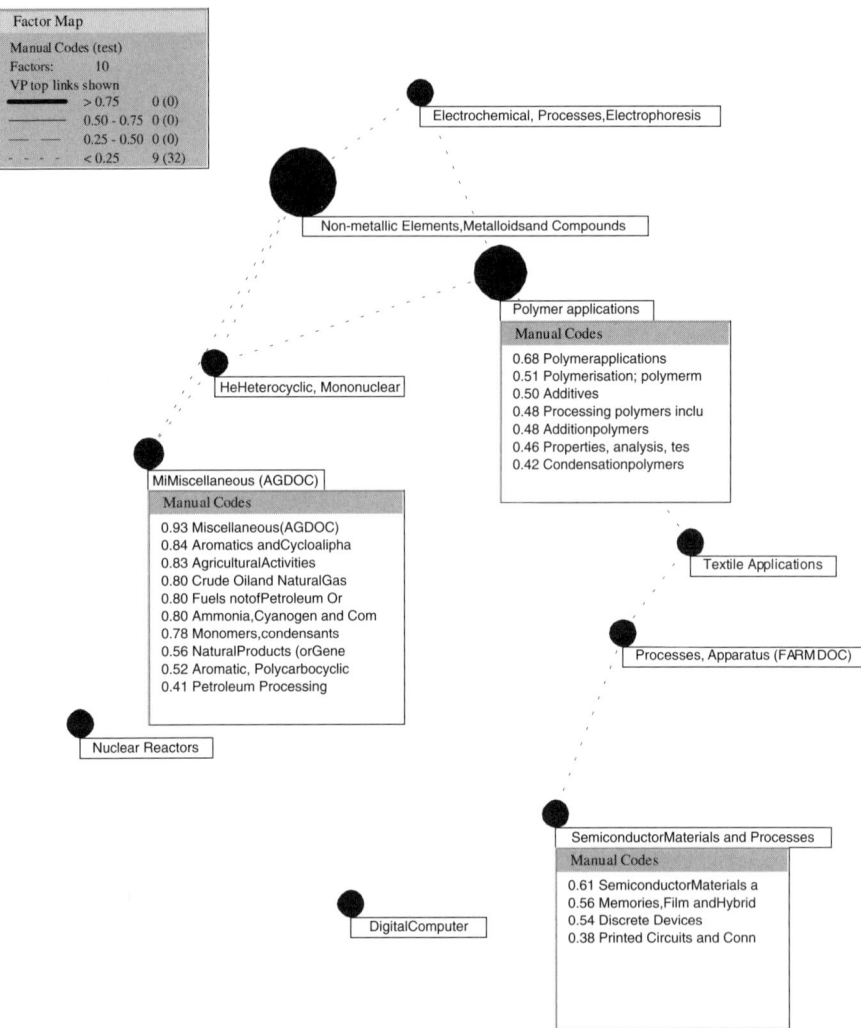

Figure 12-5. Fuel cell topical emphases

technology emphases. However, we can use *VantagePoint* to perform principal components analysis (PCA) on the Derwent-assigned "manual" codes to
map clusters of topical emphases (Fig. 12-5). Each cluster is named after its
most central manual code. Figures 16-6 and 16-7 show PCA maps based on
fuel cell *publication* keywords. Compared with Figure 12-5, those offer more
of a "technical" feel—for example, spotlighting leading fuel cell types,
advanced materials, and electrochemistry facets. Figure 12-5 tends more
toward distinguishing application areas, but also technologies (see the example
sets of high-loading manual codes shown as "pull-downs.")

TABLE 12-3. Selected Topical Clusters by Companies

| Topical Cluster | Patent Families | Individual Patents | | | |
		Ballard	Daimler	XCELLSiS (incl. DBB)	Honda
Polymer applications	1200	38	14	7	51
Misc. (fuels, etc.)	692	14	21	49	7
Semiconductors	144	1	0	4	3

We can combine "who?" with these "what?" questions. Doing so (Table 12-3), we see major differences in emphasis. For example, the polymer applications cluster (e.g., using these organic molecules as membranes to improve conductivity properties) finds strong positions for Honda and Ballard, with the overall leader being Matsushita with 82 patents. In contrast, the "Miscellaneous" (fuels, etc.) cluster finds the leading assignees to be individuals, led by Vinegar with 59, with the leading companies being XCELLSiS (including DBB), International Fuel Cells with 39, and Daimler with 21. Honda shows much less interest here with 5 families.

Curious about those 59 patents assigned to Vinegar, we look at the co-assignees. We see 7 additional individuals with at least 41 patents assigned to them and 3 others with 20 or more. Only one company appears with a single coassignment, Shell. So we look at Shell's 17 patents and find these same individuals listed as inventors on many of the patents. These may be a Shell research team and this batch of 59 patents, with a single exception, were assigned to individual technologists, not to the company. Furthermore, 58 of the 59 patents were issued in 2002, all in the United States. Next step—seek knowledgeable review of this batch of patents to figure out what Shell is up to!

Returning to Table 12-3, the semiconductor applications cluster presents a completely different picture. This is a much smaller fuel cells domain. and none of our target players appears highly active.

We could pursue "who's doing what" via additional database searches. We could search patent claims databases for fuel cell patenting by our companies to investigate their commercial interests. Also, we could search in a patents citation database to see which of these patents appear pivotal. Or we could search Derwent for any patents (not just fuel cells) filed by each of our focal companies (Table 12-3) to explore how fuel cell activities fit within their overall patent portfolio.

Ferreting Out Patent Strategies

These patent analysis illustrations reflect exploration variations. Section 12.7 ("For Whom?") suggests additional, more specific question-answering motifs. An "in-between" patent analysis assignment might profile a key competitor's patent portfolio. We might pose questions such as:

- How much do they rely on patenting, given the extent of their R&D?
- Do they identify many or few inventors (making it relatively easy or hard to identify their knowledge networks)?
- Expiration dates—What is the temporal distribution of their patents by technology (which might suggest areas they are giving up and others they are building up for a commercial thrust)?
- "Picket fence" analysis—Do they appear to be assertively building protection in certain areas? If so, can we surmise their commercial objectives?
- Portfolio valuation—What's the worth of their IP? What would it be worth to us (how does it complement our strengths and priorities)?

We can look to our empirical measures to help answer such questions. Many "patent indicators" can be constructed—cf. CHI under Chapter Resources and discussions in Chapter 13. Nils Newman offers four patent indicators of tech mining value:

(A) Concentration—mean number of Derwent manual codes (these reflect technology types and applications) per patent family for Company X's fuel cell patents compared with the mean for all the fuel cell patent families (a company with a tight concentration of codes is less apt to be blanketing or fencing than a company with a diverse code pattern).

(B) Uniqueness—frequency of manual codes for Company X that do not appear in any patent families of other companies in the fuel cells set (more uniqueness suggests possible strategic patenting or inventing around).

(C) Concept Clustering—Create a clusters map like Figure 12-5. Several distinct, loosely related clusters suggest fencing; several distinct, nearly unrelated clusters suggest blanketing; one or a few superclusters could signify the comprehensive approach; sharply defined and tightly related clusters suggest strategic patenting.

(D) Distribution—manual codes concentrated in a few areas suggest strategic patenting; many, loosely related codes suggest the comprehensive approach; a long "tail" of relatively few patents in many codes is consistent with blanketing or fencing.

The "Strategic Patenting" sidebar poses six "why?" possibilities. It then summarizes our empirical indicators relating to "blanketing" and "fencing" as pointing toward possible underlying corporate "strategy."

> ### Sidebar: Strategic Patenting—Six IP Gambits (Granstrand, 1999)
> - "Inventing around"—protecting a special application
> - Strategic patenting—a potent patent on which to build
> - Blanketing—heavy patenting on multiple facets of a technology
> - Fencing—a string of patents to block others over a range of related functions
> - Surrounding—"girdling" to cripple a key competitor patent's commercialization
> - Comprehensive—patent networks forming a comprehensive portfolio
>
> Empirical exploration of a few of the leading fuel cell companies over the past decade is intriguing, but complicated. Siemens scores low on Concentration and high on Uniqueness. Its concept clusters seem sharply defined and pretty tightly related. Distribution is relatively concentrated. Together, the indicators suggest a strategic patenting approach to secure core inventions within several specific fuel cell technologies. Honda scores low on Concentration and low on Uniqueness, its concept clusters all network together, and it evidences broad distribution. All told, the indicators suggest comprehensive and surrounding strategic inclinations. Ballard is different—fairly high Concentration, many singly occurring manual codes (high on Uniqueness), concept clustering suggests focus on both production and uses of fuel cells, with broad Distribution—interpretable together as possible blanketing. We invite you to explore these or your own indicators using the sample data file on the website.

12.6. PATENT CITATION ANALYSIS

Patents cite "prior art" to indicate their own novelty. This implies that patent citation information can help identify related bodies of knowledge. Citing publications and patents, so long as these do not obviate one's claims, provides strong defense against a competitor later challenging your patent claims based on that prior art.

Several patent databases (e.g., Derwent, Delphion, MicroPatent) provide citation information. Citing patterns inform about technological dependencies and extensions. They speak to whether companies rely mainly on their own technology or build upon that of others.

Mapping "citation trees" can elucidate intellectual relationships among patents. Figure 12-6, provided by Mogee Research & Analysis, LLC (see Chapter 12 Resources), spotlights one 1997 Lucent patent (U.S. Patent #5701152). Lucent cites seven earlier U.S. patents that its patent builds upon (the earliest being a 1978 Skiatron patent). Others cite #5701152 as a stepping-stone for their inventions—the figure shows six U.S. patents citing it as of 2001 (e.g., a 1999 Motorola patent).

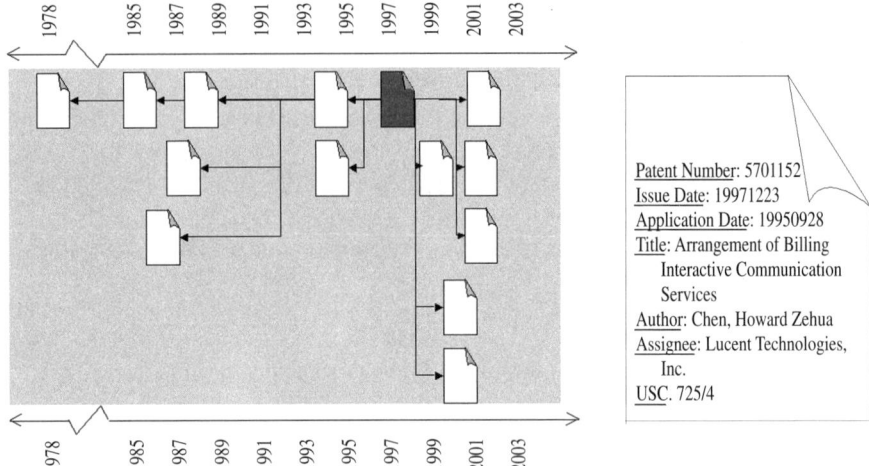

Figure 12-6. Patent citation mapping (Mogee Research & Analysis, LLC).

We can examine various aspects of the backward and forward relationships. We might note that #5701152 cites several patents in U.S. Class 725 (interactive video distribution systems) as well as two in Class 380 (cryptography)—an interesting combination. We note that the patents citing #5701152 increase in number year by year. These later patents belong to other companies, as opposed to Lucent pursuing this technology. Citation mapping can help elucidate patent patterns conveying strategic intent. Recall the "Strategic Patenting" sidebar; Granstrand (1999) illustrates six strategies and provides a variety of patent citation maps.

Mary Mogee, head of Mogee Research & Analysis, LLC, and a leading patent citation analyst, notes that such analysis can help expand a search. Also, it can suggest licensing opportunities and potential partners or competitors with related interests. Citations can reveal important inventors, vital to retain, in a firm being considered for merger.

For a target technology domain (as opposed to tracking relationships of a single patent, as just done), patent citation analyses can distinguish influential patents, prominent organizations, and/or leading inventors. One can gauge the pace of innovation by measuring median time lags from citing applications to cited patent grant dates. Tabulating the percentage of citations by an organization to itself can indicate whether it is a pioneer and whether it is building its IP in the target arena. (See "Patent Citation Reports" under Chapter 9 Resources.)

The gist of citation analysis is often to distinguish those patents receiving significant numbers of citations. Citation rates vary by year of issue (earlier patents have had more time to be cited) and by technology area. Normalization is necessary. As with most tech mining data, distributions tend to be highly skewed, with a few patents accruing many citations and most patents receiving none or almost none. In gauging a company's IP, focus on the highly cited patents and their inventors (e.g., are they still with the company?). One might identify the organizations with the most patents in the "Top 10% Most Cited" for the target technology area (corrected for year of issue).

As already mentioned, patent analysis can aid company valuation for mergers and acquisitions. CHI Research carries this notion further. They gauge the future prospects of companies based on citations to their patents (see Chapter Resources). Companies with relatively more heavily cited patents than others in their industry are promising. Those whose market valuation in relation to their book value does not reflect this technological advantage make attractive investments. Some money managers and hedge funds now incorporate patent metrics intro their models, and unit trusts have been set up on this basis.

Patents also cite scientific articles (Narin et al., 1997). CHI (see Chapter Resources) warns of the importance of cleaning these citations as they are very dirty and not standardized. Citing science is far more prevalent in some fields (e.g., biotechnology) than in others (e.g., electronics). Overall, the amount of citing of scientific publications is increasing sharply (up from under 0.5 cites per patent by U.S. patents in 1987 to about 2 cites per patent in 2000—National Science Board, 2002). This implies an increase in science-based innovation—particularly in areas such as biotechnology and pharmaceuticals. In those industries, some 90 percent of patents cite scientific papers, but other manufacturing industries also increasingly cite scientific papers. Lane and Makri (2000) have fit a growth model to the trend in patents citing scientific papers, projecting that this may continue to rise greatly. Diana Hicks updates this to suggest growth, but less pronounced (http://www.nsf.gov/sbe/srs/seind02/c5/c5s3.htm#growth). Examining science citation may help you understand key sources of ideas in an area, fundamental research worth further tracking, and collaboration networks (e.g., companies working with particular universities).

Boyack (2003) has done interesting work on scientific paper citation at the cluster level to ascertain which technology areas draw on which others. Chen (2003) shows patent "landscape" maps created by Boyack et al. changing over time. Changes in cross-domain interactions can track competitor technology interests. It is also illuminating to see who is citing your firm's patents. Such analyses help formulate a research program that will yield effective IP.

12.7. FOR WHOM?

The "Why Patent Analysis?" section arrayed rationales for doing patent analysis and distinguished broad purposes. Here, we mention a few things patent analysis can accomplish for various users.

First we point out that patents play considerably different roles in various industries. The Derwent patent "File Segments" field distinguishes chemical from engineering patents. The chemical and pharmaceutical industries rely very heavily on patenting for IP management, so patent analysis is especially important to them. Patents from the Chemical Abstract Service are valued because of the special structural (chemical formula) information set up. In other technology-oriented industries (e.g., software), the role of patenting is much weaker. Business method patents are now allowed to protect new uses for known processes, machines, or manufacturing methods. In some areas patent practices are undergoing rapid change (biotechnology). See the "Patent Scenarios" sidebar.

Sidebar: Patent Scenarios

One way to communicate competitive landscape and risk implications is through patent-driven scenarios. Imagine that General Motors were exploring fuel cell possibilities. Suppose they tallied the patent data (see Chapter 14) associated with Derwent classifications separately for electric vehicles, automotive electrics, and electric propulsion. In reviewing these data they note that three of the Top 10 companies patenting in this area are XCELL-SiS GMBH (#1), Daimler-Chrysler (#3), and Ballard Power Systems (#9)—and that these three companies share a number of these patent assignments. In contrast, General Motors is #4 and shows no co-assignments with other companies on these automotive fuel cell patents. The analysts might convey the possible risks and action options through a set of alternative scenarios. Here are two GM sketches suggesting directions forward (these are purely imaginary, not based on substantive analyses).

Scenario 1—Fuel Cell Future Without GM: Daimler-Chrysler locks up the key IP for the new low-temperature fuel cell vehicle. It's now 2015 as they introduce the overwhelmingly superior fuel cell automotive fleet, just as restrictions driven by global warming concerns kick in. We are locked out of the most promising engine innovation since the internal combustion engine. . . .

Scenario 2—Counterpunch: After careful analysis of a number of alternative propulsion technologies, we (GM) have identified and pursued two product development programs. One partners us with Siemens, the fuel cell R&D leader in a radical new bio-fuel cell. The second entails our purchase of key IP held by a small Japanese company for a combination solar fuel cell approach. We assess the prospects of each at about 20 percent to revolutionize automotive propulsion by 2020. . . .

The intent here is NOT to forecast explicit future occurrences and their likelihoods; it is to engage senior management in weighing alternative strategies given the technological environment.

Users of patent analyses include attorneys, IP managers, researchers, planners, and managers. Each has somewhat different, but overlapping, information interests.

Attorneys perform "due diligence" to learn of potential blocking patents. (Recall the sidebar on "Strategic Patenting—6 IP Gambits"; "blocking" patents mainly intend to interfere with other companies' innovation paths.) Understanding the patent landscape helps assess whether to apply for certain patents and, if the answer is "yes," to have information useful in making claims. Attorneys might also identify possible infringement of our existing patents. We have seen four patent attorneys spend a day to come up with a counter-claim against another company that was suing their client company over a different patent.

IP managers could map related patents to assess the present state of technological development pertinent to their company's interests. They also can examine patent trends to gauge competitive landscapes and prospects. This information complements understanding of our organization's strategic aims and its technological capabilities in helping formulate IP strategies. Should we exploit particular technology by manufacturing a product? Would we be better off licensing it to another company (which one)? Are certain patents not worth paying fees to maintain? Decisions bear on protecting and exploiting our IP via licensing or sale, or acquisition of complementary rights (possibly to block competitors from that arena). The sidebar illustrates tech mining being used by IP together with marketing.

Sidebar: What Might This Mean?

Merrill Brenner relates how an Air Products' technologist noticed that a competitor had seven patents in an aspect of a technology in which we and our other competitors each had one patent. We performed a signal analysis that illuminated that this competitor with seven patents had a much stronger position in one pertinent market, and therefore a different view of the value of the technology. In addition, we were concerned about the competitor taking this technology into another market area of more interest to us; in response, we made certain that our patent protected our position.

Product managers want to know which competitors are doing what in the domain. This could impact short-term product development pacing. It greatly affects marketing plans. In other words, compare a competitor's patent portfolio with their product portfolio to anticipate their future product portfolio. We have seen "expert" understanding of a competitor's interests and strengths proven wrong by patent analysis.

New product developers can scan patent distributions (who is doing what lately?) to find potential technology development collaborators with complementary knowledge. Tony Trippe (noted patent analysis authority, tony@trippe.com) relates how a company applied patent analysis to completely change plans for developing a new product. See sidebar.

Sidebar: You're Not Getting Sleepy . . .

The story unfurls as this Fortune 50 company considers developing a product to affect sleep patterns. This company emphasizes product marketing, so they began by researching which companies held highest market share. They identified three companies as the major players in the sleep adjustment space. Once again, as this company thinks first about branding and product marketing, they determined to speak to these three companies about possible partnering.

Tony had recently trained one of the team members in performing text analysis on patent documents, and this individual suggested that before the team contacted the market leaders perhaps they should analyze the "sleep" patent space. Because this project was highly technical in nature, this person was able to convince the team that patent analysis could add insights to the market analysis.

In this case the company used the Aureka Online System (formerly from Aurigin Systems, now owned by MicroPatent) to search for relevant worldwide patents based on a keyword and patent classification code search for sleep adjustment (including deprivation, enhancement, insomnia, etc.). Aureka contained a module to map documents based on shared content (co-occurrence based clustering; see Chapter 10) using 3-D surface maps (developed as "SPIRE" by Battelle, then marketed by Cartia, before Aurigin acquired it).

When the text clustering map was completed, the team noticed a concentration of documents talking about research on circadian rhythms. They were interested in pursuing additional research in this area and wondered which organizations held this IP. Somewhat surprisingly, examination showed that the market leaders did not own the IP they were representing. Instead, the IP in this space was owned by a number of smaller companies that the team had never heard of. At this point they decided to approach a few of the technology leaders, because the combination of the special IP with the marketing and branding muscle of the Fortune 50 company would produce a much better result than collaboration with one of the market leaders. The team later reported that the use of tech mining techniques on patent information made a critical contribution to the success of the project.

Strategic planners' interests dovetail with those just mentioned; obviously, these various technology management aspects require thoughtful coordination. Planners tend to address broad and long-range technology management. As such, they could appreciate 3-D topological mapping (see Fig. 16-8) of the broad patent environment. Interpretation of changing technologies together with competitor emphases and market evolution can enable better "Goal Level B" planning. Using Rouse's distinctions (1994), Goal Level A concerns short-term, fully attainable technology and product targets. Goal Level C posits ultimate targets for technological capabilities, product development, and market successes. Goal Level B is the critical in-between targeting over the next horizon. This is where understanding possibilities and the competitive environment can pay off with significantly better products (or processes or services) to beat the competition.

Managers can take various messages from patent analyses. Marketing may want to profile key customers to better understand their intents and capabilities. Human resources could use these analyses to identify prime recruiting candidates to complement your own organization's top producers. R&D managers could benchmark their efforts against leaders in their industry.

Patent analysis, reinforced with other empirical investigations (e.g., technical and business publications mining), can help answer many questions across this gamut of technology managers. Patent analysis should follow the general tech mining prescription to integrate it with expert knowledge (particularly human source intelligence about others' activities and technical domain knowledge to interpret empirical findings).

Mary Mogee (2003) provides a nice patent analysis case. A client is considering purchase of several patents to support manufacture of a medical device. Patent activity and citation analysis helps assess how important these patents are compared with those of competitors. Networking with knowledgeable persons can confirm and enrich the actual patent analysis. Imagine making such decisions without understanding how this set of patents fits into the patent landscape. Less obviously, consider how vital knowledge of what IP the various key players possess is to any decision about mergers and acquisitions, licensing, or competitive posture.

To add a cautionary note, as with all tech mining, interpreting patent activity requires substantive familiarity and thoughtful checking. This is particularly so with patents given the inherent intent of those applying for patents to disclose as little as possible of their real insights and plans. Be wary of variations in company name, subsidiaries, mergers, and filing practices. Try to assess why companies are pursuing particular strategies—for example, to commercialize a particular technology or just to block others so as to protect another technology of theirs. Check out anomalies such as the absence of patents from particular companies or nations. Check that the data sources you are using have suitable and compatible classification codes (e.g., International Patent Classifications are much less detailed than EPO or USPTO codes) and that the codes are up to date.

12.8. TRIZ

TRIZ (pronounced "trees") stands for the "theory of inventive problem-solving." It builds on work by Altshuller in the Soviet Union from 1946 onward. He and colleagues analyzed hundreds of thousands of patents to categorize invention types. They also pointed out flaws in the Soviet inventive process, earning Stalin's gratitude in the form of an extended trip to the gulag.

While imprisoned, they continued to uncover and categorize types of changes (principles) that generate worthwhile inventions.

TRIZ reflects an entirely different sort of patent analysis. One draws on certain inventive principles to creatively solve a problem at hand. TRIZ holds that problem types and solution types repeat across domains. It strives to identify essential "contradictions" by tabulating interactions among 39 system features. It compiles ideas on how to resolve those contradictions by using a set of 40 problem-solving principles (e.g., transformation of states of an object, segmentation, self-service). Extensive analysis of Soviet patents generated those principles (Terninko et al., 1998). Candidate ways to eliminate contradictions should be assessed to see whether they can boost achievement of useful functions and/or reduction of harmful functions. TRIZ thus only involves "patent analysis" as we mean it as background intelligence. We see exciting potential for an enhanced innovation management process guided by TRIZ.

TRIZ holds that about 95 percent of inventive problems have already been solved, in some other field. That's quite interesting for tech mining in suggesting that exploration in other domains can provide fruitful *analogies*. Although TRIZ typologies were formulated to address engineering problems, the approach extends to other problems. One can use the TRIZ systems perspective and checklists manually or with the aid of software, such as The Invention Machine (See Chapter Resources). Let's illustrate with approaches that specifically aim to facilitate technological innovation and new product design. You might consider this as a grand "systems" extension of our notion of "innovation indicators" (see Chapter 13).

Clarke (2000) presents a "directed evolution" approach that goes under the name of I-TRIZ. This seeks to assess where your present technological system under study stands and where it has greatest potential to advance along eight Patterns of Evolution, paraphrased as:

(1) Stage of the overall technological system in question along its S-curve evolutionary path.

(2) Ideality—ratio of the system's useful to its harmful effects; look to increase this.

(3) System Component Evolution—each component has its own S-curve against which to assess prospects for improvement and ways to eliminate conflicts among components.

(4) Dynamism and Controllability—look for ways to perform more functions, more flexibly.

(5) Complexity, then Simplification—systems tend to first increase in quantity and quality of functions, then move toward simplification.

(6) Address changes in system elements to alter their match or mismatch with each other.

(7) Getting smaller—look to advance from macro- toward microsystems.

(8) Decreasing human involvement—look to further automation of tedious functions.

Clarke illustrates with looking for a new way to close skin wounds. Known approaches include sutures (sewing) and staples. Exploration of Patterns 5 and 7 leads to consideration of a next-generation instrument to extrude liquid polymer through a pressurized nozzle through the wound against an anvil. The polymer would then be solidified to yield a continuous lattice formed to the exact wound shape to hold the tissues together. This promises gains on Pattern 2 (Ideality), 4 (Flexibility), and 5 (Simplification).

Mohrle (2000) offers a CTI application of TRIZ. Some 1400 patent abstracts relating to a not particularly "high-tech" topic—"mops"—were retrieved. The 40 judged most important were categorized with the TRIZ inventive principles. The analysts then compared their company with two key competitors in terms of which inventive approaches each used. The results informed a workshop to assess their company's R&D emphases, leading to a proposal to acquire one of the competitors for its complementary inventive approaches. We applaud using tech mining analyses in this way to stimulate further interactive solution processes.

TRIZ can be used for technology forecasting and product improvement by examining the status of a particular technology application. Plotting this on a radar chart shows how advanced the solution is along pertinent trend axes. Mann (2003) and colleagues have distinguished 35 technology trends (e.g., taking advantage of color in the application) and another 23 business trends. The features of a current capability (possibly defined by a particular patent) can be gauged (0–5 scale) as a distance away from the ultimate evolutionary potential. Figure 12-7 assesses a hypothetical product on eight dimensions recast toward business potential. One then assesses potential improvements, keying on those with the greatest remaining potential. We suggest you try out evolutionary potential assessment on an emerging technology of interest:

(1) Specify whatever 5–10 technical and/or business dimensions (trends) you consider most salient to success for this innovation.

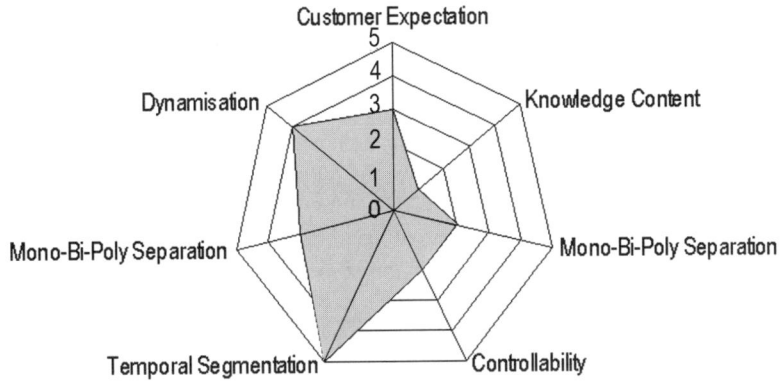

Figure 12-7. Evolutionary potential assessment

(2) For each dimension, agree on what minimum (0) and maximum (5) values make sense.

(3) Assess the current technology status on each dimension and plot on a radar chart like Figure 12-7.

(4) Looking toward those dimensions with the greatest potential for improvement, establish your technology development priorities and generate a strategy to achieve these.

12.9. REFLECTIONS

This is a long and rich chapter. Patent information provides a treasure trove of potential intelligence. Various forms of patent analysis help derive such valuable knowledge. This chapter addresses the special issues in exploiting patent information. But we would also remind readers to consider this in conjunction with the general tech mining considerations raised in the rest of the book. In particular, don't neglect to blend patent-derived insights with what can be gleaned from other S&T and contextual information resources. Also, be sure to engage suitable experts to help interpret and draw action recommendations.

Almost all large companies are pursuing more proactive IP assessment, intelligence, and management these days. Here is a sidebar to finish up with a sense of how tech mining can help a small company, too.

Sidebar: Intellectual Assets, Inc.

Paul Germeraad reported (2004) how they were able to help an enterprise initiate negotiations with 3 prospective licensees for their IP (negotiations under way as we write). They helped the company uncover prospective interest in its 6 patents through a series of analyses:

(1) Identify which companies are citing your patents. Then check who is citing those citing patents.

(2) Look for the companies that are most actively patenting in the same technologies by checking the pertinent class codes. In particular, look for companies that are increasing their patenting in the last 3 years. Also, perform thematic searches for patents based on technologies being mentioned (titles, abstracts, claims) to find active companies.

(3) Analyze your competitors' patenting.

(4) Combine the results of the previous three steps to compose a list of candidates. Examine promising ones to understand their interests. Try to ascertain their patenting strategies and objectives. Look for their probable application targets (sectors, products).

(5) Screen these to the most promising, and profile each of them. Check how extensively they cite prior art, identify their inventor knowledge networks, categorize what and where they patent. Then explore the fit between their interests and your IP. Creatively extrapolate how rights to your patents could benefit them. Determine which of your patents appear most valuable to them. Assess the consequences for your company of conveying IP rights to each of these prospective licensees. Identify promising contact persons. Summarize this information on each prospective licensee in a comparative fashion (e.g., use spider charts) to prioritize the best prospects to pursue.

(6) If initial inquiries find potential interest, prepare briefing support materials to enter negotiations well-informed about their situation. Mapping our patents together with theirs helps convey complementarities.

CHAPTER 12 TAKE-HOME MESSAGES

- Patent information is a vital component of tech mining requiring specialized skills, particularly in searching and in interpretation regarding legal IP issues (e.g., patentability, grounds for claims, infringement).
- Patent analysis contributes importantly to R&D program planning and to competitive strategy regarding mergers and acquisitions.
- Different databases provide several important forms of patent information: abstracts of "front page" information from patent applications or from patents granted, claims, citations, and full text. Match your searches to your particular tech mining needs.
- Table 12-1 offers a framework to guide use of multiple information types for four major patent analysis tasks.
- Patent analyses help profile organizations (companies) and also, with more difficulty, emerging technologies.
- Patent analysis can address "who, where, and when" questions well and "what" questions with more difficulty.
- Tech mining-oriented patent analysis can identify inventor teams—"knowledge networks"—understanding of which can help formulate "what if?" scenarios on where a competitor might be heading or the implications if they were to partner with a third organization.
- We advocate patent innovation indicators—derived knowledge based on patent information concerning technological emphasis, technology share, growth rate, patent quality, and aggregate (e.g., national) profile. These can help benchmark your technology against the leaders in particular arenas. (See also Chapter 13.)

- Quick Google searches can provide vital contextual information to flesh out issues raised by patent analyses.
- Patent analyses can generate insight into competitor technological innovation strategies.
- Patent citation analysis can distinguish key patents and significant knowledge dependencies.
- Work with your organization to diffuse patent analysis skills and familiarity among data specialists, analysts, and various technology managers to help build a robust tech mining environment.
- A good number of tech mining patent analyses will strike users as strange new ways to achieve familiar objectives (e.g., patent attorneys used to reading many patents to achieve what can now be done more broadly, quickly, and cheaply through tech mining). As such, these users are good prospects to use these analyses, but they need careful explanation of the bases for conclusions reached.
- TRIZ, based in patent analyses, offers innovative ways to solve many technology and business problems. The notion of evolutionary potential assessment can help distinguish more promising technological innovations.

CHAPTER 12 RESOURCES

There are a variety of groups and sources of information relevant to patent analysis. The Patent Information Users Group, Inc. (PIUG) is the international society for patent information. It holds multiple meetings at which patent data and software vendors interact with various users to share ideas. See http://www.piug.org/. The European Commission and the European Patent Office in 2003 hosted the EPIDOS conference along with PATINNOVA to advance knowledge and sharing of patent information.

Consulting groups active in patent analysis include Mogee Research and Analysis, LLC (http://www.mogee.com), Patinformatics (http://www.patinformatics.com), and CHI Research (http://www.chiresearch.com). Mary Mogee (of Mogee associates) publishes research in patent analysis (see, e.g., Mogee, 2003). Anthony Trippe of Patinformatics provides a thorough overview of analytical techniques and objectives in patent analysis (Trippe, 2003b). Francis Narin of CHI Research is also a leading patent researcher. Resources from Narin include Narin (2001) and Narin (1992).

TRIZ, a technique discussed in this chapter, could be pursued through many available books and articles. One software package to aid in applying TRIZ is the "Invention Machine" (2004, http://www.invention-machine.com/).

Generating and Presenting Innovation Indicators

The previous several chapters have described "doing" tech mining—that is, the analyses. This chapter applies the results of those analyses to answer technology management questions. We offer an extensive set of questions and pertinent measures to help answer them. Certain of those measures track technological development via "innovation indicators." The chapter then addresses how to present and package these tech mining findings for users.

CHAPTER CHALLENGE

- Imagine you work in Mergers & Acquisitions at "QQQ"—a high-tech multinational company. QQQ is actively considering purchase of a smaller company ("qqq") that has invented an attractive electronic technology. QQQ senior management wants you to play tech mining detective to find out what you can about qqq's technological capabilities—fast—as negotiations are heating up. What can you provide?

The "Chapter Challenge" exemplifies one of several tasks well served by tech mining—acquisition decisions—and, as it unfolds, it shows how several tech mining analyses and resulting findings can contribute. We need to continually remind ourselves that we are doing tech mining to better inform decisions about changing technologies. Thus the starting point is really the end point—what information needs prompt this analysis? These needs, of course, vary widely and can be characterized from many perspectives.

Tech Mining: Exploiting New Technologies for Competitive Advantage, Edited by Alan L. Porter and Scott W. Cunningham.
ISBN 0-471-47567-X   Copyright © 2005 by John Wiley & Sons, Inc.

TABLE 13-1. Technology Management Issues and Concerns

A. R&D portfolio selection
B. R&D project initiation
C. Engineering project initiation
D. New product development
E. New market development
F. Mergers
G. Acquisitions of intellectual property ("IP")
H. Exploiting one's own intellectual assets
I. Collaboration in technology development
J. Identifying and assessing competing organizations
K. Tracking and forecasting emerging or breakthrough technologies
   (opportunities & threats)
L. Strategic technology planning
M. Technology roadmapping

Table 2-3 set out a few technology management needs; we expand to 13 issues in Table 13-1.

In early chapters, we offered another perspective—the types of technology analyses that tech mining can aid (Table 2-2). Obviously, the distinctions in Table 13-1 could be shuffled endlessly. For present purposes, we direct attention to a level more specific—*questions*—whose answers would help resolve these issues. We noted some management of technology (MOT) questions in the initial chapters as meriting our attention but saved the full list for this chapter, where we suggest measures to answer them.

This chapter is about technology choice. How do we choose the correct technologies or innovations for our organization's purpose, based on the best possible information? Part of the answer lies in the use of expert opinion techniques, which we discuss in Section 13.1. Expert opinion techniques can be problematic: Finding the experts, and soliciting the right information from experts, requires technological intelligence to conduct correctly. We also discuss the use of innovation indicators as a way of integrating tech mining information to meet particular needs. Issues of information representation and packaging arise. The chapter concludes with a set of examples of applying tech mining to issues of technological choice.

13.1. EXPERT OPINION IN TECH MINING

This section emerged when our series editor reminded us of the importance of expert opinion to complement mining of information resources. We have mentioned expert opinion as a key counterbalance to various empirical analyses. Tech mining does not usually need to do formal surveying—that is, to compile representative sample responses to a set of questions. So, we can't just point to survey methods sources! Instead, we usually need to ask a couple of knowledgeable colleagues to:

- Assess whether our empirical findings are correct, complete, and on target.
- Draw implications from the complex of issues and empirical findings— i.e., stretch beyond the data.

This section offers some pointers.

Locating Needed Expertise

In tech mining we're usually interested in special expertise, not large numbers of responses. It is critical to define the needed expertise, which likely entails multiple perspectives. For instance, imagine we are assessing aspects of the European automotive application of fuel cells, as explored in our case analysis (Chapters 12 and 16). Depending on the decisions to be made, we should probably seek persons knowledgeable about our firm's European interests, fuel cells, automotive technology, popular acceptance of alternative energy sources, and other aspects. In other words, the nature of the expertise can address technical, market, customer acceptance, policy, and other dimensions pertinent to successful innovation.

Consider various sources of relevant expertise. Distinguish

- *External experts*—Offer richest potential to obtain diverse points of view. However, one must guard proprietary information, often key in competitive technological intelligence (CTI).
- *Internal experts*—Advantages include ease of access, common interests, and avoidance of proprietary data concerns. Folks from the same organization are also apt to share perspectives (culture, knowledge of intents), a plus. However, this can also result in the "zeitgeist" phenomenon—a shared point of view, leading to overendorsement of the party line.

A special set of key internal perspectives must be provided by the intended tech mining users. Chapter 14 treats ways to engage would-be users throughout the exercise.

Tech mining can help identify experts, particularly concerning technology matters. In the fuel cells case, we could list leading authors or inventors and profile their particular research emphases. An interesting twist is to use tech mining to locate internal, as well as external, experts. We have compiled Georgia Tech research published on fuel cells to enrich the "good old boy" knowledge of who studies fuel cells on campus. Tech mining particularly helps spot new and younger researchers in large organizations such as this. As we match candidate researcher interests to our needs for expertise, we might review their websites and ask local colleagues whether they know of their work.

Be on the alert for potential bias. From your side, be alert to avoid coloring their judgment. On their side, those knowledgeable about particular technology issues usually have a stake in them. Depending on the scale of effort and time available, it may be fruitful to seek out disparate viewpoints. An early "divergent" stage can suggest novel avenues for tech mining pursuit. Later "convergent"-stage expert contributions shift toward review and formulation of consensus interpretations and action recommendations.

Eliciting the Needed Expert Information

What do we want from those experts? It varies by phase of the study. Early on we can use help in scoping the technological and other boundaries and checking the adequacy of our data searching. In process, interim review of analyses can spot misunderstandings and gaps. At the end, review and augmentation of interpretations and action recommendations can add greatly.

Handy colleagues may be willing to collaborate. You can casually engage them whenever tech mining questions arise. Reaching out to external experts requires more care in formulating your inquiry. We suggest:

- Jot down your immediate questions.
- Deliberately step back to anticipate likely future questions.
- Prioritize and prune your questions so you can focus on the essentials in whatever amount of time the expert can give you.
- If at all possible, pilot test your questions, even if this is only asking a tech mining collaborator to check them.
- Ask whether the expert would be willing to later review your results—this bonus round can add tremendous value.

Payment issues may arise, so be prepared in terms of what you can offer, starting with the tech mining results. Determine ahead how you will capture what the expert says. Having a colleague take notes while you focus on questioning is often effective. If quite technical details must be captured precisely (e.g., chemical formulas), determine how to do this. Asking the expert to review your notes afterwards can work well.

We find that interactivity is usually key to tech mining expert opinion. Showing the expert what we have done can help trigger specific suggestions. Incidentally, providing them with the tech mining often gives them a different perspective on their field, so they may value this. Do be properly humble, because we have found experts who "know it all" and find any empirical portrayal of "their" domain automatically wrong. Keep in mind that R&D domain profiling is not generally familiar. What you are doing will strike them as somewhere between inane and inspired, but probably closer to the former.

Focusing can pose challenges. If you are asking an automotive marketing expert to review your fuel cells tech mining, most of your content will seem abstruse. Ease that expert into your study, guiding toward the aspects on which you seek his or her judgment. For instance, if you are E-mailing a hefty attachment of draft results, guide the expert to exactly what sections you'd like reviewed and what your key issues are. Interaction, in person or by phone, is highly desirable here so they comprehend their role.

If possible, we also want to interact with the experts concerning their observations and critiques. We "review" their inputs and probe with them to understand and elaborate. For instance, going over possible steps to improve the tech mining can elicit helpful pointers on where to look, how to cover gaps, and pitfalls to sidestep. Just having a one-shot E-mail from them can't provide this.

Consider group as well as individual expert interaction modes. Various forms of focus group offer potential advantages over sequential expert encounters in efficiency (multiple experts at one time) and interaction (interplay among different points of view to explore possible implications).

Offering incentives or clarifying personal stakes may greatly enhance cooperation. Sharing results is an obvious candidate "carrot," but be creative. Sometimes having the request for help come from the right person can make all the difference in securing the cooperation of a busy expert.

That is all well and good, but what about the quick study? How does one obtain expert inputs when the analysis is due in one day? As noted, tech mining can help identify candidate experts. Under severe time constraints, we need to seek the most suitable, available sources. Someone local who is somewhat knowledgeable about automotive fuel cell applications can probably meet our rush needs better than an eminent expert not close at hand. So reflect on whether true "expert" knowledge, or simply journeyman awareness, will suffice. By analogy, to statistically analyze a set of results, you don't need the world's leading statistician, just someone who understands the standard statistical treatment of such data.

Integrating Expert Opinion with Tech Mining

Figure 14-1 suggests that tech mining empirical findings enter one manager/user ear while the expert has the other ear. While "battling" intuitive judgments, tech mining really doesn't want to meet up with the expert this late in the game. Rather, we ought to have already integrated expert perspectives with the empirical findings to present a single, coherent message.

That said, we are best served by initially stepping back to sketch ahead what expertise we want to obtain at various stages of the exercise. In other words, expressly *design* which and how experts should be engaged. Early on, judgment on technological and business opportunities should contribute to scoping the exercise. Other types of knowledge should help determine the right information sources, formulate searches, interpret data, and draw implications sensitive to our organization's aims and context. In "24-hour" tech mining work, ways to tap expertise have to be greatly compressed, but not ignored.

Consider also the "last" expert. The sections that follow emphasize engaging users actively and planning how best to provide them with technology intelligence so as to inform their decision-making. This suggests a high potential in working with key managerial associates—that guy who has the boss's ear in Figure 14-1—to formulate and present tech mining findings most effectively.

13.2. INNOVATION INDICATORS

One might view Table 13-2 as the centerpiece of this book. It offers 39 questions to be addressed by tech mining. Each question pertains to certain of the technology management issues (Table 10-1)—we cross-link to these in Column 2. Because questions don't nest under single issues, we don't organize this table about the issues. We don't expect you to "read" this table—it is our grand compilation. The last column offers a "shopping list" of potential measures and innovation indicators that could help answer particular questions. The list offers both relatively straightforward measures (e.g., counting activity) and more elaborate "innovation indicators." The next section discusses innovation indicators. Note that we do not limit indicators to strictly numeric measurements.

The table is organized by question and data type. The questions are ordered as follows:

- *What?* (*questions in italics* mainly concerning content): research—development—commercialization
- Who? (nonitalicized questions concerning players): research—development—commercialization

Questions do not neatly fit categories, but they are ordered to roughly track the innovation progression from research (drawing heavily on publication data), through development (relying heavily on patents), through commercialization aspects (stretching from technical into business and popular press information resources, although we don't fully develop these).

To illustrate, Question 24 asks: Who are available experts? You might want to pose this question in addressing various MOT issues, such as: D, J, K, or L (Table 13-1). However, you might not have a need for experts in the way you are treating these issues. Conversely, for a management issue that we didn't note (e.g., in support of A—R&D portfolio selection), you might. Our associations are just suggestive. Likewise, indicators do not map uniquely to questions. For Question 24 we mention two indicators:

- Profile most prolific and most cited inventors not associated with a large company (affiliations, leading patent class codes; temporal distribution)

TABLE 13-2. Tech Mining Questions and Indicators

Tech Mining Questions: 1–23. What? questions 24–39. Who? questions	Mgt Issues	Helpful Information (Measures & Innovation Indicators)
1. What emerging technologies merit our ongoing attention?	A,J,K,M	Scorecard measures for rapid overview of multiple technologies.
		Consider separate metrics for industrial vs. information vs. molecular technology screening.
		Design alerts (what measures; one-pager type updates).
		Plot trends in publication activity, with rate of change measure, by database (with database benchmarks—e.g., multiple information technologies).
		Plot trends in patent activity, with rate of change measure.
		Overlay publication trends from multiple databases (ranging from fundamental research to development and business) and patent trends to assess technology maturation.
		Research community size—track change (publications, patents, distinct authors & inventors)—look for explosive new growth.
		Plot domain size vs. growth rate (for publications; for patents).
		Fit growth models (often S-curves) to the pertinent technologies to gauge their maturation.
2. What facets of this technology development are especially hot?	B,C,G,H, I,K,L	PCA mapping of keyword clusters to identify major R&D facets to the technology.
		"What's hot?" indicator (which leading keywords show recent upsurge in usage) Ratio of conference/journal papers (by main topics, too; ratio benchmarked against relevant research domains) (possibly sort domains on this ratio).
		3-D trend charts for topics (component technologies) over time.
		Rate of change metric compared for main topics (e.g., fuel cell types; KDD tools).
		Time slice profiles by main topics showing changing substantive emphases (using keywords, class codes, and/or alternative).

TABLE 13-2. *Continued*

Tech Mining Questions: 1–23. What? questions 24–39. Who? questions	Mgt Issues	Helpful Information (Measures & Innovation Indicators)
		Research community size—track change by main topics (publications, patents, distinct authors and inventors). Indicators of "spreading (or constricting) interest"—no. of organizations of various types engaged on each main topic, and rate of change. Plot company vs. (Academic + Gov't + Other) publication by main topic. Mark status and project tech development prospects on an S-curve (tech maturation) for each main topic.
3. What are new frontiers for this technology (fringes with opportunity)?	A,K	Use list comparison to ascertain new topics. Use NLP on titles and abstracts to generate candidate new topics (for human screening). Tally extent that patents cite scientific papers. If substantial, search hot topics and recent trends in those scientific research domains.
4. What are the component technologies that contribute importantly? significant subtypes of the technology?	A,K.L,M	Map topical clusters (PCA). Map class codes (publications). Map class codes (patents). Use ARM to help sort parent-child relationships. Map suggestive intertopic relationship flow diagram for expert review.
5. How does this technological development fit within the technological landscape?	B,C,D,I,K, L,M	White space maps. Devise density measures to distinguish "desert" from "jungle" and intermediate activity levels. 3-D surface maps. Provide normalized metrics as a function of size of the technology area. Provide time-slice versions of the maps to show evolution over time (development paths). Compare coverage of this technology across relevant publication databases (can also plot over suitable time period). Identify current alternatives and potential substitutes.
6. What is driving this technology development?	A,B,C,E,I, K,L,M	Score relative science base (% of patents citing R&D papers). Preponderance of use of terms that suggest particular drivers (e.g., regulations, market pull, technology push).

TABLE 13-2. *Continued*

Tech Mining Questions: 1–23. What? questions 24–39. Who? questions	Mgt Issues	Helpful Information (Measures & Innovation Indicators)
		Relative publishing by academic, industry, and other organizational types.
		Chart leaders in publishing (e.g., nations) and patenting (e.g., companies, nations).
7. What are key competing technologies?	A,B,C,D, G,I,K, L,M	Scan of mentions in conjunction with the target technology.
		Comparative assessments found.
		For technologies identified as potential competitors, compare their maturation, drivers, etc., with the target technology.
8. How bright are the development prospects for this technology?	B,C,D,G, H,I,K,L	Scorecard measures for rapid overview of multiple technologies.
		Consider separate metrics for emerging vs. developed technologies.
		Research activity over time (publications; patents).
		"Evolutionary potential"—Radar charts (adapt from Mann's TRIZ dimensions to suit patent and publication data in this technological domain—see Notes).
		Indicators of "spreading interest"—no. of organizations of various types engaged, and rate of change.
		Indicators of "constricting interest," too.
		"Spreading word"—no. of distinct sources publishing on this technology—trend over time.
		Research teaming—compare team size (no. of coauthors) vs. other technologies (more teaming is a positive indicator).
9. What are the likely development pathways for this technology?	C,D,E,G, H,K,L	Indicate velocity (rate of patenting and rate of change of patenting).
		Indicate velocity (publication rate and rate of change).
		Mark status and project tech development prospects on an S-curve (tech maturation).
		Mark status and project tech production process prospects on an S-curve (display together).
		Mark status and project development prospects on S-curves for critical component technologies.

TABLE 13-2. *Continued*

Tech Mining Questions: 1–23. What? questions 24–39. Who? questions	Mgt Issues	Helpful Information (Measures & Innovation Indicators)
		Perform sensitivity analyses on the growth models to assess likelihoods—i.e., depict a range of plausible future developments.
		Flag signals of regime change (chaotic transition to new S-curve).
		Possibly compile similar measures on the key variations or forms of this technology—present these comparatively.
		Distinguish differentiation possibilities (apparent drivers—present visually for human expert review).
		Generate one or more leading indicators (plot the technological maturation and commercial diffusion of an earlier technology against the target emerging technology, where we have bases for analogous growth trajectory).
		Break out application prospects (based on patent claims, class codes, publication class codes, etc.) to compare the main topics within the technology.
10. What are the component technologies that contribute importantly? significant subtypes of the technology?	A,K.L,M	Map topical clusters (PCA). Map class codes (publication). Map class codes (patents). Use ARM to help sort parent-child relationships. Map suggestive intertopic relationship flow diagram for expert review.
11. Assess the maturation of the component technologies.	B,C,D,G, I,M	List related technologies for expert screening. "Type" topics noted—technological classes. Spotlight topics relatively peculiar to this technology (vs. prevalence in the database as a whole). Generate "hot spots" maps (follow CHI model using normalized recency of patenting by component technology—set our thresholds for what constitutes a hot spot).
12. Identify technology fusion potential.	A,B,I,J,K, L	Track over time—publication topic linkage patterns. Track over time—patent topic linkage patterns. Extrapolate trends to predict possible fusion. Map time slices of high-level topic clustering (look for beginning of cross-topic association, both within our technology and extending out to others).

TABLE 13-2. *Continued*

Tech Mining Questions: 1–23. What? questions 24–39. Who? questions	Mgt Issues	Helpful Information (Measures & Innovation Indicators)
		Identify organizations that bring multiple capabilities together.
13. Should we apply for particular patents relating to this technology? What claims should we pursue?	H	Present "one-pager" to facilitate expert risk assessment. Display "hot spots" activity intersecting this patent.
14. Develop a technology-product road map.	M	Consolidate information on components and their maturation, technology development paths, production process development, and applications in the form of a visual road map. It should show for our organization: a) the integration of components to produce this technology; b) its evolution through generations; and c) products and their generations to which this technology can apply.
15. Assess the maturation of systems in which to apply this technology.	B,C,D,E, G,H,I, M	Application systems profile (identify main targets; identify status of each; present for expert inputs concerning gaps and our access to the system).
16. Which aspects (main topics) of this technology match our application interests?	A,C,D,G, H,I,M	Break out main topic by publication and/or patent claim mentions (using our own application targets thesaurus). Break out main topic by class codes (publication; patent). Plot trends in these breakouts.
17. What are our opportunities in this emerging technology?	A,B,C,D, E,H,L	Profile patent assignees (how many and how strong) in a composite visualization. Visually compare our position with the distribution of leading players. Indicate patent density. Indicate patent currency (licenses being maintained). Indicate apparent patent strategies of leading players. Indicate velocity (slope of new entrants). Indicate patent set importance with patent citations. Particularize measures (patent density, trends, etc.) for technology or application interests that intersect our own.

TABLE 13-2. *Continued*

Tech Mining Questions: 1–23. What? questions 24–39. Who? questions	Mgt Issues	Helpful Information (Measures & Innovation Indicators)
		Identify apparent "blocking" patents that could impede our intended developments.
		Benchmark with CHI patent citation data by sector.
		Supply chain—our access to necessary components + our access to systems and customers.
		Scorecard scale—relative attractiveness of alternative commercialization possibilities (e.g., developing).
		Composite indicator of height of barrier to entry based on overall extent of patenting, patent self-citation, patent concentration, overall publishing activity and maturation.
18. What societal and market needs do this technology and its applications address?	C,D,E,L, M	Scan mentions in publications (for ideas; also to count suggested possibilities and identify alternative terms). Trends in needs mentioned.
19. What applications offer promise for this technology?	D,E	Use NLP on claims to search for applications. Develop an applications thesaurus to use in screening claims for applications. Visually depict the IPC or Manual Code distribution. Visually depict the spread of IPC or Manual Code mentions over time.
20. What are the global opportunities?	E,H	Geo-plot patent assignee concentrations. Geo-plot national patenting by various assignees (as indicator of market potential). Geo-plot research publication by author nationality. Delimit important barriers and impediments by region or country (e.g., competitors, standards, regulations, restricted access). Denote significant assets and advantages by region or country (e.g., partners, local capabilities). Risk scorecard metric (devise suitable risk assessment for global opportunities for our organization).
21. What is changing in the competitive environment?	H,J	Indicate new entrants (first patents). Indicate changing rate of entrance of new patenters.

TABLE 13-2. *Continued*

Tech Mining Questions: 1–23. What? questions 24–39. Who? questions	Mgt Issues	Helpful Information (Measures & Innovation Indicators)
		Identify companies "exiting" (no new patents in N years; not maintaining patents).
22. Does this technology offer strong commercialization prospects?	A,B,C,D, G,H,L	Gauge the technological infrastructure (track required component technologies' availability; track development of systems to absorb this technology).
		Scorecard measures for rapid overview of multiple technologies.
		Benchmark patent activity against comparable technologies.
		Benchmark spread of patent class code and publication activity by industrial sectors.
		Tabulate applications mentioned by industrial sectors.
		Keyword richness indicator—Plot this measure of R&D specialization over time.
		Technology decomposition (keyword type spread)—Using keyword type thesaurus, show trend for spreading development interests (categorizing relative emphasis on fundamental science, process, technology development, materials, applications).
		Pie chart—% of R&D publication by industry (vs. academic and gov't or other) (alternatively, plot % by industry over time).
		Ratio of (business and trade) to technical papers (benchmarked).
		Tabulate applications mentioned, over time (from patent claims, publication abstracts).
		3-D trend plot of publication class codes over time (a spreading focus indicator).
		Identify substitutes for this technology (real or potential competitor technologies).
		Nominate scale score on "ease of adoption"— request expert refinement.
		Compare correspondence of technological development and market opportunities.
23. Assess the competitive environment.	B,C,D,E, G,H,J	Score the 5 Michael Porter competitive forces: – Rivalry (none; few; many competitors) – Suppliers (lean to rich) – Potential entrants (threat low to high) – Buyers (dry to juicy marketplace) – Substitutes (threat low to high)

TABLE 13-2. *Continued*

Tech Mining Questions: 1–23. What? questions 24–39. Who? questions	Mgt Issues	Helpful Information (Measures & Innovation Indicators)
		Chart recent (e.g., 1/3 of time period) slope for no. of class codes appearing annually, benchmarked against earlier period slopes. Map dispersion of the technology: – Diffusion of patenting across countries, over time – "Fringe" patenting (no. of companies with 1 or 2 patents, vs. time) Depict our organization's role vis-à-vis competitors.
24. Who are the available experts?	D,J,K,L	Profile most prolific and most cited inventors not associated with a large company (affiliations, leading patent class codes; temporal distribution). Profile most prolific and most cited authors (leading keywords, affiliation, web address, possibly disciplinary background).
25. Which universities or research labs lead in this technology —overall or in particular aspects?	A,B,I	Profile "Top N" publishing organizations in recent years (topical emphases, temporal activity distribution, core team strength). Profile "Top N" publishing—for particular types of organizations and/or for particular locations (e.g., U.S.). Compare publication class codes (visualization). Map leading research organizations' publishing on 3-D surfaces, showing evolution over time. Longevity indicator—temporal activity pattern. Map authors (coauthorship teaming; topical emphases).
26. What are the strengths and gaps within our own organization?	A,B,C,D, F,G,I,M	Tabulate publication and/or patent activity in related technologies. Map who collaborates with whom inside our organization. Identify our gaps in required capabilities to achieve innovation objectives.
27. Which companies lead in particular aspects (main topics) of this technology	E,F,G,H, I,J	Profile "Top N" patenting companies by main topic (or use class codes). Profile "Top N" publishing companies by main topic (class codes, keywords). Concentration indicator (% of patents held by top companies).

TABLE 13-2. *Continued*

Tech Mining Questions: 1–23. What? questions 24–39. Who? questions	Mgt Issues	Helpful Information (Measures & Innovation Indicators)
		Focus indicator (proportion of a company's patents on this main topic vs. related vs. unrelated).
		Map leading companies' patents on 3-D surfaces, showing evolution over time.
		Map one company at a time to understand emphases.
		Provide easy rating one-pagers to facilitate combining expert judgment with tech mining outputs.
		Graphically depict outlier companies (those with distinct patenting patterns).
		Matrix of leading companies by main topics.
28. How strong are the leading companies' R&D teams?	G,H,I,J	Map inventors (coinvention teaming; topical emphases).
		Map authors (coauthorship teaming; topical emphases).
		Map team activity over time (by topical emphases—spot dead-ending; fusion or fusion potential).
		Identify inventors associated with highly cited patents (note degree of concentration).
29. Which companies lead in this technology?	G,H,I,J,L	Profile "Top *N*" recent patenting companies.
		Graph patent citation distribution for most-cited companies.
		Graph patent citation distribution for high-citing companies.
		Profile "Top *N*" recent publishing companies.
		Plot publications and patents over time for each leading company.
		Longevity—Compare duration of engagement (publications; patents).
		Indicate patent self-sufficiency (reliance on other patents).
		Focus indicator (proportion of a company" patents on this technology vs. related vs. unrelated).
30. How do leading companies' development emphases compare to ours?	D,E,H,I,J	Compare IPC or Manual Codes (visualization) against ours.
		Compare publication class codes (visualization) against ours.
		Identify their hot spots and their citing of others' hot spots vs. ours.

TABLE 13-2. *Continued*

Tech Mining Questions: 1–23. What? questions 24–39. Who? questions	Mgt Issues	Helpful Information (Measures & Innovation Indicators)
		Spotlight rates of change in our vs. their R&D activity patterns (e.g., recent years vs. earlier years).
		Scorecard as potential collaborator; develop suggestive indicators pointing to joint R&D, IP acquisition, etc.
31. What other technological strengths does each leading company have?	H,I,J	Profile their related patenting by IPC or Manual Codes.
		Map their patent emphases.
		Compute a concentration measure showing % of patenting in this technological area.
32. Characterize a company's IP relating to this technology (competitor analysis, or collaborator analysis).	F,G,H,I,J	Profile leading assignees (companies and others) comparatively (how many patents each and how strong) in a composite visualization.
		Indicate concentration for each (what % of their patenting is on this technology; what % relates; what % is unrelated).
		Indicate publication concentration (on this and related technologies) for key publishers, over time.
		Track concentration over time.
		Indicate patent currency (licenses being maintained).
		Indicate apparent patent strategies of leading players.
		Indicate each assignee's patent set importance with patent citations.
		Profile this company's R&D strength (core researchers and inventors, longevity, emphases).
33. What smaller companies or individuals have attractive IP relating to this technology [potential acquisitions or hires]?	G,I	Profile assignees (how many and how strong) in a composite visualization.
		Indicate patent currency (licenses being maintained).
		Indicate apparent patent strategies of leading players.
		Indicate velocity (slope of new entrants).
		Indicate patent set importance with patent citations.
		Identify companies receiving SBIR R&D support, and from which Federal agencies.

TABLE 13-2. *Continued*

Tech Mining Questions: 1–23. What? questions 24–39. Who? questions	Mgt Issues	Helpful Information (Measures & Innovation Indicators)
34. Who's partnering with whom (competitive environment)?	D,E,F,H, I,J	Identify coassignees. Identify coauthors. Profile timing of these collaborations. Plot organization interests; flag intersections as potential future collaborations. Profile candidate company-company partnering (status; mode—e.g., joint development, comarketing; capabilities of each). Profile candidate company-other (university, lab) partnering (status; mode; capabilities of each; emphases).
35. Competitor Profiling	J	Profile their capabilities pertinent to developing this technology (scan across tech development stages from research to sales and maintenance and life cycle issues) (frame for expert judgment with our "hints"). Profile their resources (financial, knowledge, technical, marketing, customer base). Scorecard scale—this company's innovativeness in various regards: technology, products, services, production processes, delivery systems (pose to invite expert inputs or refinements). Scorecard scale—goodness of fit of this technology's potentials with their business strategy.
36. What companies should we place on watch?	J	Scorecard measures for rapid overview of multiple companies. Profile candidate companies (their hot spots; their areas of citing hot spot patenting—i.e., patents cited much more heavily in the last year or so). Design alerts (what measures; one-pager type updates).
37. Who might be prospects to license our IP (or partner in some way)?	H	Identify organizations with complementary technologies under development as candidate licensees for our technology.
38. How entrepreneurial is the competitive environment?	E,J	Indicate velocity (slope of new small business entrants in patenting). Trend in companies receiving SBIR awards (and rate of change). Scorecard item: availability of venture capital.

TABLE 13-2. *Continued*

Tech Mining Questions: 1–23. What? questions 24–39. Who? questions	Mgt Issues	Helpful Information (Measures & Innovation Indicators)
39. Assess each key competitor.	G,H,I,J	Chart competitors' recent vs. earlier publishing and patenting rates (looking for activity and rate of change). Assess competitors' technology strategy (identify probable targets and uncertainties in our knowledge of same). Assess competitors' SWOT position (strengths, weaknesses, opportunities, threats). Assess competitors' product positioning. Assess competitors' likely time to market with target new products or services. Assess competitors' new product development effectiveness. Construct a rough technology roadmap for each key competitor (identify likely hurdles for them with risk estimates). Scorecard "them vs. us" across these and other pertinent factors. **Pose all these for expert input or refinement of our rough, initial estimates.

Note: SBIR = Small Business Innovation Research; KDD = Knowledge Discovery in Databases.

- Profile most prolific and most cited authors (leading keywords, affiliation, web address, possibly disciplinary background)

These profiles might also help answer other questions. On the other hand, you may figure out better ways to identify the experts you want. For instance, perhaps you seek a couple of Japanese professors with complementary interests to each other to participate in a special workshop. Tailor your tech mining accordingly.

Use Table 13-2 to stimulate your thinking. In using it, we find that scanning the questions is the best way to focus. Then consider whether any of the suggested measures, as is or modified, would help answer your questions. Browse nearby questions for additional ideas for measures. Derivation of some of the indicators is not fully explained, but space and sanity preclude treating each. Exhibit 13-1 introduces basic indicator notions.

EXHIBIT 13-1	*Explanations of Selected Indicator Notions for Table 13-2*

Indicator	Explanation
Scorecard measures	Reduction of various numerical scales to simple 0–100 stacked bar presentations (or similar) so that a user gains a general comparative sense
Alerts	User-tailored, periodic updates that show changes over time on certain key metrics
One-pagers	Composite presentations of multiple indicators to give a quick "heads-up" on complementary intelligence to help answer a particular question
Benchmarks	Comparisons presented against one or more targets
Growth models and S-curves	Basic technology growth models, including S-shaped growth curves (e.g., Pearl or Gompertz curves), exponential, or linear trends
PCA mapping	Clustering of keywords, or such, to show major activity concentrations and their interrelationships
NLP	Natural language processing to break titles or abstracts into noun phrases for further analyses
Map authors	Show who publishes with whom
"Top N"	Lists cut off to show the most interesting leading entities (e.g., Top 10 university research centers)
Profiles	Usually take the top "N" from a list of interest and provide auxiliary information on each of those leading entities
White space maps	Depictions of activity in a domain so as to identify topics not heavily addressed (see Chapter 12)
3-D surface maps	Plots of records grouped by their common use of terms with concentrations depicted as "peaks"
Spreading interest	Indications of increase on a metric of interest over time
Velocity	Tracking change over time, and also rate of change over time (often by comparing recent vs. prior activity)
Map class codes	PCA used to map co-occurring classifications (analogous to keyword cluster mapping)
IPC	International patent classification codes
Time slices	Time broken into multiyear periods to help discern patterns
Patent density	Concentration of patenting activity in certain topics
Hot spots	Patent (or other) activity concentrating in the most recent years (using certain criteria, especially citation rates)
Geo-plot	Show activity overlaid on a geographic map
Keyword richness	Extent of keyword specialization, possibly categorized by type of terms (e.g., materials)
Fringe	Activity at the R&D domain boundary, usually low frequency

CHAPTER CHALLENGE

- Pursuing our acquisition issue, you scan Table 13-2 to identify some 15 questions pertaining to "G." These trigger your thinking as to what concerns you most in assessing qqq as a potential acquisition by QQQ. You seek IP, so you need to carefully gauge how valuable qqq's patents are (i.e., Questions 26 and 27). You also need to know about their R&D team and assure yourself that the key players remain in place (i.e., Question 8). You might then check how well qqq's technology complements your technology via a gap analysis (Question 9—to assess what capabilities you need to strengthen). Consider extending this to think through how the combined (QQQ + qqq) technology could meet your application targets (Question 19).

- Once you have sharpened the questions for your tech mining work, you again scan Table 13-2 for ideas on which measures to generate for each. These will depend on the data and time available to provide your recommendations to QQQ senior management.

- Reflecting on the Chapter Challenge,
  - Could you answer all the questions (15) associated with acquisition issues in Table 13-2? Theoretically, yes; practically, no.
  - Should you answer all those questions? No.
  - Would you generate all the listed measures and indicators for given questions? No, tailor to your needs.

Consider Table 13-2 as a combined questions and indicators idea list to help you decide what questions to answer, using what indicators. The next section develops our rationale for innovation indicators.

In undertaking to answer particular questions, many suggested measures are quite straightforward (Table 13-2). Take Question 27: Which companies lead in particular aspects of this technology? One measure tabulates the companies that have the most patents relating to the target technology. Other measures elaborate somewhat—"profile the 'Top *N*' recent patenting companies." By "profile" we imply compiling additional relevant information for those "Top *N*" companies. "*N*" is typically 2–10, depending on the breadth of the competition and our particular technology management concerns (Table 16-9 gives an example profile). However, other measures can become complicated. We see high potential for a particularly appealing set of these, the *innovation indicators*, conceptually derived from our understanding of innovation processes and patterns.

Technological Innovation Processes

The key distinction between tech mining and data mining is that tech mining begins from a conceptual model of technological innovation processes. In contrast, data mining is theoryless. The goal of data mining is to first find empirical regularities in the data that hopefully offer business value. Only later (if ever) does one seek theoretical explanation for the findings.

Innovation processes take place through highly complex, socio-techno-economic systems, so any model will be approximate. These aspire to capture key cause-and-effect relationships—what drives successful implementation of new technologies?

We have found the "technology delivery system" conceptual model helpful in understanding what translates an idea into an effective innovation. This model points toward a broad complex of influential factors:

- Consider dependencies both on component technologies and on the larger systems into which our target innovation must fit (infrastructure considerations).
- Address internal organizational capabilities (institutional and technical resources).
- Attend to external socioeconomic forces affecting the innovation in question (e.g., emergence of standards, regulatory environment).
- Identify applications and factors affecting customer acceptance (does implementation demand behavioral change?).
- Assess intended and unintended effects and likely acceptance of the innovation by all the stakeholders.

A rich body of research and experience documents factors that promote (or inhibit) successful technological innovation. Studies of new product development, technology substitution, technology transfer, and other aspects of innovation abound. We scavenge "innovation success factors" from multiple perspectives. Intriguing notions include technological generations, the innovation funnel, champions, sponsorship, substitution, early adoption, technological fusion, acceleration, and competitive and contextual forces. Others have identified innovation indicators for emerging technologies and for looking within organizations (see Chapter Resources).

These conceptualizations help focus tech mining on generating results that speak to the prospects of attaining technological innovation. We also draw on our experiences in the conduct of over 100 tech mining studies of various types (see Chapter Resources). Feedback from users of *VantagePoint* as to what works and what doesn't also contributes to this growing collection of innovation indicators.

Three Types of Innovation Indicators

Innovation indicators exploit empirical information to estimate factors that affect technological advance and successful commercialization. Indicators can be formulated to measure the implications of R&D (or other) activity with respect to a technology under scrutiny. Or they can be framed to assess one or more competitor organizations' several technology development activities. You need to select indicators that help resolve particular managerial issues.

As tech mining matures, it should build a repertoire of indicators validated by series of analyses.

Watts and Porter (1997) argue for interpretation of publication, patent, and business information with regard to implications for successful technological innovation. They distinguish three types of innovation indicators:

(1) Technology life cycle status
(2) Innovation context receptivity
(3) Market prospects

Technology Life Cycle information keys on determining how far along the development pathway the technology has advanced, its growth rate, and the status of technologies on which it depends. These indicators key on technological *maturation*. The dominant model is S-shaped growth (slow takeoff, followed by rapid rise, then asymptotic approach to a limit—cf. Porter et al., 1991; Glenn and Gordon, 2003). Activity over time provides the key data. Gathering such information from multiple sources that range from fundamental research to business attention can prove enlightening.

Contextual factors begin with measures of whether supporting technologies are sufficient. Often these are not fully ready, so indicators may project likelihood and risk considerations. Competing technologies affect innovation prospects, so they merit indicator development. The extent of involvement in the technology by other organizations is vital intelligence. Tracking establishment of standards and regulations warrants inclusion, too. Contextual indicators should draw on sources beyond R&D publication and patent data, namely, popular press and business compilations.

Market Prospect indicators address the potential commercial payoffs of the technology and products to which it contributes. These indicators aim to gauge requirements to attain such payoffs. Identification of potential applications suits such indicators. Competitive assessment of IP and market strength of other companies is vital. Tracking the dispersion of commercial activity (e.g., new product announcements, international patenting, other market estimations) is important.

These three types of innovation influences interlink and overlap, so separation is somewhat arbitrary. We use the three types to help conceive candidate indicators. We don't follow through to categorize the resulting specific indicators by these types because that seems unnecessary.

Innovation indicators are measures with a purpose. We offer the image of a large set of candidate indicators on a "palette" from which to choose. Assessment of a small company target for acquisition, for instance, warrants very different technological intelligence than comparing proposed R&D projects. Hence, Table 13-2 is only a starting point.

Interim Indicators

One view of tech mining is that of an activity flow. The tech mining activity flow:

- Exploits data,
- Uses tech mining software,
- Generates technology information products (TIPs), and ultimately
- Supports decision-making.

This is another way of looking at our tech mining process introduced in Chapter 2. We will return to this process and activity flow again in Chapter 14, where we consider utilization. The aspect of note for this section is that tech mining serves multiple parties—information professionals, researchers, and technology analysts, as well as end users. Table 13-2 concentrates on answering managers' questions. Before this payoff, we may well want to resolve needs of the other players along the way. Other chapters go into relationships among these players and their distinctive perspectives, so we won't belabor those here.

Tech mining players may want to set up their own categorizations and assign records to them. These categorized documents can then be examined further to develop more specialized targets, possibly leading to query refinement. Some analysts prefer examining outputs in other formats, particularly while processing preliminary results. *VantagePoint*, for instance, can hand off outputs to *MS Excel* or *Bizint SmartCharts* for further analyses.

In Section 13.5, we advocate "one-pagers"—composite, highly visual TIP representations. Versions of these can be generated from initial search results to help refine understanding. The point is to consider formulation of useful interim tech mining results that will likely never be presented to senior managers.

Indicators Relying on Additional Information Resources

If Table 13-2 doesn't offer enough candidate indicators, we can generate more! In fact, many more could be derived by extending the prime information sources beyond publications and patents. Note that nearly all of the measures and indicators of Table 13-2 result from mining these sources. We focus on those as particularly rich, but that is not to say high value cannot be obtained from business databases, popular press compilations, market information reports and databases, and Internet scanning. Our knowledge is greater on the publication and patent side, and this book's scope accentuates these. A consequence of our focus is that the innovation indicators we offer are strongest for technology life cycle, intermediate on contextual influences, and weakest on market prospects. We invite suggestions to enhance the innovation indicators set.

To whet your appetite, here are some candidate indicators based on these wider ranging information sources:

- Patent infringement activity (using other patent information resources)
- Market assessment (market size and attribute sources)
- Market forecasting (using new product announcements among other sources)
- Competitor resource profiling (drawing on corporate financials)
- Environmental issues posed by the innovation (environmental databases)
- Customer acceptance (test and evaluation with current and new customer groups)
- Supply chain assessment
- National absorptive capacity assessment (incorporating country information sources)
- R&D funding (assessment and projection of available resources based on funding agency announcements, project databases, and other intelligence)
- Organizational change profiling (human resources information plus cultural intelligence)

Patent Indicators

Obviously, Table 13-2 uses patent information heavily in the generation of many indicators. Chapter 12, especially Section 12.4, offers special considerations for patent analyses and patent-based indicators. In this section we suggest some additional patent-based indicators.

Ernst (2003) presents a rich array of possible patent indicators oriented toward actionable technological intelligence. Possibly his greatest contribution is to point us beyond basic patent activity tabulation toward derived knowledge. We build on his repertoire to offer some appealing indicators (and, in fact, you can find threads of these across Table 13-2). Adapt these ideas to generate indicators based on patents, plus other information, to help inform your organization's key issues.

(A) Technological Emphasis: Compare an organization's patenting activity (applications or patents granted) among "fields" (e.g., use International Patent Classes or your own markers of domains of special interest).

(B) Technology Share: For a given field, compare various organizations' extent of patent activity.

(C) Rate of Technological Growth: Compare recent versus earlier activity levels.

(D) Patent Quality: Adjust the amount of patenting to take into account the quality (see discussion below).

(E) National Profile: Examine a country's indigenous patent activity and extent of patenting by foreigners.

Let's illustrate possible applications.

(A) *Technological Emphasis* could be used to benchmark your organization (Us) against a key other (Them) to flag areas where you compete and areas where you complement each other's IP for collaborative opportunities. Instead of just tabulating organizations' patent activity relating to the target technology, you might tally:

- Patent activity in one or more particular patent authorities (e.g., EPO or South American countries)
- Current patents (using INPADOC—the International Patent Documentation Centre of EPO—to ascertain which patents are being maintained, or those being dropped)

(B) *Technology Share* could be based on patents granted and/or on patent applications in one or more key patent locales (i.e., EPO, JPO, USPTO). One might compare all firms active in certain domains of interest, or just selected key firms. Consider using relative measures to facilitate comparison. The leader is scored "100," and others are scaled against this. For instance, if the leader has 30 patents in the chosen field over the pertinent time frame its score is 100; another firm with 10 patents would score 33.

(C) *Rate of Technological Growth* can be calculated for selected time slices. These should roughly accord with the rate of change in the domain under scrutiny. For fast-changing domains, we want to compare recent activity levels. Suppose we have decided to compare the two most recent 5-year periods. Firm U increases from 5 to 20 patents (ratio of 4.0), while Firm T increases only from 10 to 12 (ratio of 1.2). For areas with many players over many years, calculation of an average rate of change using moving averages to smooth out lumpy patterns may facilitate interpretation.

(D) *Patent Quality* can involve various measures. Ernst (2003) notes four: (1) ratio of patents granted to applications, (2) international scope (extent of filing in multiple locales, especially the "triad" of EPO, JPO, and USPTO), (3) technological scope (coverage of pertinent areas to the interests in question), and (4) citation frequency. These should be normalized to the relevant domain (e.g., citation frequency compared to other patents in the same area, calibrated by age).

(E) *National Profile* can be derived from Derwent patent families for patenting in given countries. Patenting activity by natives indicates indigenous R&D capability. Patenting by foreigners provides an indicator of that market's appeal. Together these can indicate the viability of the country's IP protection. These indicators could contribute to decisions on investing in that country. Although WIPO data nominally indicate which countries, many applicants designate many countries when filing a PCT application. They are truly interested in those for which they actually pursue national IP protection (information available in databases like INPADOC, as well as Derwent WPI—see Chapter 12 on patent analyses and Appendix A on data sources).

13.3. INFORMATION REPRESENTATION AND PACKAGING

What do we do with all these indicators? We use the term technology information products, or TIPs, to get ourselves thinking about the importance of effectively communicating our "product." That tech mining product is value-added information derived from raw empirical technology information—that is, indicators. We need to think hard about how best to deliver this to the intended users.

Multiple Indicators

Indicators can be combined to provide richer, comparative insights. Let's illustrate with the patent indicators just discussed. Figure 12-1 shows how variants of (B) Technology Share and (C) Rate of Technological Growth can make a 2×2 chart to help benchmark competitive positions. Suitable combinations of these, and other, metrics can generate informative breakouts. For example, plot Technological Status on the X-axis versus Rate of Change on the Y-axis. Depending on how many organizations we want to compare and their concentration, we might do a four-quadrant chart, as in Figure 12-1.

Publication activity measures complement the patent tallies, if the organizations in question both publish and patent. You might display the comparison with dual-column charts ranging from the areas in which "Us" and "Them" compete most strenuously to those areas where only one participates.

R&D "input" measures, such as expenditures and number of scientists and engineers in R&D, juxtapose against patents and publications—R&D "output" measures. You might, cautiously, examine various divisions within your organization in terms of output-to-input ratios, or compare countries or organizations for which such data are of interest.

Another extension is to compare patent measures with "downstream" economic activity. For instance, profile a set of competitors based on patenting, revenues, and market share, in a target area (cf. Hall et al., "Market Value and Patent Citations," http://emlab.berkeley.edu/users/bhhall/). You can further benchmark them against the median (or mean) values for their sector.

Multiple indicators give a more complete picture of technological innovation prospects. Another example—classic investment portfolio analyses seek to maximize potential gains while keeping risks tolerable. This approach adapts well to R&D project "portfolio management." Indicators can help weigh gain versus risk. Furthermore, tech mining provides a vehicle to integrate external R&D prospects into the determination of internal R&D project selection. For example, structured technology forecasting can help NASA

(U.S. National Aeronautics and Space Administration) focus scarce support on R&D concerning technological components rated least likely to otherwise be ready when needed for a target technological system (see Chapter Resources).

Information Visualization

Discussion to this point concerns the content to be presented. We now add a different dimension—consideration of how to present that information. Information can be represented in many ways. Information visualization is the aspect receiving the greatest attention because enhanced computing capabilities present much richer visualization possibilities than ever before. See Chapter Resources.

Three Cognitive Styles

However, visualization is only one worthy approach. Effective communication draws on (1) linguistic, (2) analytical, and (3) spatial intelligences. Each of these styles helps communicate different "TIP" aspects. The challenge is to succinctly summarize information patterns in ways that best speak to the target audiences. Options may include one or more of *text* (our mainstay), *tables* (providing more details conveniently), or *figures* (more visual).

Tech mining can use at least four *linguistic* techniques to facilitate written communication. First, presenting *sample records* can help the user gain in-depth insights. Second, *hyperlinks* allow the reader to follow chains of reasoning to assemble patterns of content by following on-screen links between documents.

The third strategy relies on *automatic document classification*. Clusters of related documents are more easily browsed to discern content patterns (Cunningham, 1996). The fourth strategy is the most advanced, relying on *automatic abstracting* to create short summaries of new developments with natural language processing (NLP) techniques. Automatic abstracting techniques vary, but they all seek to offer quick digests of bodies of information. *VantagePoint* can select sentences from a PCA cluster of abstracts based on coverage of the high-loading keywords that define the cluster.

Analytical techniques count activity to be presented as lists, matrices, or arrays. Lists lend themselves to tabular presentation. Matrices combine two lists and are even more effective interactively, allowing the user to explore subtopics. Arrays, or "profiles," provide multidimensional reports. For instance, a profile of leading authors might indicate organizational affiliation of each, prominent keywords used, and temporal publication distribution.

Spatial techniques communicate patterns and interrelationships best for many users. *Maps* can convey emerging issues in a landscape of scientific and technological (S&T) discovery. Distances between items depicted communicate relatedness. Items may be documents, or index terms including keywords, classifications, or research institutions. A related approach uses the metaphor of *networks* to communicate relationships. Nodes represent items or collections of items. Linkages show relationships among nodes. Indirect links may be investigated with the aid of software such as *Pathfinder*. However, don't assume that your target users like spatial presentations; they are not intuitive for everyone—ask. Also, spatial techniques encourage a gestalt form of understanding that may or may not fit the tech mining needs.

Reducing to Practice: Scripting

A key benefit of scripting (i.e., automating routine processes using MS Visual Basic commands) standard tech mining measures is that analyses can then be done dramatically quicker. We know of instances in which the time to generate particular technology information products has been reduced from months to a day or so. That makes such analyses viable decision support in technology management arenas where time is of the essence. This is a qualitative change whose importance cannot be overestimated. Tech mining advocates need to demonstrate such capabilities to their organizations to build demand.

Scripting enables semiautomated generation of desired TIP components. It can also support development of selected *palettes*. Table 13-2 suggests this approach—for a given question, we offer the analyst a shopping list of indicators from which to choose. As the next step, a set of ready-made templates can expedite composition of those indicators into composite TIP representations—"one-pagers." Section 13.6 pursues these possibilities, including the notion of standardizing business decision processes to take strong advantage of them. *VantagePoint* can be scripted in conjunction with *MS Office*. "Wizards" can guide users in simplified analyses ("one-button" options) or through more elaborate options.

One-Pagers

Indicator packages offer strong appeal to address a given issue in a concise, informative way. The sidebar on "QQQ—the Resolution" suggests consolidation of the pertinent information into a single composite presentation—ideally on one page. Then decision-makers and stakeholders can quickly digest this to gain insight into the prospects.

Information compilations differ greatly between technology questions and organizational questions. Figure 13-1 illustrates a one-pager for fuel cells on an organizational question—Question 7 from Table 13-2—Which companies lead in particular aspects of this technology? We adapt this question toward the patent scenario (Section 12.7) in which General Motors is hypothetically exploring fuel cell possibilities. Our modified question to address becomes: *"Which companies lead in fuel cell development in the automotive arena, and how do we (GM) compare?"*

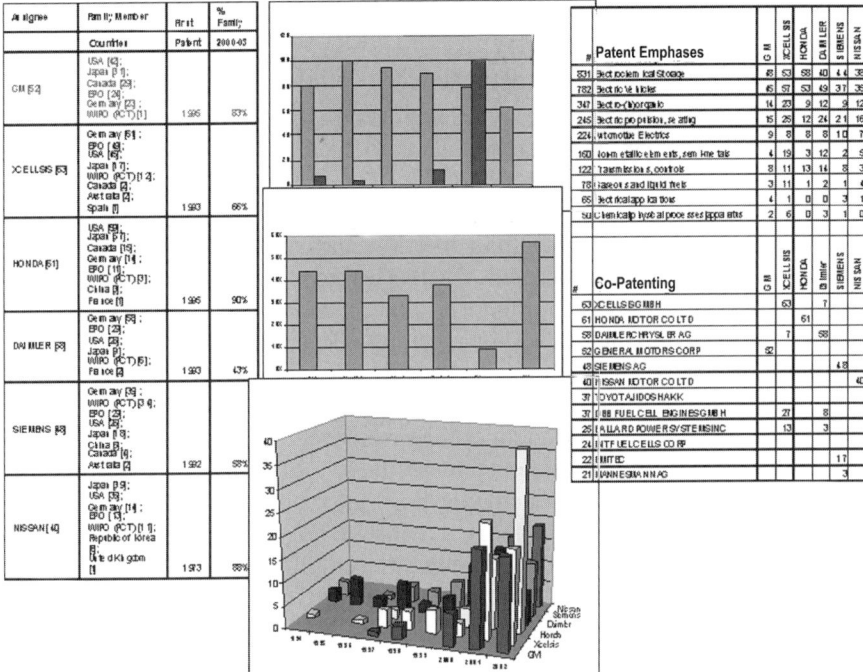

Figure 13-1. "One-pager" on Which companies lead in automotive fuel cells

As the tech mining analyst tackling this question, one would review the suggested indicators for Question 7 (Table 13-2), but adjust to our target users' expectations. For instance, the one-pager focuses on the leading five other companies, plus GM. You can see that including fewer entities unclutters the presentation but risks missing important competitor activity. This points toward the benefits of standardization, where all involved share expectations on what is being presented.

Obviously, the one-pager should be backed by more detailed breakouts and it requires interpretation. This initial try would need refinement to tailor to user needs. It presents "at a glance":

- Us vs. Them—GM alongside five leading automotive fuel cell patenters.
- All of us show comparable levels of patenting in this domain.
- Siemens is by far most actively publishing R&D; the two Japanese firms publish nothing.
- Trends show all of us, but especially Honda, energetically pursuing automotive fuel cells.
- Interestingly, we (GM) and the two Japanese auto companies emphasize patenting in the United States and Japan, whereas the three European

companies emphasize Europe; Nissan also has six Korean patents, whereas no one else has any.

- Patent emphases show general similarity among all six firms.
- Copatenting suggests extremely important collaborations are underway. As an auto manufacturer, we need to find out more about Daimler-Chrysler's partnering (see Chapter 12).

This one-pager is less graphical than others we have generated. This reflects the focus on showing how six companies' activities compare, in some detail. In contrast, in focusing on an emerging technology, general visuals often work well. Maps of intersecting themes within a technological domain pack a lot of information. Color can be used more aggressively (e.g., contrast GM vs. the others), but we had an eye toward black-and-white publication here. Pie charts show the extent of industrial vs. academic and government R&D well (an innovation indicator). Trends and projections convey how hot an area, or subtopics, are. For other technology management questions, information on individuals can be spotlighted (e.g., experts, knowledge networks). For some purposes we like to show "outliers"—players with distinctly different interest patterns.

The "scorecard" notion of simplified, comparative presentations of selected overview indicators offers great promise. The one-pager presents charts showing relative publication and patent activity in the area and showing concentration in automotives. The latter flags Siemens as patenting the most on fuel cells associated with other applications.

In technology-oriented one-pagers, scorecards are a great way to get a sense of technological maturity, competitive climate, science base, and interdependence on other areas. We are exploring scorecard effectiveness in an on-going research project.* Initial trials with Air Products and Chemicals, Inc., finds managers liking such high-level abstraction, but analysts fearing superficial decision-making could result.

The "QQQ—the Resolution" sidebar wraps up our chapter challenge, with a nod toward one-pagers.

Sidebar: QQQ—The Resolution

So, how do you judge this small company's (qqq's) capabilities? Patent analysis, plus. Start by identifying all of qqq's patents. Then highlight all the patents that relate closely to the electronic technology in question. Extract these as a cluster to make a new sub-data set of relevant patent records. List the inventors and map them to identify who has been most active in

---

*"QTIP"—Quick Technology Information Products—Search Technology, Inc., supported by the U.S. National Science Foundation as Small Business Innovation Research (SBIR) Project # DMI-0231650.

this domain, and who works with whom. Use this to tentatively identify the core team working on this electronic technology. (See examples in Chapters 9 and 14.)

Now perform additional searches in patent and R&D publication databases (e.g., INSPEC) to track the activities of the core team members. Reach out to capture qqq annual reports, financial records, product announcements, and so forth. Extract the information into succinct presentations. Show the results to knowledgeable persons and ask them to help interpret.

Then consolidate the empirical evidence with your understanding to tell the qqq story. This could incorporate "one-pager"-type compilations with interpretation. The story should close with its punch line—to buy or not to buy?

"The Rest of the Story"—The QQQ case is based on a real corporate case that took place in 2002. What the tech mining analyses suggested, and experts confirmed, was that the electronic technology development team was no longer intact at qqq. qqq's capabilities to advance this technology were now marginal. Successful tech mining ended QQQ's plans to acquire qqq.

Technology Information Products

The bottom line for tech mining is that we do not know how to represent the information we generate in ways that technology decision makers (R&D leaders, senior researchers, engineering project directors, technology transfer professionals, etc.) accept. We do not think this is because we are stupid, but rather it requires a significant behavioral shift. A colleague at one company using *VP* laments that they have 40 analysts assessing the technological environment, but what sways senior managers is the "Sunday *Times* Technology Section." We need to provide the right content, in the right form, delivered in a friendly way.

Figure 14-1 assays the situation. Our manager listens to that expert by his side. Expertise fits our comfort zone—it is familiar and quick. At present, technology managers rely far more on intuition and expertise than on hard data and sound analyses. Tech mining strives to balance this through relevant TIPs that reach that manager's "other ear." The aim is not to base decisions solely on either empirical intelligence or expert opinion, but to rely on a balance.

An image embedded in our psyches is the cubicle of our DARPA U.S. Defense Advanced Research Projects Agency program manager on a 1995 research project leading to *VP*. He oversaw $40 million in research from within mountains of paper (we suspect largely unread). He got his technological intelligence from the network of researchers and colleagues who had his ear. Although he endorsed the concept of exploiting electronic information, he did not do so personally.

This chapter is really about providing the right information. The first part of the chapter concerned the right content to answer the technology manager's questions. Section 13.2 offered ways to package that content to have maximum punch. This section ponders ways to deliver those information packets most effectively. Chapter 14 pursues process issues in communicating TIPs more effectively.

Alternative Delivery Modes

Dissemination of tech mining findings can be approached in many ways. Researchers performing tech mining at Air Products expressed the desire for easy portability. That is, they want to output information from tech mining software in highly modular form for easy cut/paste into MS Office applications for ease of presentation and reporting.

Interactivity presents another dimension. Various users (recall Figure 14-1) want to take different information from tech mining analyses. The capability to "dig down" to specifics, just as needed, can amplify findings powerfully. For instance, VP analyses can be shared with users via *VP Reader* software that enables them to explore what interests them hand-in-hand with digesting a report (we invite you to try this out on the Wiley website).

Another appealing route is to develop an intranet website for technological intelligence and foresight. Combining two key concepts could amplify the power:

- Standardized innovation indicators to allow easy assimilation
- Multiple technology and/or competitor profiles, supporting comparisons

Such a web-based technology information forum encourages periodic updating to keep the technology profiles current, as a potent organizational resource.

Integration into Business Decision Systems

Suppose we could standardize how our organization presents tech mining findings. Then, when such tech mining findings are presented, the users will recognize them. An organization should develop sets of "standard" indicators. Although these should not preclude experimentation with new measures, having well-recognized sets of indicators will greatly aid their acceptance as they become familiar and understood. In addition, such sets lend themselves to scripting to automate their generation.

Air Products & Chemicals ("APCI") has shared their systematization of their intellectual asset management process. This stage-gate process poses a dozen major technology management concerns, including (paraphrased):

(1) Who has important IP in this technological domain?

(2) For a given company with important IP, what are their emphases?

Each of the major APCI concerns opens out further into several vital questions. For instance, Concern 2 breaks out into questions including (paraphrased):

(A) Is Company X focusing its R&D on particular technologies?

(B) Do they publish actively? Patent actively?

(C) Do they partner with other companies? In which areas?

(D) When did they enter particular technological arenas? Exit any?

APCI requires particular analyses and specific outputs to answer these questions. For instance, at the initial stage of a project, the researcher might need to address Concern 2, but only to spell out Question a by providing a table to show the "top 5" International Patent Classifications for patents in Company X's portfolio. If the project passes the first-stage screening, the researcher might later be required to address all the questions within Concern 2 in depth. APCI prescribes which S&T databases, and which analytical software, to use at every stage.

Standardization does not obviate the need for interpretation. Furthermore, it must not preclude experimentation and refinement of what information is used. That said, we see systematization as "the" way to improve technology management.

13.4. EXAMPLES OF PUTTING TECH MINING INFORMATION REPRESENTATION TO USE

Table 13-2 arrays a host of possible measures and indicators. This chapter goes on to suggest several ways to package TIPs and alternative delivery routes. We close the chapter with examples of successful, and unsuccessful, TIP uses.

One failure illustrates the sensitivity to misunderstanding about dealing with large bodies of information. This can lead to a researcher wanting to know why someone he or she knows appears to be misplaced, or missing, in a map of the R&D domain. Unless the tech mining response to such challenges is compelling, the credibility of the whole analysis can collapse. This happened to us in a study for the U.S. National Institute of Occupational Health—NIOSH. A researcher in a high-level Institute briefing perceived herself mislocated in a map. The NIOSH program manager presenting the findings was unable to convince her of the legitimacy of the analysis from the podium, and his personal credibility suffered a serious blow. It turned out that the mapping was statistically quite proper and why she was positioned in the "wrong" topical domain could be explained step by step. But that was a moot point; the damage was done.

Interpret this cautionary note to mean that tech mining principles and outputs are not generally familiar. As with any innovative approach, the burden is on the innovators to convince the establishment of their legitimacy and value. In presenting tech mining "answers," do your homework thoroughly. Know how familiar your intended users are with these sorts of analyses. If they are relatively unfamiliar, explain what has been done with extra care and thoroughness. Check tech mining results with knowledgeable persons—anomalies need not mean "failure"; they can be fruitful stimuli to further exploration. Chapter 14 suggests ways to enhance the utilization of tech mining analyses.

We next describe several experiences in which tech mining analyses were used successfully, in different ways. These illustrate possibilities in directing selected tech mining outputs to particular target applications.

Gap Analysis for the Vice President for Strategic Planning, Georgia Institute of Technology

Georgia Tech was determining whether to develop a new center of excellence in an applied artificial intelligence (AI) area. The Vice President for Strategic Planning needed to ascertain what complex of skills was needed for the proposed center to succeed, and how we stacked up on those skills.

Tech mining helped by mapping the ten pertinent subtopics that contribute importantly to this defense AI application. Note that the Vice President did not request such a "map"—he had never seen such a representation. Tech mining then helped identify active players within Georgia Tech.

The Georgia Tech AI self-profile suggested weakness in a couple of areas. The resulting profile was presented to a group of Georgia Tech research managers and AI researchers for review. None of them was used to using tech mining. That meant that presentations had to be "transparent" so they could understand and judge whether the tech mining "answers'" were credible. There were doubts! They quickly corrected the tech mining profile to note an active Georgia Tech Research Institute group who were highly able, but rarely published because they primarily did defense contract work. We were still left with a gap in another area.

At that point, the external research domain profiling helped identify potential sources of that expertise. Such expertise could be tapped by recruiting new graduates or faculty from universities with prominent research centers working the gap area. Alternatively, expertise could be accessed by forming a collaborative relationship with such research centers. Another option was to rethink the proposed center of excellence more narrowly to better match current Georgia Tech strengths.

Note that this case does not fall neatly within a single technology management issue (Table 13-1). Although the questions asked by the Vice President in this case involved setting R&D priorities, other important decisions were involved. These included evaluating other academic research enterprises and assessing the potential for R&D funding in the target area. So this study actually addressed a mix of MOT issues.

Using Tech Mining to Identify Research Thrusts in Data Mining on Large Datasets to Scope a Ph.D. Dissertation

Figure 13-2 illustrates one R&D profiling view (Porter et al., 2002). The domain in question concerns the use of data mining on very large data sets (e.g., sets of images generated by satellite cameras passing over a target region). The figure shows the relative attention to three different data mining aspects: data splitting/segmentation (smallest segment), data reduction/analysis (largest segment), and data integration. It also indicates that the overlap is considerable [e.g., only 41 of the 146 data splitting records fail to also address either data analysis (68 records), data integration (20), or both (17)].

This activity profile helps understand certain major research thrusts in large dataset mining. It also sets the stage for further analyses. For instance, we might identify researchers interested in both data splitting and integration. In this case, we were pursuing how "neural networks" contribute to large data set mining, seeking to focus a PhD dissertation. We hypothesized that neural

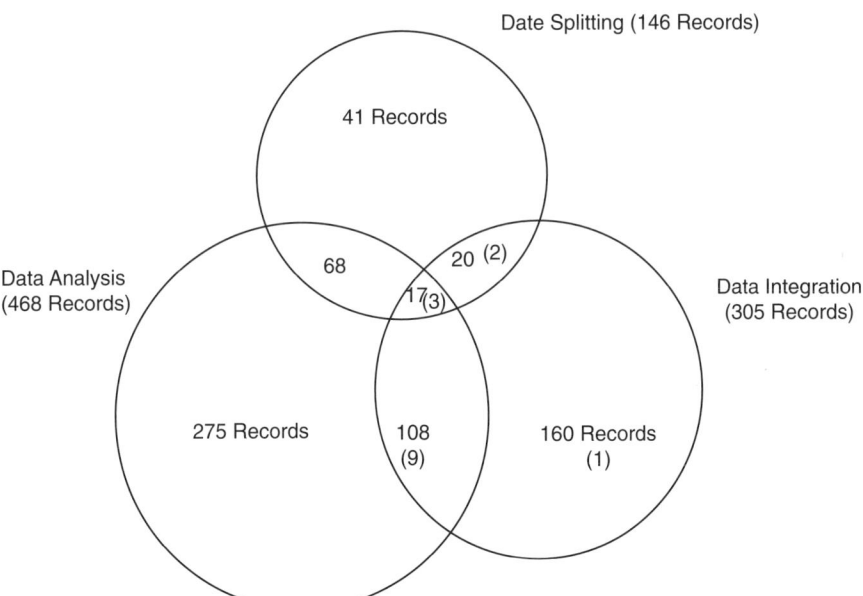

Figure 13-2. Profile of a research domain: data mining for large data sets. [Note: The numbers in parentheses represent neural networks associated with data integration.]

nets could play a major data integration role. The search set of 991 records from INSPEC contains 86 that address neural nets, of which only 15 concern data integration (add the numbers in parentheses in the right circle of Fig. 13-2). This alerts us to the possibility that neural nets might serve multiple needs in mining large datasets. We go further to pull out the 86 neural nets records as a separate subset and map them various ways (not shown here). We find promising data splitting and data analysis roles to pursue.

This example demonstrates a very specific TIP—providing "bird's-eye views" of data. Chapter 16 explores such "research profiling" further. The tech mining results gave a fresh perspective on the research opportunities. The case ended with an interesting twist. The PhD student involved became so intrigued with text mining in support of technology management that she switched her dissertation to a tech mining topic.

Reassessing Ceramics for Tank Engines

At the U.S. Army Tank and Automotive Command (TACOM), a researcher wondered whether ceramics could improve automotive engine performance (Watts and Porter, 1997). Previously the Army had supported ceramics research but had abandoned it as a failure in the late 1980s. Watts hypothesized that the technology might have matured sufficiently by the mid-1990s to warrant renewed Army investment. As you might imagine, convincing "once-burned" senior R&D managers to change their mind about a technology takes some doing. Tech mining did the job.

Figure 13-3 provided the clue to ceramic maturation. The back row shows that ceramic R&D publication relating to engines peaked in the 1980s. Through subject matter experts we learned that, at that time, sponsors lost confidence in this overhyped area and funding dropped precipitously. By the mid-1990s, research activity had risen modestly. The middle row In Figure 13-3 shows a similar pattern in the involvement of research organizations. The number involved had dropped dramatically, showing a modest upsurge in the mid 1990s. The front row tells something different. It indicates that the complexity of the research in the mid-1990s has jumped dramatically—extending way beyond the nature of the 1980s topical coverage. The interpretation: R&D had indeed progressed beyond general fascination with ceramics to specifics. This pattern of specialization implies pursuit of real applications.

Figure 13-3 led to further analyses. We went on to identify the change in topical coverage from the 1980s to the mid-1990s in specific areas. Our paper tabulates keyword coverage, by time period, for particular topics such as materials, process engineering, etc. The trends showed that "research" appeared to be moving into "development" aspects. Critically, ceramic experts examined these indicators and confirmed the interpretations. The combination of empirical evidence and expert evaluation led to the Army Tank Command authorizing Watts to pursue ceramic possibilities.

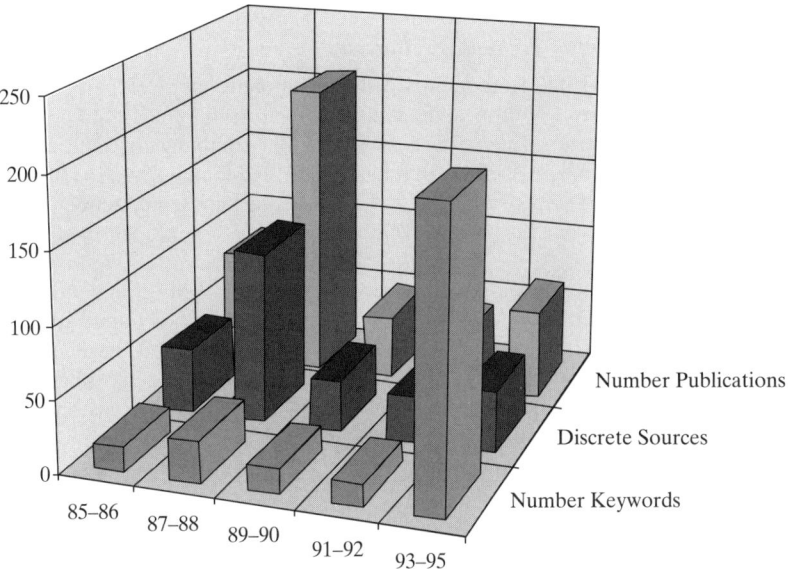

Figure 13-3. Alternative technology maturation trends in 3-D

The case continued to showcase tech mining capabilities. Watts next explored which organizations were doing the cutting edge R&D in ceramic applications pertinent to automotive engine conditions. Somewhat to the surprise of his colleagues, this R&D was not being done by those they associated with structural ceramics application. For instance, one interest concerned coating metallic surfaces with thin ceramic films to enhance performance and engine life. They identified the most interesting thin film ceramic research in the microelectronics sector! There, researchers were working to improve performance of semiconductors with ceramic coatings. Anyway, the Army decided to partner with one national lab and one company to see whether these ceramic films would perform well in the automotive engine environment. And they did. As we write, an engine processing facility is going into operation to coat military engine parts for extended life and improved environmental performance as well.

This example demonstrates the utility of innovation indicators—"keyword richness," in this case.

Bonus Uses of Technology Analyses

Technology information products (TIPs) can be used beyond answering the immediate, internal question. The sidebar describes an imaginative extension.

Sidebar: Bonus Use of Technology Analyses

KPN, the leading Dutch telecommunications company, has been performing scenario analyses to better understand changing technologies and contexts for its businesses. They use these within the company for multiple purposes. For instance, KPN Research has steered R&D initiatives to meet priorities based on scenario analyses. That is, certain telecommunication products and services were developed to meet targets identified in scenarios.

Many companies develop scenarios but keep them highly confidential. In contrast, in the late 1990s KPN began to share its scenarios (alternative futures pertaining to telecommunications). Notably, KPN used these as a marketing aid. As their representatives visited current or prospective clients (e.g., a bank), they would share the KPN scenarios to help that customer explore its future telecommunications needs under various contingencies. Sharing the KPN scenario information enhanced the credibility of the KPN representatives and generated goodwill—and sales.

13.5. SUMMING UP

In several respects, this chapter is the heart of tech mining. By focusing on technology management issues, we can craft technology information products (TIPs) that answer the key questions being posed. To accomplish this aim, we offer a range of candidate measures and innovation indicators (Table 13-2). These can be synthesized into composite representations to answer particular tech mining questions. Those representations can be delivered in different ways to maximize their utilization.

However, this is a nascent area of study. Our innovation indicators are just an initial set of ways to tap publication and patent abstract data resources. They need to be fire-tested to see what measures and representations thereof "work" for technology managers and other users. They need to be enriched by tapping other data sources—patent and publication citations are lightly mentioned in Table 13-2. R&D inputs (e.g., project databases, funding information) warrant probing. Business, marketing, public policy, and popular press compilations need to be mined. Table 13-2 is richest in technology maturation indicators, moderate on contextual influences, and weak on market indicators. We have also not investigated best ways to integrate these empirical indicators with human intelligence. We hope scholars and practitioners will tackle these opportunities to develop tech mining.

In sum, tech mining consists of:

- Addressing technology management issues (Table 13-1 offers 13) by
- Answering particular questions (Table 13-2 offers 39) through

- Empirical indicators (see Table 13-2, including both straightforward measures and conceptually based innovation indicators, generated from compilations of R&D publication and patent abstracts, and other sources)
- Composed with the aid of various computer processes including: text mining software, scripting of routine steps, wizards to guide analyses, drawing on palettes of question/indicator choices and templates to facilitate information presentation, particularly in the form of
- Composite information visualizations—"one-pagers" (Fig. 13-1).

CHAPTER 13 TAKE-HOME MESSAGES

- Tech mining begins with technology management issues (Table 13-1) to be resolved, not with data to be massaged.
- Table 13-2 offers an extensive list of questions that tech mining can help answer. This is a key resource pointing to candidate measures and indicators to address those questions of interest in a given tech mining effort.
- We sort Table 13-2 into "what?" and "who?" topics. You can subdivide further by innovation stage: research (publication-oriented indicators), development (patent-based indicators), and commercialization.
- Explore which particular tech mining outputs resonate with your target users in helping them to make decisions.
- "Innovation indicators" relate empirical information to technological innovation processes to help assess prospects of success.
- Innovation indicators speak to a technology's maturation, context, and market prospects.
- Tech mining also provides useful interim information to the parties involved in the analyses to refine queries and enrich perspectives.
- This book concentrates on mining publication and patent abstracts, but we note extensive complementary information resources pertinent to the business environment.
- Derive your own innovation indicators that speak to your organization's keen concerns, whether technological maturation, contextual support, and/or market prospects.
- Go for multiple, not single, measures.
- Devise "one-pagers" that compile the information to best answer the central technology management questions confronting you.
- Organizations should develop a system of preferred indicators to build user familiarity.
- Such indicators should be incorporated into systematized business decision processes to best leverage tech mining power.

- Document-based delivery of technology information products (TIPs) should facilitate easy incorporation into end uses (e.g., into MS Power-Point presentations).
- Web-based TIP delivery offers good prospects for comparing tech mining findings across technologies.
- TIPs have proven very effective in informing various technology management concerns, but be sure to set user expectations carefully to avoid misunderstandings.

CHAPTER 13 RESOURCES

Our earlier book, *Forecasting and Management of Technology* (Porter et al., 1991) offers a chapter on expert opinion relating to technology analyses. This previous work is oriented around how to perform expert opinion analyses; in contrast, the current chapter discusses how and why technology mining should be integrated with these techniques to gain a complete perspective.

The literature relating to technological change and innovation processes is extensive and diverse. Some concepts to illustrate the breadth of considerations are: competitive forces (M. Porter, 1985), the innovation funnel (Dunphy et al., 1996), technology generations (Mahajan and Muller, 1996), substitution (Smith, 1992), infrastructure (Modis, 1993), industry interplay (Anderson and Tushman, 1990), design standards (Dror, 1989), diffusion (Metcalfe, 1988), tech transfer (Cohen et al., 1979), sponsor and adopter (Souder et al., 1990), and acceleration (Millson et al., 1992). We note other approaches to innovation indicators (Ortt and van der Duin, 2004; Tidd et al., 1997; Dundon, 2002; Gaynor, 2002).

Georgia Tech's Technology Policy and Assessment Center website (http://tpac.gatech.edu) encapsulates much of our experiences leading toward "innovation indicators." See the "Papers" posted, especially those relating to "Technology Opportunities Analysis." Case explorations of candidate indicators appear there for Knowledge Discovery in Databases (KDD) and Fuel Cells. "HotTech" illustrates a web-based information delivery approach to answer 15 technology management questions—see the Intelligent Agents example.

Technical risk management has its own literature (cf. Michaels, 1996). Blending information on internal and external technology development efforts can help plan and allocate resources (Kirby et al., 2001).

Information visualization is exploding with new approaches, ideas, and software. Especially pertinent to tech mining, we note the work of colleagues (cf. Borner et al., 2003). Special journals include *Information Visualization*, edited by Chaomei Chen. His most recent book (2003) focuses on visualizing scientific information.

Chapter **14**

Managing the Tech Mining Process

This chapter offers ways to enhance the utilization of tech mining in an organization. Successful tech mining must do more than analyze science and technology (S&T) information. It must manage an array of personal and organizational processes. This chapter begins by posing tough challenges facing those who would manage tech mining successfully. We offer a checklist to enhance the prospects for an organization to make use of these studies. This leads to considering how tech mining is institutionalized by the organization. The chapter concludes by considering ways to facilitate the learning of tech mining.

14.1. TOUGH CHALLENGES

Tech mining is an emerging area without an established track record. It consists of an amalgam of analytical approaches, carried out by a loose cadre of professionals, to contribute to diverse technology management ends. This chapter and Chapter 15 grapple with how to manage such an ill-formed field.

Does tech mining infrastructure appear to be coming together smoothly? The answer is, unfortunately, "no." Would-be tech mining practitioners face difficult hurdles in getting the raw *data* they need. Neither database providers nor information services managers have fully grasped the desirability of providing sizable sets of text records at moderate cost for mining (i.e., unlimited-use licensing to readily retrieve thousands of abstract records). The software

Tech Mining: Exploiting New Technologies for Competitive Advantage, Edited by Alan L. Porter and Scott W. Cunningham.
ISBN 0-471-47567-X   Copyright © 2005 by John Wiley & Sons, Inc.

tools to exploit these text resources are not fully integrated. Tech mining users confront distinct search engines, retrieval mechanisms, analytical software, visualization software, and end user presentation aids. Potential tech mining *users* don't fall neatly into single organizational units. In sum, those who would put tech mining to use have their hands full.

Those tech mining infrastructure weaknesses can be overcome by commitment and hard work. Then tech mining can derive potentially valuable knowledge. But that is not enough. Our experience, reinforced by colleagues, is that the tech mining process often breaks down at transferring the knowledge into the decision-making processes (cf. Hansen et al., 1999). Communication between analysts and users needs hard work.

So why bother? Throughout the book we make the case that the potential to exploit rich technology information resources is so strong that it can't be stopped, indeed, that essentially "all" technology managers must utilize tech mining in the near future to compete successfully. Chapter 1 set forth some of the "Information Age—Knowledge Economy" considerations that offer tremendous tech mining opportunities. That's why we think the tech mining potential warrants confronting the severe management challenges to make it work well.

Furthermore, on the upbeat side, the authors believe that we have useful things to do with tech mining. Tables 2-1, 2-2, and 2-3 laid out real needs that call for tech mining. Chapter 13 posed substantial sets of *technology management* issues and questions and proceeded to show how tech mining can address them effectively. In various places, we have raised additional uses of tech mining, not limited to managerial concerns, including:

- Intelligence—understanding internal and external knowledge networks
- Technology foresight—alerting to potentially disruptive technologies
- Knowledge discovery—potential instigation of new research
- Research profiling—helping researchers relate to S&T developments within and adjacent to their R&D domains
- Public policy analysis—such as national technology capability assessment and research funding prioritization

These share a theme of more fully exploiting external information resources. This chapter expands the discussion by examining how the value from such analyses can best be captured within an organization.

14.2. TECH MINING COMMUNITIES

In composing this section, we began by distinguishing two communities—*analysts* and *users*. Chapter 2 made the case that these are usually quite different folks. A third community we are nominating consists of *information profes-*

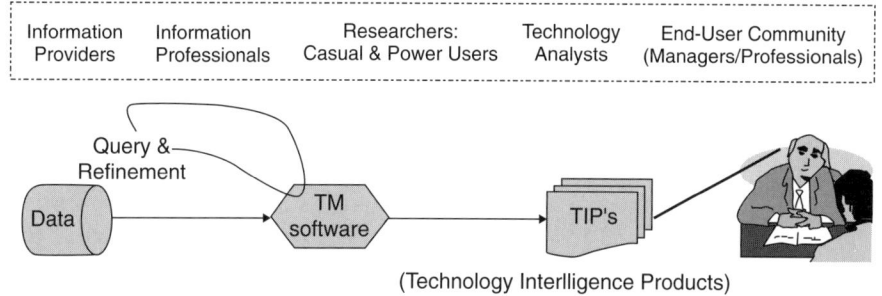

Figure 14-1. Tech mining process and players

sionals (searchers, librarians). We also have come to see *researchers* as a cadre involved with tech mining with distinct needs. Figure 14-1 points to five communities meriting attention if tech mining is to be successfully delivered.

The tech mining process whereby S&T and other (e.g., business) information resources are exploited through the use of tech mining software is addressed elsewhere in the book. A keen interest of this chapter is ensuring that the resulting derived knowledge reflected in "TIPs" (technology information products) does serve user needs. Section 14.4 concentrates on that. This section raises considerations for the particular player communities.

Over time, we keep finding more tech mining players. Information providers include patent office personnel. They not only process patent applications but increasingly involve themselves in enhancing information provision and working to promote information utilization. Likewise, database providers' roles are moving beyond the traditional stopping point of "here's the data." They are extending interests into analytical aspects. For instance, two major patent database providers, MicroPatent and Derwent, provide various text mining capabilities. In particular, MicroPatent offers Aureka and Thomson Scientific offers Derwent Analytics (a version of VantagePoint tailored to exploit Derwent World Patent Index, Delphion, and Web of Knowledge, including Science Citation Index). (See Appendix B.)

Somewhat surprisingly, many *information professionals* have not yet "gotten it"—the idea of gaining metaknowledge from analyses of collective bodies of S&T information (e.g., 10,000 fuel cell technical publications and/or patents). At the 2003 Patent Information Users Group (PIUG) meeting in Chicago, one of us (ALP) listened to successive, contrasting presentations. A technology analyst-manager posed farsighted possibilities for applying tech mining as a key component in one multinational's newly systematized, strategic business decision processes. Then a corporate information professional, still anchored in the paper world, returned to the old standby of how to locate the right few papers.

We believe that exploitation of electronic text information entails a paradigm shift, possibly as momentous as the shift from handwritten to printed text information. Both open entire new realms of information uses, so we shouldn't be shocked that certain interests resist change. Imagine back about 1455 when Gutenberg started printing a few bibles. (Incidentally, checking that date took about 20 seconds with Google—a phenomenon of this electronic information age.) Were the clergy pleased at this newfound capability to share the Christian message widely? Certainly not—it threatened their power enough to occasion burning bible translators at the stake. Over a few centuries, the clergy have adapted to provide new services to those who can read the bible messages themselves.

Today's information professionals, likewise, have to redefine their profession. They need to take into account users directly accessing electronic resources. Information professionals can (and are) developing new opportunities for themselves to contribute to information assurance and analyses. Today's technical librarians are electronic data specialists. Increasingly, many are learning tech mining tools (under whatever name). The exciting opportunities lie in their bringing these skills to bear in new ways. We suggest that these include being the trainers who teach researchers and other professionals how and when to apply tech mining tools. We also want information professionals to become full team members in research and technology development projects (Newman et al., 2001).

Another tech mining community should be distinguished—*researchers*. We mean to include scientists, inventors, product designers, and R&D staff. Scientists and engineers need to perform some tech mining as part of their job requirements. Many scientists and engineers implicitly perform competitive technological intelligence ("CTI") as part of their job function. Technology "hot spots," such as Silicon Valley, draw on concentrations of knowledgeable personnel. These personnel share experiences in keeping on top of new technologies and figuring out how to combine them to best advantage. High job mobility ensures that interorganizational learning of successes and failures occurs. In the coming years on-line communities will reduce the "critical mass" of knowledgeable personnel needed to create "learning communities" about technology.

In a pilot tech mining project with Air Products & Chemicals, we learned that researchers emerge as explicit "doers" of tech mining as well. Air Products' stagegate process (wherein projects pass through a series of checkpoints to assess their merits) entails early screening. "Researchers" (including inventors and technical project managers) perform tech mining at their desktops, having unlimited-use access to a few key S&T databases. They examine the R&D domain and patent landscape to provide prescribed TIPs that contribute to the decision to proceed, or not, with the project in question.

Researchers doing tech mining may or may not become expert at it. Most will be casual tech mining doers. For them to successfully accomplish this a couple times a year requires that the tech mining software be very easy to use and embed considerable analytic knowledge. Chapter 13 probed these emerging tech mining capabilities to expedite analyses and information representation of the findings generated. On the other hand, Air Products' experience indicates that some researchers will do more. They will appreciate tech mining potential and enjoy becoming adept at it. These "power tool users" can help the spread interest in tech mining and provide vital distributed support helping others to do it. The involvement of these industrial researchers in doing tech mining encourages us that faculty and student researchers will pick up tech mining interests and skills in the near future as well.

Returning to the electronic information paradigm shift, we note that this impacts a key end user contingent—*managers*. The availability of pertinent information presses them to shift from intuitive to empirically substantiated decision-making. That will only come "over the dead bodies" of many traditional managers. Dave Grossman, a General Motors veteran, shared his perception that executive selection processes at companies like General Motors favor those acculturated into "business" over managers with technical backgrounds. The resulting lack of managerial familiarity with technologies reinforces tendencies to rely more on marketing and financial information in reaching decisions.

We surmise another selection process at work that leads toward executives having undue faith in their own intuition. Enough intuitive decision-makers, making enough choices, will hit a proportionate number that work out right. If you are lucky and make the correct choice on some high-visibility decisions, your star ascends. Quite likely, your ego does too—you come to believe in your intuition. We assert that managerial intuition is inherently inferior to the *combination* of empirical evidence with experiential judgment (intuition). As tech mining contributes to good decisions, we forecast a landslide in this direction over the coming 5–10 years. In the meantime, expect tensions between tech mining advocates and traditional managers.

In an ideal organizational world, information professionals, researchers, technology analysts, and senior technology managers would unify in support of tech mining initiatives. In the real world, anticipate and counter likely multiway tensions. But also look for interest and support from those attuned to the importance of knowledge exploitation across these communities.

14.3. PROCESS MANAGEMENT

Tech mining emphasizes logical resolution of issues based on conceptual models, information capture, and empirical analyses. Most tech miners are "analytical types" more than "people persons." But good analytical work in itself rarely accomplishes anything. A colleague at one company using *VP* laments that analysts and managers speak different languages. Analysts love to detail the complete story, replete with quantification. Senior managers are always in a rush to get the unadorned bottom line, preferably conveyed as a personal story. And pictures are really nice too. We need to provide the right content, in the right form, delivered in a captivating way.

Decision-makers routinely disregard analyses, even when well done, timely, and pertinent to the issues at hand. Figure 14-1 hints at competition for that manager's ear. That other person with the manager represents an expert. Expertise is familiar, quick, and to the point; empirical analyses are often unfamiliar, slow, and obtuse. What is to be done? Consider "process" in addition to "analysis." Andy Hines, veteran of futures research at Kellogg and Dow, suggests that such professionals reallocate their energies significantly. He recommends 70 percent of your time be devoted to process with only 20 percent to content (analysis), and the remaining 10 percent to organizational culture change!

The "Technology, Policy and Management" Faculty of the Technical University of Delft, where one of us (SWC) serves in the Policy Analysis Department, includes another department, that of Process Management. This faculty studies mechanisms to facilitate decision processes that involve multiple stakeholders. This book is mainly about analysis, but here we challenge ourselves to attend to process as well, if we are to get the several tech mining communities to work together effectively. The sidebar offers a possible analogy.

Sidebar: The Education of a Poker Player

The title of this aside is that of a classic book on poker written by Herbert O. Yardley (1957). Yardley was a World War II intelligence agent who spun charming tales of his poker playing experiences in backrooms from Indiana to China. A key message was that to be good at this game, one had to study the players, as much as knowing the cards and the odds. At the extreme, really good players can identify situations where they can expect to win no matter what cards they are dealt. The analogy fits the technology analyst— it is essential to understand the players as well as to handle the analytics.

Two of us have applied process management concepts (de Bruijn and Porter, in press) to explore the bedeviling question—"Why don't technology managers use our technology analyses?" (Porter and Newman, 2001). Imagine a hypothetical situation—our firm is deciding whether technology innovation T is worth pursuing to improve service S. We commission a tech mining study. What are the factors that determine whether the tech mining analysis influences our firm's decision? Certainly, the quality of the analysis is a necessary condition, but it is not sufficient.

One needs to consider several factors in gauging the likelihood of tech mining impact and, moreover, in figuring out how to enhance the prospects of real influence. First, is the problem at hand relatively structured? That is, does it lend itself to an authoritative solution? With respect to our hypothetical case, it's not hard to imagine some early stage innovations (T) that lack determinant information to gauge their prospects. Or our firm's executives may not agree fully on decision criteria. For tech mining utilization, such uncertainties point toward a contested issue. Here, any finding may prompt questions about tech mining assumptions, system bounding, data, and analyses. Expect conflict.

A second factor is that most tech mining studies serve a network of interested parties, not a single decision-maker. This implies that multiple stakeholders are involved. They are likely to have differing interests in technology T and service S. This portends a contested decision process, likely to generate challenges to the tech mining approach and findings.

A third factor is the importance of the issue at stake. On unimportant issues, the decision-makers may not care about tech mining analyses. The opposite situation is that the subject is of great importance to many participants. Although this is likely to generate interest in the tech mining results, it is also apt to mean that certain players will have strong incentives to criticize the tech mining study if it generates findings with which they are uncomfortable.

A fourth factor is urgency. If time is of no essence, very limited countervailing reasons may be sufficient to undermine tech mining-based recommendations. A shared sense of urgency will enhance interest in the analysis, if there is enough time to perform tech mining and comprehend it.

In the paper just mentioned (de Bruijn and Porter), we work through a decision tree with the possible situational combinations. Compressing this, we suggest that tech mining proponents perform a situational analysis (see sidebar).

Sidebar: Situational Analysis

Questions to ask about how your tech mining fits into the situation:

- Does the decision process involve one actor or multiple actors?
- Are the interests of the actors closely aligned or not?
- Is tech mining generally accepted by these actors or not?
- Is the issue at hand considered important?
- How urgent is the decision?

Simplifying somewhat, we pose five alternative tech mining process management actions based on this situational analysis. These largely address the extent of user involvement, but they extend to other considerations that may be suggested by certain situational factors:

(1) Perform tech mining analysis, clean and simple—when minimal discordant views are to be expected, just perform a quality, timely study and communicate results clearly (see next section).

(2) Work on communication especially to initiate the tech mining exercise—strive to engage all the parties in the issue at hand (information professionals, analysts, intended users, others). Getting the players involved tends to add insights and obtain "buy-in" to the study and its eventual results. When either (a) stakeholders differ somewhat on values but probably share confidence in the tech mining or (b) stakeholders differ on values but don't consider the issue very important, use this strategy. Limited "process management" here aims not to involve the stakeholders actively in the tech mining but to learn their interests well so as to generate clear and understandable tech mining findings for them.

(3) Actively involve stakeholders in the tech mining process throughout—when either (a) the actors don't have confidence in tech mining or (b) the actors have strongly different interests. Interaction should include consideration of the "rules of the game" regarding which stakeholders will be involved in the analyses and when, how to reach decisions at key tech mining steps, and what stakeholders must commit to a conclusion.

(4) Focus on the decision-making space—for important, urgent issues with discordant stakeholders, who don't trust tech mining, target a "no-regret" decision. Get stakeholders to point out potential bases for regrets to analysts, who then provide findings that address prospects and influences relating to these bases. Analysts offer sensitivity analyses, trade-off analyses, "what-if?" scenarios, and such for stakeholders to weigh in arriving at a decision that retains flexibility. Think "negotiation" and how tech mining can catalyze innovative resolutions.

(5) Don't attempt tech mining—we note the possibility that situational analysis may imply that tech mining is so unlikely to be used that it is unwarranted. Factors to weigh include stakeholders so hostile that process agreement is unlikely, widespread suspicion about tech mining validity on the issue, or an issue that is considered unimportant.

Tech mining practice can learn from other decision support approaches. For instance, risk assessment has expanded its attention from analytical concerns, to effective communication of results, toward proactive involvement of the different stakeholders (Fischhoff, 1998). Other technical fields are also working to incorporate broader considerations and process aspects into their scope.

14.4. ENHANCING THE PROSPECTS OF TECH MINING UTILIZATION

The preceding discussion reminds "analysts" to attend to process as well as analysis. This is half the battle, and it will go far toward boosting utilization

TABLE 14-1. Tech Mining Utilization Checklist

Tech Mining Utilization Indicators		In Good Shape:	Needs Work:
User-Analyst Environment: (1) Analysts: Know thy users! Managers: Know thy analysts! (Share your expectations for TIPs)	1.		
(2) Analysts: Get your users involved. Managers: Get involved. • in formulating the analysis • in the analytical process	 2a. 2b.		
(3) Managers & Analysts: Check & enhance the organizational climate • Are your organizational unit(s) receptive to new information? • Get other users' support for the study. • Reduce the extent that the TIP threatens some users. • Budget the TIP appropriately.	 3a. 3b. 3c. 3d.		
Information Characteristics (the TIP): (4) Work together to build credibility • Ensure that the analyst's credentials are properly appreciated • Get the TIP endorsed by well-respected person(s) • Ensure that TIP methods are familiar and accepted by users	 4a. 4b. 4c.		
(5) Encourage vivid **#@!#*!! (memorable) reports	5.		
(6) Ensure that the TIP is on time for decision making	6.		
(7) Be clear on what TIP content is needed: • The right blend of information for taking action • Answers (as opposed to raising more questions)	 7a. 7b.		
(8) Ensure best possible communication • Present the right level of detail for each user • Analysts personally communicate with each user • Present findings interactively (with 2-way feedback) • Provide information in readily digestible amounts • Customize TIPs to match all key users' demands	 8a. 8b. 8c. 8d. 8e.		

prospects. A good portion of the previous section pointed toward involving tech mining users with the doers. Table 14-1 consolidates advice on ways to accomplish such engagement gained through our studies of utilization of tech mining.*

*We appreciate research support from the U.S. National Science Foundation (NSF—Project DMI-9872482, 1998–2001) Management of Technological Innovation Program and from the Center for Innovation Management Studies (CIMS—2001–2002), presently located at North Carolina State University, Raleigh, NC, USA.

Table 14-1 provides an easy-to-use checklist for analysts and their counterpart end users (nominally, "managers") to use in assessing prospects and then taking action to bolster utilization of a given tech mining exercise. Depending on who is performing tech mining for whom, you might want to involve information specialists and researchers as well. Eight factors serve as reminders to check stakeholder needs and whether these are being addressed. The first three factors key on interactions between analysts and managers. The last five factors key on the "technology information products (TIPs) that convey findings.

We suggest that the principal persons involved in a tech mining sit down together and check off whether each factor is under control or needs work. Determine jointly how best to follow up on those that need work.

One might restate Table 14-1 as "eight ways to enhance communication"! It starts with knowing your audience and what they need to know (indicators 1 and 7) and moves to engaging that audience (indicator 2). We will pursue organizational issues (indicator 3) in the coming section. Indicator 4 ("Work together to build credibility") does not go without saying; for many analysts it goes against the grain to tout credentials. But how else does management know that tech mining works and that you are good at it? Call attention to successes.

In our CIMS study (see last footnote) we interviewed 32 colleagues about the utilization of technology analyses. Most compelling was "Al," a fellow who started as an information scientist and moved up to become the analytical right-hand man of the Chief Technology Officer (CTO) at a major multinational. Al did things purposefully—the sidebar illustrates. The salient message is that if you want instant credibility for your tech mining method with the Executive Suite, go to Al, their gatekeeper. If he endorses your approach, the executives will be positively inclined.

Sidebar: Process Management in Action

One of Al's tales to us actually concerned reorganization of the R&D function at his company. He was doing the background research to prep the executives for a workshop to determine the best way to go forward. Al decided that a certain paper really laid out important alternatives, but the paper was not "executive length"—i.e., it was on the order of 40 dense pages. He took as his mission seducing the executives into reading this in preparation for the workshop. He attacked these challenges like a battlefield commander. He mustered weapons (endorsement by the CTO), made personal appeals to particular executives ("Charlie, you need to read this before the meeting because . . ."), and added his own summarization that pitched the importance of reading the source document. Of course, he provided it in convenient form. And it worked—a majority of these extremely busy managers did read the paper and it contributed to a productive decision process.

Paul Germeraad, former Vice President of Aurigin, producers of Aureka tech mining software (now offered by MicroPatent), adds a keen insight. Analysts tend to tell the "whole story," possibly answering the key question prompting the tech mining, but burying this with 18 other observations, caveats, further questions, and general distractions. Those busy executives are far better served by sharply pointed reporting that gives THE major conclusion, and recommendation, if suitable. The "rest of the story" should be appended in case an "AI" desires to trace through how the conclusion was reached. Paul advised that vivid presentation with a clear take-away message is vital to put tech mining to use (Indicator 5, Table 14-1).

Indicator 6 ("Ensure that the TIP is on time") seems obvious. But it warrants explicit attention. "We" analysts tend to place thoroughness, rather than relevance, at the top of our priority list. We can attest to the resulting discomfiture when proudly presenting that profound tech mining based on months of work to really get it right, only to have the requestor blankly stare as (s)he tries to recall—Did I ask for this? Why? The course of events can move ahead faster than the analysis gets done. So, do the best tech mining you can in the time available, even if that's one hour; don't do the best tech mining however long it takes.

Likewise, Indicator 8 ("Ensure the best possible communication") cannot be taken for granted. It takes attention and serious effort to communicate effectively. A successful tech mining analyst-manager told us to rebalance the allocation of effort from the typical 95 percent on analysis and 5 percent on communication toward a 50-50 split. Fantastic tech mining content is worthless unless it is delivered effectively.

The 2002 World Future Society spotlighted an unexpected speaker for a meeting of futures researchers. He was a Walt Disney "communicator." His message was to "tell a story." Again, this goes against the grain of well-trained tech mining analysts who want to detail their repertoire of empirical data and show off their thorough statistics. But the vivid case, presented "up close and personal," provides a hook on which listeners can attach the take-home message of the data-based analyses.

14.5. INSTITUTIONALIZING THE TECH MINING FUNCTION

Centralized, Decentralized, or Outsourced?

Tech mining can support a wide range of organizational functions, as explored several places in this book [e.g., R&D management, intellectual property ("IP") management, strategic planning]. It also has much to offer to new product, business, and marketing strategy development. It involves information specialists and technology analysts and strong interaction with target users (managers and professionals in many corporate units). So, where should the tech mining function be located? We don't have an easy answer, but consider key options: decentralization, outsourcing or centralization (and if centralized, in what type of unit).

We have worked with a few organizations that have pursued distributed tech mining activities. In particular, technology monitoring and CTI assignments can be allocated among a large contingent of researchers, engineers, and analysts. However, such work tends to be perceived as an extra burden, not core to the individuals' job descriptions. In one case, we recall lots of effort invested in setting up files, allocating topics, and arranging a reporting process. We do not recall much carry-through. On the positive side, distributing CTI and tech mining functions takes advantage of existing topical interest and expertise. It can direct attention to emerging technology opportunities. Distributed tech mining can foster carryover from ideas to implementation by directly engaging "the players" themselves. We suggest that those initiating distributed tech mining devise clear incentive structures for this work. Buckman Labs has shown that demonstrably serious corporate commitment to knowledge sharing can work (http://www.knowledge-nurture.com/). As their tale goes, initial incentives of vacation trips for effective knowledge sharing transitioned to "apply your knowledge to solve others' problems or lose your job."

Outsourcing also presents strengths and weaknesses. Small and large consulting firms offer technology monitoring and assessment. That expertise can pay off nicely, offering fresh perspective (breaking from the customer organization's own conventional wisdom). Consultancies can sometimes lower your tech mining costs through multiclient studies. Downsides include lengthening of the information chain, making rapid response on immediate decision issues almost impossible. Also, the outside perspective may be less well suited to generating workable technology solutions for your organization. Andy Hines has worked both as an external consultant (at Coates & Jarratt) and as an internal futurist at Kellogg and Dow. On balance, he sees the internal role, including a lot of process management, as outperforming the external aura of expertise, in effecting action.

Although the "centralized vs. decentralized" argument always has two sides, we believe that some centralization of tech mining makes good sense. External R&D publication and patent abstracts tend to be costly information resources. Capturing benefits from these databases gains from having a core of knowledgeable persons who build up experience in handling those data. In addition, one doesn't want the organization paying multiple times for the same data. A centralized unit can bargain for the best deal in licensing databases and help ensure data quality. Centralized operations also facilitate sharing and retention of tech mining results via intranet sites, report series, etc. In addition, centralization facilitates sharing of tech mining experiences to support continued learning, effective training, and mutual reinforcement. On the proverbial "other hand," some decentralization can facilitate interaction with key tech mining users.

Having just lauded the potential for centralized information processing, we now argue against tech mining being situated in an Information Services unit. One reason is that we do not believe the current information professional culture is fully cognizant of the emergent tech mining potential (e.g., metaperspectives on R&D activity of entire fields—see Section 11.2). Nor are information specialists usually the right persons to interact closely with senior managers (see previous section). When *VantagePoint* tech mining software has been purchased by information services (library) units, results have often been poor; they are unable to develop convincing tech mining outcomes and rarely renew the software license. In contrast, sales to various technology analysis units fare far better, although some of those peter out if champions are transferred, units are reorganized, or management does not appreciate the value of results.

We suggest that tech mining should be based at an appropriate technology analysis unit. Such units can be located in many branches of the organization. You might look to any of the following: the CTO's office, R&D management, engineering, IP management, strategic intelligence or planning, and so forth. Favorable attributes include solid ties to information services and to key technology managers. The "Tech Mining Software & the Organization" sidebar offers an insight.

Sidebar: Tech Mining Software & the Organization

Over the past few years, Search Technology (providers of *VantagePoint*) and colleagues involved with Aureka, our main competitor for the patent analysis tech mining market over this period, have informally compared notes over a few beers at the trade shows. The simplified story goes that technology analysts often favor *VantagePoint* as an analyst tool, but struggle to convince senior management of the value of the tech mining findings that they generate. In contrast, executives love the 3-D visualizations of the patent landscape from Aureka, but analysts are less sanguine about what the software does. The message to both companies is that you really need analysts and managers on board, plus cooperation of the information specialists, for enduring success. As we go to press, MicroPatent and Aureka are to be taken over by Thomson Scientific, which licenses use of Derwent Analytics (VantagePoint)!

It is also important that tech mining not be "captured" by a single business function so as to ignore other applications. Effective networking to build relationships with multiple users is one key to developing a robust tech mining capability, able to survive personnel shifts, changes in management, and economic downturns. The "Organizational Tech Mining Synergy" sidebar shares one instance of value gained through multiple linkages.

Sidebar: Organizational Tech Mining Synergy

Merrill Brenner relates how Air Products had several "voices in the wilderness" for a number of years talking about a new technology that could be both an opportunity and a threat. In fact, it could be a problem for five Air Products businesses, but it did not appear big enough to any one of those five to generate action. The Technology Intelligence group developed a background package and brought together a team for a SWOT (strengths, weaknesses, opportunities, threats) analysis. Although Air Products was too late to participate in the core technology, they did determine to get involved through a partnering strategy.

Innovative Institutionalization

In 2003, a tech mining breakthrough occurred. We viewed this through our "*VantagePoint* sales" lens. A major customer that had used our tech mining software for several years came to us to obtain a much broader corporate site license. The reason was that this company was developing a systematized process in which specific tech mining analyses contribute to the company's strategic business decision processes. Tech mining analyses thus become standardized and routine. This makes a dramatic difference in their familiarity to technology professionals and managers in the organization. That familiarization leads to a jump in the utilization of tech mining findings, in turn providing positive feedback for "more" tech mining to be undertaken.

This is not a single instance, as we also know of another corporation moving toward a similar sort of business process being established with Aureka. In both cases, the organization is prescribing that certain software (not just the two tech mining packages) be used to provide well-specified technology information products (TIPs) for each stage of new product development, intellectual asset management, and other decision processes. This standardization also allows for scripting of major portions of the tech mining information retrieval, cleaning, analysis, and representation stages. That scripting drastically speeds up and reduces the cost of tech mining.

Within one of the companies, we were surprised to learn that finance is playing a central role in this systematization—that is, that technology intelligence is being integrated with financial information in the decision processes. We see this as important because it moves technology management a big step toward reliance on empirical knowledge (i.e., tech mining) as a balance to managerial intuition (not to replace tacit knowledge, but to complement it). This is an analogy to manufacturing process managers shifting from intuition to statistical quality control measures as key input to improve product quality. "Six Sigma" quality would not be possible without empirical measures that underpin process control. We feel that empirical knowledge offers the potential for corresponding major advances in technology management.

We, at Search Technology, are developing indicators of technological innovation tailored to provide "one-page" TIPs to address particular technology management decisions (Chapter 13). We are working to provide "palettes" of indicators derived from R&D publication and patent data. These data are transformed to address specific questions that address one issue. For instance, this might be a quick profile of small companies leading in a target technology for possible acquisition. Or, for competitive intelligence, we would generate a comparative profile of what our key competitors are emphasizing in the target arena. We believe such indicator options, together with information visualization templates, will enable other companies to systematize their tech mining. This offers major competitive advantage. A colleague has included tech mining analyses in the proverbial "pick 2 of the 3"—quick, cheap, or good. We think this is changing; systematization promises all three!

Leon Hermans quite correctly pointed out for us that systematization poses concerns as well. To begin, this is likely far more suitable for large organizations. Standardization could sometimes require tech mining activities that are not truly useful. Of greatest concern might be the generation of "almost, but not quite, useful" information forced into a standard format that does not best address the issues at hand. On balance, we believe that pursuing systematization of technology management processes will prove highly worthwhile, as long as we keep our eyes open to avoid the pitfalls.

14.6. THE LEARNING CURVE

Tech mining is a nascent area that needs to learn "what works" through experimentation. The field needs to become committed to interchange of ideas to improve its performance.

Intraorganizational learning requires some centralization of tech mining function to nucleate interest. Centralization also enables accumulation of experiences. To the extent that these experiences are subjected to explicit evaluation, learning can be tremendously enhanced. Ideally, tech mining evaluation would be designed into all, or a representative sample of, tech mining exercises. Design elements to address (Chapter 15) this include:

- Specification of objectives for the tech mining activity in question (provides a bonus advantage in fostering discussion of what those objectives are among analysts, information specialists, and managers and professionals)
- What criteria spell success
- Who would compile data on those criteria—i.e., explicitly measure tech mining results
- How evaluation results should be shared and incorporated into tech mining process improvement

Cross-organizational learning poses further challenges. Certain professional meetings offer good venues at which to share experiential learning and fresh conceptualizations of how to advance these activities. We note:

- SCIP—Society of Competitive Intelligence Professionals—brings together a strong contingent of technology intelligence professionals
- PIUG—Patent Information Users Group—and other gatherings of patent analysts
- ACS—American Chemical Society, Division of Chemical Information
- PDMA—Product Development Management Association
- PICMET and IAMOT—venues focusing on management of technology
- AUTM (Association of University Technology Managers) and other venues focusing on technology licensing and transfer
- ASIST—American Society for Information Science and Technology— and other venues focusing on S&T information (e.g., Science & Technology Indicators Conference)
- MATI—Management of Accelerated Technology Insertion—and other consortia concerned with technology forecasting, roadmapping, and assessment practices

At this stage, it seems more desirable for tech mining advocates to interact with such intersecting professional interests through organizations such as these than to set out on their own.

Should there be explicit training programs for tech mining skills? We see clear value in short workshops on specific tools and methods. These might be offered in conjunction with meetings like those noted, as stand-alone offerings of commercial or university continuing education units, or on site. Perhaps most compelling is the development of electronically delivered training modules. These could be amplified through scheduled web meetings to present and discuss practice issues.

However, we argue that most learning of "add-on" skills such as tech mining should be job-centered. The best time for me to learn a particular tech mining skill is when I want it to contribute to a project on which I'm working now. The best place for me is at my computer, using the actual software applied to the data I really want to analyze.

We aspire to develop computer-based tech mining modules. In conjunction with software (*VantagePoint* or *Derwent Analytics* in our case), we want a tech miner to be able to quickly (e.g., in 20 minutes) generate useful results. Consider the word processor—one wants to type a simple document right away, then augment skills as needed. Likewise, we think tech mining software should enable easy start-up. A series of training modules would be available at the tech miner's "fingertips" to allow acquisition of a particular skill (e.g., generating a special thesaurus) or generation of a specific output (e.g., a knowledge network map).

Are there prospects for university level courses? Yes. These could be offered in conjunction with masters programs in management of technology ("MOT") or technical tracks within an MBA. They would complement information science and related programs. We could imagine good fit and interest within S&T policy analysis, futures research, data mining, and "information engineering" programs. Distance learning offerings would hold special appeal, given the argument just made for on-the-job tech mining training.

Courses offer the potential to help build a tech mining community. Just the fact that I took a course with others interested in this field would strengthen my sense of collegial support for such activities. Courses would also provide an indirect "imprimatur" for tech mining—that is, the fact that there is sufficient organized knowledge to formulate a course may help convince practitioners and customers that there is substance of value. Instructors also can be called on to make presentations to dubious colleagues. Course alumni could someday achieve sufficient initiative to build networks that could evolve into critical mass for a professional organization.

To wrap up, this chapter concentrates on the elements of tech mining beyond the analytical stream. It starts by posing infrastructure needs (concerning data access and tools), leading into addressing the needs of all the players involved in tech mining. Tech mining poses notable challenges for information professionals, researchers, analysts, and end users. Under the rubric "process management," we suggest ways to make tech mining useful. A vital element of process management is ensuring that analysts and users share information well—use of our checklist can ensure this happens. We then turn to issues in institutionalizing tech mining. Within an organization we favor a degree of centralization of tech mining function, but we also suggest extraorganizational exchange of knowledge on how to do tech mining.

CHAPTER 14 TAKE-HOME MESSAGES

- Assess to what extent the five potential communities that play prominent roles in tech mining—information providers, information specialists, researchers, analysts, and target users (managers)—are on board.
- As you undertake a given tech mining exercise, assess the situational fit on 5 dimensions—whether multiple stakeholders are involved, whether their interests align, if tech mining is generally accepted, issue importance, and urgency.
- Then determine what process management actions optimize your utilization prospects: tech mining analysis alone, good communication, interactive process management, no-regret options development, or no analysis.
- Step through the Tech Mining Utilization Checklist (Table 14-1) with the information professionals and technology managers involved to assess specific actions to bolster utilization.

- In establishing tech mining operations, explicitly consider alternative organizational options; we favor basing tech mining in a key technology analysis unit, with suitable outreach.
- Consider ways to systematize tech mining components in your organization's strategic decision processes.
- Participate in interorganizational exchanges on tech mining practices.
- Look to on-the-job modular tech mining training and more formalized educational opportunities.

Chapter 15

Measuring Tech Mining Results

How effective is tech mining? This chapter addresses ways to assess that effectiveness to help improve tech mining. The chapter discusses what should be measured, how measuring is performed, and steps to ensure that the measures used are effective and valid. The last section of the chapter reminds us that sometimes measurement even points out our successes! The chapter begins by first asking "Why measure the success of our tech mining efforts?"

15.1. WHY MEASURE?

Chapter 14 argued for attention to a tech mining learning curve and bootstrapping ourselves up the curve. To do that, we need to know how we're doing to enable us to improve. That means we need to measure tech mining results.

We use "measuring" as the chapter focus. We might instead have called this "tech mining evaluation" or "tech mining assessment." Those correctly suggest that the activity has a reflective element. For some, "evaluation" connotes serious attention to design and execution of fairly elaborate exercises in their own right. For others, evaluation carries a pejorative aura. For present purposes, we need not pursue these aspects deeply. This chapter makes the case for measuring tech mining results and suggests how to do so.

Why wouldn't we measure tech mining results? The incentives for measurement and evaluation tend to be much weaker than those for doing "anything else." There are other pressing needs with more immediate payoffs.

Tech Mining: Exploiting New Technologies for Competitive Advantage, Edited by Alan L. Porter and Scott W. Cunningham.
ISBN 0-471-47567-X   Copyright © 2005 by John Wiley & Sons, Inc.

And evaluation conveys threat. Measures may indicate that our present practices are less than optimal—who wants to hear that? Furthermore, assessment may suggest that we ought to change what we're now doing, and change is uncomfortable. So, while we definitely see upbeat aspects to measurement in helping demonstrate the efficacy of tech mining, we doubt that much will get measured without explicit prodding. We urge tech mining managers to mandate and support measurement of results.

Measurement payoffs can follow various routes. Most of the time we would use results internally. Measures inform those who direct tech mining practice to improve its effectiveness. They also can aid related business functions in determining appropriate roles for tech mining in their units. But we also want to propose the value of providing tech mining assessments for external use. More or less formal assessments can be shared in conferences or other venues to advance the state of the art.

15.2. WHAT TO MEASURE

Tech mining is complex. It involves multiple doers, users, outputs, and purposes. Measuring tech mining must therefore be multidimensional. To begin, the would-be assessor needs to specify the unit of analysis. In most situations, this will not be routine. Tech mining often builds upon earlier related analyses, passes through iterations, and generates results used in more than a single application. We suggest determining tech mining bounds based on the primary objectives of the evaluation. For instance, if a key concern is to determine the value of a particular database, we want to measure the effectiveness of multiple analyses that incorporated data from this source. Or if we want to contrast the usability of alternative analytical approaches, compare tech mining using one tool set with tech mining using different tool sets. Or if one organizational unit is assessing the decision support value of tech mining efforts, it would want to look across several exercises performed on its behalf.

We focus on a few overlapping measurement dimensions:

- Direct vs. indirect measures—in particular, "asking" the participants and users about aspects of the tech mining exercise (e.g., surveys, interviews) vs. "observing" (e.g., tallying patents or new products, comparing balance sheets)
- Factual vs. opinion data—However gathered, decide on the suitability of empirical tallies (how much of certain attributes) vs. opinions (as to efficacy, appropriateness, etc.).
- Players vs. observers—Do you want to "hear it from the players themselves" or get perspectives from interested bystanders?

What balance of measures help answer the key evaluation questions? You want valid and reliable measures, and you want them cheaply and easily. Your evaluation design needs to weigh these essential trade-offs. Should these be quantitative or qualitative? Most empirical data gathering lends itself to counting. Even for opinion gathering approaches, simple scaling of query responses expedites data collection and comparative analyses. However, open-ended queries allow respondents to explain the essence of what was really important. We recommend that the tech mining measurement system designer explicitly consider:

- Survey instrument—what questions to pose to whom
- Mode of measurement—can questions be posed via E-mail or do they warrant more personal interaction (in person, phone)?
- Whether there are secondary data sources that can be used to avoid having to ask people some or all of the questions (if you can find the answer yourself, don't ask)
- Suitable blend of closed-form, simple measures and open-ended opportunities for key players to expound on what mattered most
- Suitable mix of respondents—i.e., ask the tech mining analysts about process, the end users about satisfaction, and the subject experts about accuracy.

One's measurement objectives should determine what one measures. We will not discuss the possibilities in detail, but instead we array a number of candidate measures in Table 15-1. Suitable subsets of these or similar factors could be targeted for measurement. Within an organization, *standardization* of the measures used builds familiarity and credibility. Some factors are trickier to measure. For instance, "mistakes or costs avoided" has the tenor of an ephemeral double negative but nevertheless could be highly important.

Note that these measures, particularly those concerning tech mining processes, are largely qualitative. This does not imply loss of rigor, but it does demand attention to measurement quality. Survey questions must be formulated with care and should be pilot-tested. These should be posed to minimize biasing responses. In general, we advise drafting your tech mining evaluation protocol, then reviewing every item for relevance to the key issues targeted. Throw out every "nice to know" item, retaining only the "need to know" measures. This will simplify data collection and analysis. But most importantly, reducing the burden on colleagues bolsters the prospects for high response rates.

Auxiliary information may need to be gathered, too. Demographic information about the analysts, target users, information professionals, and others involved helps understand "who" contributed. Knowing educational background and job responsibility could prove relevant in understanding why certain findings proved valuable. Tracking the extent of prior familiarity with particular tech mining elements would help discern needs for training and familiarization.

TABLE 15-1. Factors of Interest and Candidate Measures of Them

Factor	Measures
Focus/content	User judgment of relevance to the issues at hand, completeness, proper inclusion of contextual factors, etc.
User needs	What their tech mining needs were and the degree to which each was met
Tech mining outputs	Value placed on each output
Process management	Relative effectiveness of various participatory actions, including time/cost trade-offs
Credibility	User judgments concerning credibility of (1) analysts, (2) the tech mining generally, and (3) particular findings
Communication	User judgments on effectiveness overall and of particular media used
Information access modes	Extent of use of alternative modes (e.g., workshops, websites) and user judgments on the effectiveness of each
Timeliness	User judgment
What users liked	User judgments on comprehensibility, vividness, etc.
What else users want	User opinions on what additional information would be valuable, possibly contingent on available resources for the tech mining
Information quality	Analyst and information specialist views on the adequacy of information inputs (collectively; individually)
Tech mining resources	Analyst and information specialist judgments on data adequacy and value of particular software tools
Tech mining practices	Analyst and information specialist assessments of the adequacy of tech mining steps and ways to improve them: search, retrieval, data cleaning, analyses done, technology information products (TIPs) generated, information visualizations, and communication mechanisms used
Tech mining validity	Views of various analysts on whether the best available information and analyses were performed, and on the validity of conclusions reached; suggestions on ways to improve
Tech mining costs	Direct costs of the tech mining activity ($); opportunity costs
Tech mining payoffs	What value was obtained from the tech mining activity? Direct benefits? Secondary benefits? Unintended benefits?

The sidebar gives the flavor of a research project involving technology managers and professionals at Air Products in 2003. An initial survey inquired as to what they wanted from an analytical process called "QTIP" (Quick Technology Intelligence Process). QTIP set out to mine topical searches in EI Compendex and MicroPatents to profile emerging technologies in one day or less. This effort relates to "user needs" (Table 15-1). It illustrates the range of possible tech mining-related facets to consider measuring.

Sidebar: User Interests in Tech Mining—the "QTIP" Case

An E-mail survey asked some 30 senior technology managers for their preferences concerning a number of factors. Highlights include:

- More interested in profiling particular technologies than in comparing companies across technologies
- Quite interested in screening unfamiliar technologies
- Want to see trends in R&D activity for target technologies, and for subtopics
- Would like tech mining to help forecast technologies' development
- Interested in identifying related or extension IP
- Like graphical outputs
- Generally would rather not be heavily involved in performing the tech mining analyses

After generating five sample tech mining analyses, we reviewed these with Air Products researchers working on those technologies. They expressed a high priority for hands-on involvement in working with VantagePoint files and preliminary tech mining outputs. They sought to personally refine, interpret, and present suitable findings to senior technology management.

Note that this experimental arrangement involved an additional level of players beyond the "three communities" that we initially anticipated. Here, we had (1) information professionals (helping with searches and data cleaning), (2) tech mining analysts (applying *VantagePoint* to generate basic tabulations and "innovation indicators"), (3) *researchers* assigned to assess some aspect of a changing technology and related business opportunities, and (4) senior technology managers as end users of the findings.

15.3. HOW TO MEASURE

Tech mining measurement is akin to research evaluation. A tech mining study is done and we want to document *what was done and how effective it was.* Did tech mining result in the desired outcomes? If not, why not? If so, can we capture why? We want to build our knowledge of how to perform effective tech mining by evaluating our tech mining exercises. Research evaluation has its own body of knowledge that we will not delve into deeply (e.g., journals such as *Research Evaluation*, see http://www.prism.gatech.edu/~sc149/reseval/html/past.html). We share a few basic considerations concerning the nature of evaluation and evaluation design here.

The distinction between formative and summative evaluation helps align measurement priorities. Formative evaluation obtains feedback on how an activity is going, to inform improvement—in our case, to do tech mining better. Summative evaluation, in contrast, assesses the merits of a completed activity. This contributes to judgment on its efficacy, and, perhaps, on whether it should be continued or not. The tech mining manager setting up a measurement program needs to decide which (s)he wants.

When should tech mining review procedures be designed? It is always best to design the measurement system in advance—a priori. This enables collection of the needed information and establishment of suitable baselines and comparisons. When a priori evaluation design is not possible, post hoc can be attempted.

What constitutes a good measurement system? The magic answer is suitable, fair comparisons. The notion of "threats to valid inference" has proven an extraordinarily valuable guide in designing interpretable evaluations (cf. Campbell and Stanley, 1963; Cook and Campbell, 1979; and Section 10.2). Those designing measurements should keep in mind four general concerns:

- Internal validity—Can you discern whether the activity of interest (tech mining or a related activity) made a difference? Do you have good bases for comparison?
- Statistical conclusion validity—Whether the differences observed could reasonably be attributed to chance? This is enhanced by highly reliable measures that minimize measurement noise.
- Construct validity—Can we ascertain the underlying variables (constructs) at work (to help generalize our findings)? Have we operationalized our understanding of cause and effect factors well enough to assess?
- External validity—Can the tech mining outcomes be generalized to other situations of interest? Could the results be generalized to other persons, in different contexts, at different times?

To meet these four tests of valid inference, we need to set up solid comparisons. Contrast this with the more typical "case study," in which a singular activity is reported, often in glowing terms. In the case approach, we might have performed the coinventor analysis only; then reported that "it worked great." In general, such stand-alone case studies are good only to generate hypotheses for further assessment; they are not generally interpretable regarding the effectiveness of tech mining. The Klavans team, in contrast, compares coinventor analysis with four other approaches (see sidebar "Comparative Tech Mining Design"). Those comparisons might be made along a single measure, but would be even better with multiple measures. That would help us see in what ways coinventor analysis was, or was not, more effective than the other tech mining approaches tried.

Sidebar Case—Comparative Tech Mining Design

Dick and Judith Klavans described (American Chemical Society Annual Meeting, San Diego, 2001) an intriguing tech mining approach and, what is really quite unique, a careful validation of it. They worked with a pharmaceutical firm to determine what patent analyses would most accurately portray that firm's competence in a particular technology. They compiled over 750 patents filed by the firm over a 5-year period. They then compared five methods:

(1) Coinventor analysis (cluster based on which individuals appear together on particular patents)

(2) Cocitation analysis (cluster patents based on their tendency to be cited together)

(3) Coterm analysis (cluster patents based on similar language used in the abstracts)

(4) Patent categorization (cluster based on International Patent Classification categories)

(5) Science categorization (cluster based on the scientific journals cited by the patents in question)

They determined that coinventor analysis best characterized the technical competencies of the firm. The intent, of course, is to apply such methods to learn about other firms. They also note that methods (4) and (5) help probe for other organizations with related technology interests.

To repeat, we recommend that measurements be set up to provide relevant comparisons. Some basic comparisons are (Georghiou and Meyer-Krahmer, 1992):

- Before and after—"This time we applied tech mining; last time, for a similar issue, we did not"—here's how the results differ along several dimensions (e.g., decision quality, timeliness).
- Control group—"Two units faced a similar issue; our unit applied tech mining and theirs did not"—here's how results differed along several dimensions.
- Comparison groups—extend the control group notion to multiple instances. These might apply different versions of tech mining, as well as no tech mining, or other decision support activities.
- Pre- and posttreatment comparison groups—Ideally, an experimenter would like to control many factors that could distort comparisons; random assignment of subjects to treatment really helps do this. This almost never would make sense for tech mining measurement. Best would be to

combine the before-and-after type comparison with the comparison groups notion in a "quasi-experimental design."

Given the natural disinclination to perform evaluation, we need to devise tech mining measurement schemes that balance simplicity and cost-effectiveness with interpretability. Campbell and colleagues have generated a helpful, limited list of about 50 specific factors that can jeopardize valid interpretation. Guarding against a few of the key factors that seem to bear strongly on tech mining measurement can yield more informative results:

- Subject selection—Are the tech mining doers and users reasonably similar in the units being compared (e.g., in their familiarity with tech mining, overall subject knowledge)?
- External influences—Can we rule out extraneous influences as causes of observed differences (e.g., this decision supported by tech mining worked out better than another decision without tech mining, but could the difference be attributable to the change in CEO? to the upturn in the economy?)?
- Other explanations—Play detective!—Can you come up with alternative plausible explanations for the observed differences (e.g., we put a lot more resources into resolving Issue A than into Issue B)?
- Reactance—Are the players involved aware that they are being evaluated? If so, could they distort the measures to convey the desired results (e.g., the manager reports that tech mining made a tremendous contribution in order to boost chances of future tech mining resource allocations)?
- Robustness—Do we observe similar effects over multiple instances (e.g., be wary of one-time findings that tech mining did, or did not, contribute)?

Two "who" issues deserve mention. First, we need to be clear as to whose judgments we want to obtain. Some tech mining measurements (e.g., data quality) are best provided by information specialists; others (e.g., usability of tech mining software) by analysts; and still others (e.g., were particular tech mining outputs influential in reaching decisions?) by end users.

Second, we need to decide who will take the measurements. Tech mining analysts are most familiar with the work done and the various players, but they have a strong stake in coming up with favorable findings. Sometimes, it may be worthwhile to get neutral observers to assess the tech mining results and practices. It's a trade-off whose resolution depends on the intended uses of the measures.

15.4. ENABLING MEASUREMENT

What is needed to initiate tech mining measurement? Here are some factors in addition to those raised in the previous sections. First, someone needs a

TABLE 15-2. Documenting Tech Mining

Tech Mining Action	Document
Sources used	Which were considered; why chosen; how accessed
Resources invested	Funds, personnel, special materials available for the tech mining exercise
Tech mining actions taken	Activities log (indicating who did what, when)
Process steps	Log significant steps (e.g., contacts, meetings, key inputs, how users were engaged in tech mining activities, feedback mechanisms used, communications received).
Outputs (TIPs) generated	File draft reports.
Reactions	Compile responses to tech mining presentations, reports, requests for additional information.
Follow-up	Compile information on what happened regarding the issue at hand, reports on how tech mining information affected decision processes, hindsight assessments of how well tech mining had gone and the issue outcomes.

clear assignment to perform tech mining assessment. That needs to be bolstered by definitive organizational backing for the measurement effort. Otherwise, it won't happen.

In practice, we have found that *documentation* of tech mining activities does not come naturally. Yet it is invaluable for measuring tech mining results and ascertaining the merits of particular practices. Being able to track "how" someone performed that search, why they used particular sources, what prompted generation of that chart, and so forth, facilitates many aspects of tech mining management. It also fosters quality assurance awareness and actions (if necessary). Table 15-2 provides a starter checklist to stimulate thinking on what should be documented. Once you have decided on the measures to maintain, it is desirable to make record-keeping easy and convenient, but also accessible and understandable by others.

It helps to hold a wrap-up meeting at which the players in the tech mining exercise reflect on who did what, and what worked well. This gives analysts a chance to reflect on data issues, how they determined analytical approaches, and the nature of interactions with users. It offers interested users a chance to enhance their understanding of what lies behind the tech mining results. It can help information specialists see the "rest of the story"—what was done with the information they helped provide and refine. Enhanced mutual understanding helps build one's tech mining community. The sense that others share an interest in performing these analyses fosters resolve to overcome hurdles. Networking with each other builds a more robust environment so that tech mining structures don't collapse when one key person leaves.

15.5. EFFECTIVE MEASUREMENT

So far, we have set forth a number of considerations in generating measures. Let's now focus on what we intend to gain from this effort. In so doing, the first facet we emphasize concerns measures of success. When all is said and done, what do we want from tech mining measurement? We distinguish two main criteria: *validity* and *utility*.

Validity concerns whether the tech mining was done as well as it could be. Did it make the best possible use of the best available information? Did the analysts apply the most suitable tools to generate the most meaningful indicators? Were those tools applied correctly?

Note that validity does not key on whether tech mining forecasts prove correct. For instance, suppose we generate competitive technological intelligence ("CTI") asserting that competitor C could undercut our position in market M by acquiring certain intellectual assets. As a result of this warning, our company aggressively counters C's moves and our market position holds firm. The threatened bad outcome thus does not come about, but that does not mean the tech mining was wrong. Indeed, effective tech mining ought to lead to actions that alter projected futures in our favor.

Other tech mining addresses inherently uncertain situations. We may identify several alternative courses of action and assign relative probabilities to each. Our test of validity bears less on what proves out than on whether we find flaws in the underlying analyses. In the face of uncertainties in forecasting emerging technologies, we want robust tech mining. That is, we seek results that are replicable by others. We recommend explicit expression of tech mining assumptions as a starting point for robust analyses.

Determination of validity relies on critical review of tech mining activities by well-informed professionals (the participants or more neutral observers). Reviewers may be able to suggest alternative information resources or new analytical tools besides those used in the study in question. Useful review can be quite simple or more elaborate.

Utility concerns whether the tech mining proved useful.* This can be multifaceted, considering:

- Findings directly contributed to the focal decision processes; the target users used the results.
- We already knew what decision we were going to make; tech mining helped in justifying that decision.
- The findings contributed to resolving other issues, possibly by other users than those initially targeted (i.e., secondary benefits).

---

*Utility has other meanings, too. Economists distinguish variations on the "greatest good for the greatest number." Others argue that overconcentration on economic benefits downplays other important human interests (e.g., equity, transcendence). There is a loose parallel for us from such debates—the utility of tech mining can take many forms.

- The tech mining processes enhanced "everyone's" understanding of the issues relating to the topic, possibly leading to redefinition of the critical issue.
- The tech mining processes enhanced appreciation for the use of empirical knowledge in technology management, leading to better decision processes generally.

Measurement of utility can become involved. At the first tier, we may ask the target users whether the tech mining influenced their decision-making. Pursuing further, we might explore how tech mining outputs affected people's thinking on the issues. Also, we can try to understand more about the processes through which tech mining was applied. For instance, we might pick up that tech mining saved X amount of time in coming to a decision (or conversely, that the decision was improved a little bit, but it sure was slowed by tech mining!). Or we could seek to find out reasons why certain players appreciated the tech mining and others did not. The sidebar "Keeping Awake at Night" illustrates utility of a special nature.

Sidebar: *Keeping Awake at Night*

(an Air Products experience related by Merrill Brenner)

Merrill asked a technology manager in an emerging area what was keeping him awake at night, and he replied with three "A vs. B" technology choices that could undermine his proposed product offerings. The Technology Analysis unit gathered substantial background information on these three topics and did analyses to determine corporate paths. That is, they played out alternative development strategies and their relative payoffs for Air Products. From these, senior management chose a favored strategy. The technology manager was very pleased to stop the "recycling"—he was asked often to change directions based on the latest external developments. He now has a set direction, with a documented basis, and he only has to worry about the basis changing.

One danger in focusing measurement or assessment on single tech mining activities is losing sight of cumulative benefits. Compilation of tech mining results across many studies of different technologies and organizations (competitive intelligence) can prove very useful. A given technology profile can now be contrasted with that of other technologies. Recognition of similarities and differences can aid technology forecasting. Other technologies' development patterns may make good analogies. As such, they could be "leading indicators" for the target technology (e.g., VCR market penetration as a model for DVD emergence) (See Chapter Resources). Similarly, a given competitor profile can be benchmarked against other companies (or countries or whatever).

Moreover, cumulation of tech mining results over time can build up valuable indicators. Particular analyses can be updated regularly (e.g., semiannually). Updated information of this sort can flag important changes, acting as an early warning system. Time series on R&D and business activity over time may lend themselves to fitting of growth curve models and projections. The combination of tech mining outputs over multiple topics, over time, can be a powerful information resource. Consider sharing this information via intranet websites to provide a base for organizational knowledge sharing and application.

15.6. USING MEASUREMENTS TO BOLSTER TECH MINING

And last, a word in flagrant tech mining self-interest! Although measurement of tech mining results entails threats to its practitioners, it also offers great opportunities. As a new enterprise, tech mining does not come with ready-made credibility. We need to document that tech mining works, that it generates really valuable information, most of which is not available by other means. We also need to show that the tech mining benefit-to-cost ratio can be extremely high. As advocates, we believe these positive results are within our reach. That does not mean that all tech mining efforts will be unqualified successes. Rather, we sense progress up a sharp learning curve that will make tech mining indispensable someday. So, measure tech mining results and broadcast the results to build support for the endeavor.

Recalling a message of the last chapter, personalized stories provide a compelling, vivid way to convey messages. So, in addition to generating credible measures and tabulating the advantages accrued from tech mining, we need to collect success stories. These have their first use within organizations—where well-respected colleagues relating value received from tech mining exert bona fide influence. We also hope that somewhat sanitized story versions can be shared outside, helping to build the larger tech mining enterprise. To be even-handed, we need to also share tales of failure, along with putative remedies.

While we are considering how to promote tech mining, we suggest thinking about ways to lower the barriers to tech mining use. One key seems to be provision of cost-effective infrastructure. For tech mining, this begins with access to data—that is, unlimited-use licenses for a core set of science and technology databases. Database providers and gateways should explore ways to expand such access. For one, data provider agreements with trade associations could provide affordable data access to small and medium enterprises (SMEs). Integrating data access with tech mining software to streamline analyses and business arrangements would help a lot. Allowing technology consultants to resell data from R&D and patent abstract databases could expand tech mining an order of magnitude. We ought to target universities so that students graduate with an awareness of what tech mining can contribute and how to make it happen.

Within large organizations, the infrastructure breakthrough is beginning. The inclusion of tech mining elements in strategic business decision processes elevates tech mining to a new platform. Making it part of the decision system means that a much wider user cadre becomes aware of its potential and familiar with its technology information products (TIPs). That should provide needed positive feedback to develop more powerful tech mining tools and more effective results. It also promises a robust base for tech mining, in contrast to the situation of a lone champion whose departure results in collapse of the endeavor.

One final boost—were there a visible research program in place, this would have broad impact. In the early 1970s, the U.S. National Science Foundation set up modest funding for technology assessment (TA). That helped spawn research and publications thereof, academic courses, books on TA, and a community of practice. Recent European Community research initiatives support the extension of technological foresight methods and practices (e.g., Workshop on Future Technology Analyses, Institute for Prospective Technology Studies—IPTS—Seville, May, 2004). Tech mining would benefit tremendously from having a research program focused on its practices. That, in turn, could stimulate a peer-reviewed literature to develop practice.

CHAPTER 15 TAKE-HOME MESSAGES

This chapter has addressed ways to assess a tech mining exercise. Exhibit 15-1 digests our experiences to offer more informal rules of thumb.

EXHIBIT 15-1 *Ten Tech Mining Commandments*

(1) Focus your tech mining exercise with care; the right scope is critical.

(2) Get with your users; process management makes a huge difference.

(3) Be clear on the main questions to be answered and what information is desired, when.

(4) Gain access to suitable data sources.

(5) Search and run preliminary analyses for review by knowledgeable persons to refine the search.

(6) Clean your data.

(7) Pick multiple indicators that address the target management questions most effectively; address contextual and market considerations, as well as technological maturation.

(8) Run the necessary analyses (not all conceivable ones); check results; run sensitivity analyses.

(9) Be alert to surprises—activity at the fringe of your target technologies; don't become a captive of the conventional wisdom.

(10) Invest serious effort in representing your findings so they convey the punch line effectively, but also the uncertainties and risks.

We suggest regarding tech mining evaluation:

- Develop a systematic strategy to measure, or evaluate, your tech mining results.
- Decide whether formative and/or summative evaluation is wanted.
- Design tech mining evaluations with suitable, fair bases for comparison.
- Address the main threats to drawing valid inference from your evaluation.
- Be clear on whose judgments are most suitable on each facet of the evaluation.
- Decide who should take the measurements and make this a clear responsibility.
- Require tech mining documentation.
- Hold a wrap-up meeting to mark the completion of a tech mining exercise to spell out lessons learned.
- Measure the validity of the tech mining in question.
- Measure the utility of the tech mining in question.
- Gather tech mining findings and share them via the intranet.
- Use measured results and success stories to promote tech mining in your organization.
- Look to systematize tech mining as part of your organizational decision or policy processes.

CHAPTER 15 RESOURCES

There are many technology forecasting and assessment methods. Glenn and Gordon's (2002), *Futures Research Methods* (http://www.acunu.org/millennium/FRM-v2.html), is an excellent resource. See also Technology Futures Analysis Methods Working Group (2004).

Chapter **16**

Example Process: Tech Mining on Fuel Cells

How do you do tech mining? This chapter shows the way through an illustrative analysis of fuel cells. If you wish, you can step through the tech mining actions with us by using software (VantagePoint Reader) and sample data provided via the Wiley website (ftp://ftp.wiley.com/public/sci_tech_med/technology-management).

16.1. INTRODUCTION

Read this chapter in conjunction with Chapter 4. It introduces fuel cells and presents selected results for those interested in using tech mining, but not necessarily doing analyses themselves. This makes for some cross-referencing of results presented there, for which we apologize.

We select *VantagePoint* ("VP" for short) as the software to illustrate the analyses. You may have come across versions of the software under other names. *TechOASIS* is the version of VP available without charge for U.S. Government users.* Another version of VP, *Derwent Analytics*, is tailored to work easily with Derwent patent data (and also Delphion and Web of Knowledge data).

You might question whether this choice is biased—it is. We are intimately involved with VP development. But this is the software we know best and have

*Development was supported by the Defense Advanced Research Projects Agency (DARPA), the U.S. Army (particularly TACOM—the Tank-automotive and Armaments Command), the U.S. Navy (ONR—Office of Naval Research), and the National Science Foundation (NSF).

Tech Mining: Exploiting New Technologies for Competitive Advantage, Edited by Alan L. Porter and Scott W. Cunningham.
ISBN 0-471-47567-X Copyright © 2005 by John Wiley & Sons, Inc.

collectively used in over 100 tech mining analyses of various sorts. So discount our enthusiasm, but judge for yourself the value of the sorts of outputs we generate. Helpfully, using a single software package enables us to provide a "reader" version of it so that you can get a hands-on feel for tech mining. You may want to visit the website and track the tech mining steps on sample files containing about 10 percent of the fuel cell records.

This chapter aims to present the steps leading to outputs of tech mining, rather than the mechanics of using this particular software. Hence, we mention some of the dirty work—such as data cleanup—but without detailing it. We want you to gain a sense of the requirements and capabilities of tech mining.

VP works in conjunction with Microsoft Windows programs. This makes tech mining results easily available in the Microsoft Office suite for further analyses, reports, and presentations. How, other than using VP, could you do such tech mining? You have several major alternatives, including:

- Do it yourself using the capabilities of a database search engine and other common software. For typical Boolean search engines, you can generate activity counts by varying your search terms. For instance, through the Engineering Village website, you could search INSPEC for "fuel cells." Then restrict that search year by year (e.g., "fuel cells" and "2002") to jot down annual activity. You could then enter those counts into general analytical software, such as MS Excel, to generate a trend plot over time.
- Apply search engines that offer an analytical option. For instance, Chem Abstracts' search engine, SciFinder, can directly generate activity lists in the form of nomographs.
- Create your own tech mining scripts with special purpose languages such as Perl.*
- Use other text mining or statistical software. Possibilities are noted in Appendix B, "Text Mining Software."

Stepping through this chapter should give you a good sense of what tech mining entails. For tech mining veterans, you may also gain some new ideas on alternative ways to generate useful outputs. We follow the nine-step process for carrying out tech mining, used throughout the book—Table 16-1.

This chapter treats each of these steps in turn, in greatly varying degrees of detail, for selected fuel cell aspects. In particular Steps 5 and 6, basic and advanced analyses, are illustrated in depth.

16.2. FIRST STEP: ISSUE IDENTIFICATION

The goal of this step is to establish a clear set of questions for performing the tech mining study. In the present case of fuel cells, the motive is simply illus-

---

*See http://www.perl.org or the excellent O'Reilly books on the topic (Wall et al., 2000).

TABLE 16-1. The Nine-Step Tech Mining Process

1. Issue identification
2. Selection of information sources
3. Search refinement and data retrieval
4. Data cleaning
5. Basic analyses
6. Advanced analyses
7. Representation
8. Interpretation
9. Utilization

tration. Additional motives in performing an analysis of fuel cells might focus upon:

- Competitor analysis—profiling and interpreting what one or more organizations are pursuing in fuel cells
- Particular subtechnology analyses—exploring what is happening in one technology (e.g., solid-oxide fuel cell advances) or comparing multiple technologies (e.g., alternative bio-fuel cell approaches)
- Target applications (e.g., the relative merits of certain fuel cells, compared with each other and with other energy sources, for electric vehicle use)

In the present case we explore various application topics in a casual manner, for example, profiling R&D for "microgrids."*

16.3. SECOND STEP: SELECTION OF INFORMATION SOURCES

The goal of this step is to find the most authoritative sources available for mining information about the topic. In the present case of fuel cells, we selected the following three databases.†

- *Science Citation Index* (*SCI*—from Thomson ISI's Web of knowledge website)—an extensive source for fundamental research
- *INSPEC* (from IEE, the Institution of Electrical Engineers)—an excellent source of engineering research
- *Derwent World Patents Index* ("Derwent" for short)—a leading source of patent documents published by multiple patent offices

---

*"Microgrids" are multifunctional electrical networks with distributed energy production.
†We accessed data via Dialog, a leading gateway to over 400 different databases. We expressly thank IEE, Thomson Derwent, Thomson ISI, and Thomson Dialog for their permission to utilize these data.

Furthermore, other attractive fuel cell publications sources include *Energy Science and Technology* and *Chem Abstracts* (coverage noted in Section 3.2). Also, Internet resources are used to identify more current activities and contacts. Eventually, these empirical studies should be complemented with expert opinion.

16.4. THIRD STEP: SEARCH REFINEMENT AND DATA RETRIEVAL

The goal of this step is to formulate queries for extracting the records (whether patents or papers) from the databases.

In October 2002, we searched for abstracts of papers relating to "fuel cells" in SCI and INSPEC. In March 2003, we updated these searches and also searched for patent abstracts mentioning "fuel cell" in Derwent. Alternative search schemes might have restricted searches to particular data fields (e.g., titles, keywords) or certain years.

In real tech mining we would iterate these searches to sharpen the focus. We would also continue to grow our knowledge of fuel cells by expanding and refining our queries as our awareness of the technology deepened.

We are not seeking to retrieve a handful of records that precisely match a target interest—as a good librarian might do for a researcher who wants to read a few state-of-the-art papers. Nonetheless, we do need both to ensure that the search results reflect the scope of the field of fuel cells and to reduce the truly extraneous noise. We usually do this by downloading 500 recent records from a simple search on the identified issue. We then use VP to generate a list of leading keywords. We show the most frequent keywords to the target user or others knowledgeable about the technology and identified issue, requesting feedback on the following:

- Synonymous terms to add to the search algorithm
- Clearly wrong terminology, possibly to suggest phrases to exclude from the search
- Example "right on target" items

This sometimes serves to open the door for the user to alter the original tech mining request!

16.5. FOURTH STEP: DATA CLEANING

The goal of this step is to eliminate redundancy and unnecessary variations in the data. Although the search begins with nearly 24,000 patents, and nearly 12,000 publications, these are reduced by cleaning. Table 16-2 helps track the data consolidation processes.

TABLE 16.2. The Data Cleaning Process

	Databases	Aggregations	Focus	Quantities
1	Derwent	Patents	All records	23,899
2	Derwent	Patent families	All records	<23,899
3	Derwent	Patent families	Excluding solely Japanese families	9,724
4	Derwent	Patents	Excluding solely Japanese families	31,559
5	SCI, INSPEC	Publications	All records	>11,764
6	SCI, INSPEC	Publications	All unique records	11,764

In the case of patents, some patents were defined too narrowly and were only of the most narrow specialty interest. In the case of publications, duplications were introduced as the two different database companies indexed the same article. This section considers how to find and eliminate these excess records.

The Derwent search on fuel cells yielded 23,899 records. Many of those patent records reflect patents applied for only in Japan. We focus on 9724 patent family abstract records (a subset of the 23,899 records) including only records whose patent family includes one or more patents in a country other than Japan. (VP enables this by designating a "group" made up of every country other than Japan for the patent family field. We then form a new dataset based on this group.) Unless one has a particular interest in Japanese patenting, this process yields a better perspective on world patenting of fuel cell technologies.

The 9724 patent family abstracts reflect 31,559 patents (i.e., 3.3 patents per family). However, we see 34,073 instances because some patents appear in multiple families (closely related patents).

Generating a single research papers file of 11,764 publication abstract records presented other challenges. We consolidated searches in two databases (SCI and INSPEC). We also blended the October 2002, and March, 2003 searches that contained major overlaps for the year 2002.

To do this, we first applied VP's "data fusion" function, then its "remove duplicate records." The latter can be done simply—as we did, just matching on article titles to remove one if there were another paper with the same title. For more critical applications, it can be adjusted more finely to use information from multiple fields (e.g., author and source) and to apply fuzzy matching. The right cleanup process will depend on sensitivities given the intended uses.

Data fusion was necessary for two reasons. First, where duplicate records were found, we needed to delete one of the two records and create a single uniform record containing all available information. Second, for analysis reasons we need a single uniform record. Where fields differ by name or content between our two databases, we must fuse the data to create a common record type for analysis. Note that we chose to not fuse the publication and patent records together. The two kinds of data are qualitatively different, and we deem it best to view these as complementary, rather than comparable, indicators of technological progress. What follows is a more detailed example of this data fusion process.

We needed to combine "the same" fields from SCI and INSPEC records (e.g., "author" from SCI records with "author" from INSPEC records; "date actual" from INSPEC with "source publication year" from SCI). We likewise combined the nonidentical "affiliation" (of the first author) from INSPEC with "affiliation" (of all authors) from SCI. Such nuances suggest careful review of your tech mining objectives before fusing records from different sources.

For some fields, combination requires trickier choices. For instance, SCI and INSPEC each provide *keywords*, but not with identical meanings. We searched INSPEC through Georgia Tech's Electronic Library wherein author- or journal-generated keywords are consolidated with INSPEC-assigned keywords in a single field. On the other hand, we have SCI "keywords plus" (keywords using cited title phrases distinct from the article itself; these differ from INSPEC's standardized index terms), separate from SCI's field for author-generated keywords. For present purposes we combine all of these as "keywords-combined."

Data cleaning can dramatically improve the quality of tech mining analyses to follow. One way to expedite the process is to script the basic cleaning steps you routinely use. We have several "generic cleanup" macros for VP. You can select a suitable one and let the computer crank through the steps—often fairly time-consuming for files containing thousands or tens of thousands of records. Cleanup tasks differ by information source (e.g., markedly different for publication and patent databases) and intended tech mining uses. Section 16.6 deals with additional cleaning aspects. Do clean your data!

16.6. FIFTH STEP: BASIC ANALYSES

The goal of this step of the nine-step process is to begin the analysis of the data. These analyses may target specific information you know you want and exploratory purposes, to find out about the topic. As you read the following materials, note that we describe four distinct tasks:

- Exploratory Analyses
  This task aims to explore both content and players. This can help scope available sources of information and confirm that the query used is actually getting the data intended.
- Additional Data Cleaning
  This task aims to explicitly reduce unwanted noise, variation, and redundancy. It involves identifying sources of noise that must be cleaned before even basic analyses can be conducted.

- Producing Lists ("first-order" analyses)
  This task aims to identify the top areas of research, as well as the top research participants, via lists. Lists identify the major components of the data, whether they be leading keywords, leading research institutions, or whatever.
- Combining Lists as Matrices ("second-order" analyses)
  This task helps to understand complexities of the data—second-order effects, if you will. It can help produce research profiles.

Note that the tasks are usually iterative. For instance, in browsing a list we notice extraneous or duplicate information, which prompts us toward further cleaning. We discuss first-order and second-order analyses in Chapter 9.

This section combines distinct analyses of publications and patents to get a robust picture of activity in the field of fuel cells. We therefore discuss the two sources of data side by side in this tech mining example.

First Basic Analysis Task: Exploratory Analyses

In the exploratory analyses we examine the fields available to us in our patent and publication abstract records. We get a sense of the general scale of activities by examining the numbers. Where warranted, we probe further by using our tech mining software to conduct more detailed queries against the data. We then look at specific records and patents to make sure we have downloaded relevant information. And, having identified sources of noise in the data, we use tools such as algorithms and thesauri to clean the data. This section indicates specific steps taken to clean up names of various research institutions and countries.

First, we begin with the examination of general fields and headings. Table 16-3 lists the available headings from the Summary VP sheet for our combined SCI & INSPEC publication abstracts file. The reference "Cleaned" mentioned below indicates that lists have been processed using "list cleanup" and/or the-

TABLE 16-3. Available Publication Abstract Fields

Fields	Numbers of Items
Raw record	11,764
Affiliation (cleaned)	2,522
Authors (cleaned)	12,720
Class codes	793
Country	342
Keywords	8,929
Record type	22
Source (conference)	465
Source (journal)	1,367
Title	11,764
Title phrases (cleaned)	18,962

sauri operation(s). The VP Summary page provides background information (number of records, databases searched, date). It also gives field descriptions, which are not shown in this table. These include the following.

- Whether the field is "as is" or is derived from the raw record
- The nature of the data in that field
- Whether it has been tagged as a certain type of data such as "year"

As shown in Table 16-3, we see how many items from each field appear in the data set of 11,764 fuel cell R&D publication abstract records.

Table 16-3 conveys the scale of research activity on fuel cells (e.g., do we find 10 or 10,000 papers?). It also provides additional information that can be surprisingly useful. For instance, note there are 12,720 authors for these 11,764 papers, or just over 1 author per paper. We can go further by tallying the number of authors with a single paper (6700), plus 2 times the number with two papers (2472), on through the one author with 492 fuel cell-related papers. This yields a total of 42,534 authorships for the 11,764 papers—an average of 3.6 authors per paper. This large degree of teaming suggests a field actively pursuing technology application.

Next we examine our patent mining counterpart to Table 16-3. Thus Table 16-4 lists the available headings from the Summary VP sheet for our

TABLE 16-4. Available Patent Abstract Fields

Fields	Numbers of Items
Raw record	9,724
Abstract advantage	5,948
Abstract novelty	4,227
Abstract use	3,619
Basic patent country (cleaned)	35
Basic patent year	39
Derwent accession number	9,724
Derwent classifications (cleaned)	278
Family member countries (cleaned)	42
Family member years	39
Family member years (most recent)	37
File segment	3
Inventors (cleaned)	10,112
Patent assignees (cleaned)	3,311
Priority countries (cleaned)	41
Priority years	44
Priority years (earliest)	44
Tech focus	1,892
Titles	9,631

Derwent patent abstracts file. The VP Summary page provides background information akin to that for the R&D publications of Table 16-3. Table 16-4 tallies how many items from each field appear in the data set of 9724 fuel cell records.

The actual summary will vary according to which fields you import to analyze. And, as already noted, databases differ in the numbers and types of fields they provide. As an example, Derwent rewrites patent abstracts to better convey content and claims (note the several Abstract fields in Table 16-4), and assigns its own classifications (Derwent Classifications shown here, as well as others not shown—File Segment classification, Manual Codes, Chemical Fragment Codes, Keywords Indexed, etc.).

The data summarized in this table present the three following patent types:

- "Basic"—a patent application cited in the search report of a patent application filed later, whose invention is an improvement upon that of the former application, which is sometimes called "basic invention"
- "Family"—a group of patent applications and patents directed to the same invention, i.e., associated with the same priority patent application
- "Priority"—the first patent application within the meaning of the Paris Convention, sometimes called "priority patent application," upon which other patent applications are subsequently filed claiming priority, thereby forming a patent family

Here we have only imported country and year information for these types; useful information varies according to your purposes. For instance, competitive technological intelligence ("CTI") can garner insight from plotting the distribution over time of a company's priority vs. family patents (overall or on a specific topic). The lag between priority applications and issued patents can tell about that company's patenting skill, global market interests, and strategies (e.g., "submerging" patents, in the U.S., by ongoing modification of a patent application to delay the patent to be issued)—Chapter 12 pursues patent analysis issues further.

We have now performed a basic survey of the data. Before concluding the survey, let's first examine the data for a given topic of interest. VP supports our surveys with a "find" function. We mentioned interest in "microgrids"— the "find" function applied to the Keywords list of our publications locates none. So, we search the Title Phrases list and there we indeed find one. Reading that abstract suggests that we expand our search to look for the phrase "distributed power," yielding ten hits. Searching "Raw Records" expands the set to 45 hits. These could be examined further to weed out unrelated items and to identify additional worthy search terms. Such probing may well suggest you go back and refine your search.

We shouldn't neglect the straightforward aspect of "mining"—locating "gems" in the form of specific key articles or patents to be studied in depth. Here again tech mining software supports our needs in getting a quick survey over the data. For instance, given our interest in microgrids, we certainly want to read the publication abstract entitled: "Microgrids [distributed power generation]." Our data set of abstracts does not contain the full conference paper, so obtaining that requires external search and retrieval. Some science and technology databases do provide full texts and others provide links to full papers or patents. This depends on the licensing arrangements made by the database providers, as discussed in Chapter 6.

Second Basic Analysis Task: Additional Data Cleaning

Note that we have a suspiciously large number of countries in our publication data—Table 16-3 shows 342 distinct countries publishing in the area of fuel cells! Inspection of the "countries" shows that the file has considerable noise in trying to extract country name from the author affiliation field. We'll keep this in mind and correct the totals for countries of interest.

This illustrates an essential tech mining principle—don't pursue "perfection' in your searching or cleaning unless it directly serves a tech mining objective. You can expend (waste) gobs of resources seeking the unattainable goals of capturing every related paper, with no unrelated ones, and getting every author, institution, and term just right. BUT—you must set your users' expectations to match. Tech mining depicts "big picture" activity patterns that do not depend on supreme accuracy. Yet it also probes deeply, where accuracy can become essential. Our message is to decide how important recall (getting all relevant records) and precision (no irrelevant ones) is to each tech mining subtask as you proceed. For instance, if your broad-brush examination suggests that one company is a promising acquisition candidate, then go back and refine your search and data cleanup for that company's R&D activity records.

So, continuing with our data, what about the institution names? We also recognize that the names of institutions are noisy and redundant. Cleaning these names is important to get accurate counts of the leaders, or organizations about which we have particular interests, in research. Don't fritter away tech mining resources trying to get every last publishing organization of our 2500 precisely right. Who cares?

To make a better tally we find and group variations on the leading institutions. For example, we have VP find any affiliations containing the string "julich." Examination of the 9 variations suggests these all pertain to "Inst. fur Werkstoffe und Verfahren der Energietechnik, Forschungszentrum Julich GmbH," so we combine them (second item in Table 16-5 below). We usually consolidate affiliations at the company, university, or research institute level. However, for some purposes, one might need to distinguish particular locations of a company, or a single research lab within a research organization or university (see sidebar).

TABLE 16-5. Top Organizations in Fuel Cell Publication (Raw)

Affiliations (Cleaned)	Number of Records	Revised Numbers of Records
Univ London Imperial Coll Sci Technol & Med	148	158
Res. Centre Julich KFA, Germany	115	241
Tokyo Elect Power Co Ltd	87	102
Lab. of Materials Science, Delft Univ. of Technol., The Netherlands	83	90
Los Alamos Nat. Lab., NM, USA	72	128

Sidebar: Who's Who?

Affiliation cleaning example—which of these three organizations that VP nominates as being the same are truly the same?

(1) Adv. Energy Technol. Grp, Korea Electr. Power Res Inst., Daejeon, South Korea

(2) Dept of Mater. Sci and Eng, Korea Adv. Inst. of Sci and Technol., Daejon, South Korea

(3) Korea Electr. Power Res. Inst., Taejon, South Korea

We deem the first and third as the same organization—overriding the nominally different city and the different organizational level. Cleaning the data often entails making somewhat uncertain choices. For some uses, one prefers a conservative approach to avoid clouding the case. For most, however, aggressive cleaning offers net gains.

We repeat this process of finding possibly common entries, inspecting those candidates, and tagging the ones we deem to be the same institution, forming a *group*. (If you have access to the Wiley website, you might want to follow this discussion by using *VP Reader* on the 10% sample fuel cell data sets.) We then sort the list on this group (e.g., University of London) to view all the tagged items together. In the case of the University of London, searching on "london" finds a number of entities that are not the University, so we untag them. But it also suggests that "Imperial College" is affiliated with the University, so we find the string "imperial coll" too, and add tags to the University of London group. Some knowledge of the organizations involved obviously helps in this process as it might be preferable to retain separate identities for these colleges. In VP, we can *"create thesaurus from groups"* to indicate these are all to be taken as variants of the particular institutions. Then we apply the resulting thesaurus to generate a list with those organizations consolidated as we desire. The last column of Table 16-5 indicates a revised ordering of the top organizations publishing research on fuel cells. You need to decide the most suitable organizational level (e.g., you might want to separate departments within a university, consolidate units of one campus, or, as in this case, combine colleges of one university system).

VP keeps getting smarter—"learning" by continually enhancing given thesauri. Software "out of the box" may come with premade thesauri. Nonetheless, it may require additional learning to adapt these thesauri to the terms, authors, and organizations most commonly met during your targeted studies. We continue to enhance thesauri for certain prominent multinational companies and American universities (e.g., cumulating variations on "MIT"). Others might keep building suitable thesauri covering keywords, authors, and organizations of enduring interest.

We often want to distinguish specific types of research organization. We have been growing a thesaurus that categorizes affiliations into industry, academia, and government. We do not, for our purposes, make a distinction between nonprofit or other forms of research organizations—others might wish to do so, however. Bibliometric indicators demonstrate that we often encounter "80/20 rules"—the first 20 percent of all organizations will probably encompass more than 80 percent of all references we encounter in practice. This is both a blessing and a curse—a thesaurus can be made quickly and effectively. But if we aspire to perfect cleanliness in the data, it is going to take a lot of work, and many, many more references to fully complete.

This organizational type thesaurus uses general cues (e.g., "univ" is usually university and "ltd" for limited is usually corporate) plus specific identities (e.g., "Georgia Tech" is a university). Figure 4-2 illustrates the distribution of the three types of organization publishing on fuel cells, namely, academic, corporate, and government. This coarse analysis has only captured about 70 percent of our fuel cell organizational identities. But that's OK—it gives us the sense of what's going on. Of the 8418 organizations so classified, 1799 are taken as corporate—21 percent. This is quite high for technical publication, implying serious corporate interest in fuel cell development.

Third Basic Analysis Task: Producing Lists

Making lists that tally the content in particular fields over a large set of records sounds simple, and it is. Nonetheless, this is often the most valuable tech mining step. In a matter of minutes, a tech mining novice can gain understanding of "who's doing what" in a research domain. One can then concentrate on the aspects of most interest for further analyses.

Tech mining adds value to a bibliographic search beyond finding particular "gems." We can make lists of any of the fields of Table 16-3. We could list leading journals covering fuel cells. We could spot intersecting research domains by applying a thesaurus to the "Class Codes," and on and on. Let's create a "Top *N*" list for affiliations (research organizations). We don't really

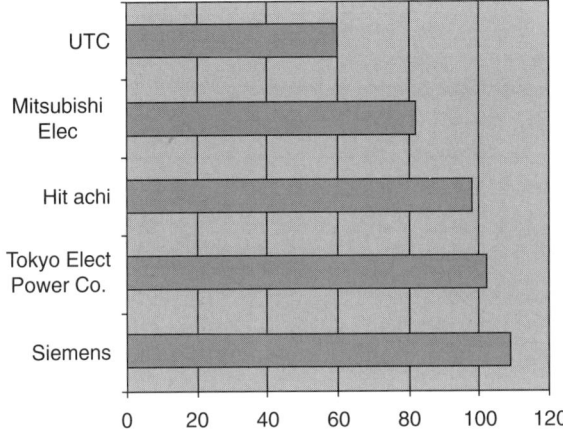

Figure 16-1. Top five firms publishing on fuel cells

want to list all 2522 affiliations; Table 16-5 offers the five leading ones based on number of records.

Figure 16-1 illustrates the activity levels for the five top publishing firms. Note that none of these appear in Table 16-5, even though some publish more than TU Delft. Comparing the table's "number of records" with the "revised number," we find considerable increases. These resulted from searching and tagging variations on these leading affiliations to consolidate their publications better. In doing this, we did not check all the affiliations, so we missed that Siemens belonged in the table. (Then, we left the table and figure as they are to illustrate the foibles of something so simple as counting!). Other issues also arise—in some cases we were not sure of all variations on a company's name; for Siemens, we decided to include Westinghouse, and we combined Siemens research from Germany and the United States. For Mitsubishi, we don't include Mitsubishi Materials or Mitsubishi Heavy Industries. We'll return to these data in conjunction with patenting later.

Another example—what are the leading American universities in fuel cell publication? We again use the find; tag; and group process to consolidate variations. Figure 16-2 depicts these leading American universities. We'll refine this tally in the next section when we combine these data with additional information to produce "matrices." Note also that we could chose "Top N" as we like—in this case we stopped with Top 8, at ALP's alma mater (Caltech).

Universities tend to be the most frequent publishers, whereas corporations tend to be the most frequent assignees for patents. We can probe finer levels in many ways. Suppose we're examining Western European activity in fuel cells oriented toward automotive applications. To illustrate use of a data subset, we select (1) automotive-oriented fuel cell patents, (2) dating 2000–03, for (3) assignees in Western European countries—yielding 278 records. Table 16-6 gives the "Top 8" companies (chosen as the cutoff because there is a drop

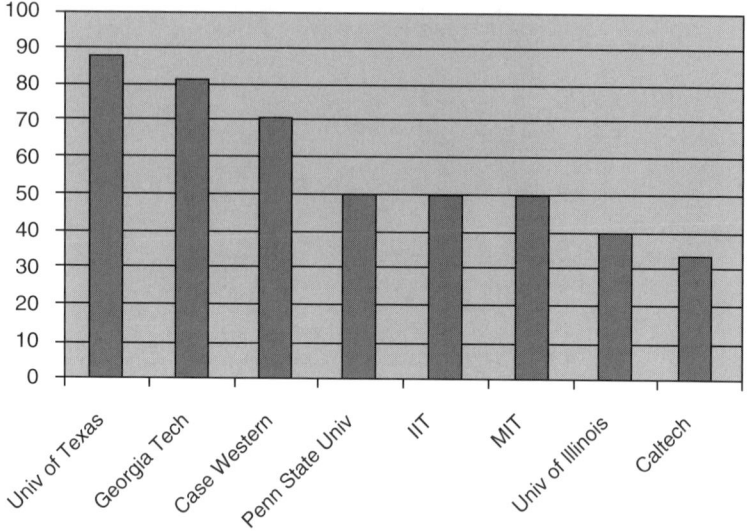

Figure 16-2. Top eight American universities publishing on fuel cells

**TABLE 16-6. Leading Automotive-Oriented Fuel Cell
Patent Assignees in Europe, 2000–03**

Patent Assignee	No.
Xcellsis	49
Daimler-Chrysler	40
Siemens-Westinghouse	22
Mannesmann	20
Volkswagen	18
Emitec	17
Renault	16
DBB Fuel Cell Engines	15

to 7 patents for the ninth company). Note that many are automotive companies, but others, such as Siemens, have related fuel cell patents in this domain
as well as in other areas.

Table 16-7 shows the family member distribution for priority patents from
these top European automotive fuel cell companies. This table indicates the
patent protection behavior for European firms. For instance, 47 of Xcellsis' 49
patents show German priority. Xcellsis has then gone on to file with EPO on
35 of these, and for U.S. protection on 34, but only 11 in Japan. Keep in mind
that bibliometrics cannot substitute for knowledge of the field and the data.
We will explore Xcellsis further shortly.

TABLE 16-7. Distribution of Priority Country by Family Member Countries for European Assignees

#	Patent Assignee	224 DE	101 EP	83 WO	80 US	41 JP	37 FR	24 AU	12 CA	5 GB	4 CN
					Family Member Countries						
49	XCELLSIS	47	35	8	34	11	1	1	2		
40	DAIMLERCHRYSLER	40	12	4	13	5	3			1	
22	Siemens-Westinghouse	19	10	20	8	5		2			2
20	MANNESMANN	20	5	14	1	2		3			
18	VOLKSWAGEN	18	2								
17	EMITEC	17	4	17	4			6			
16	RENAULT		5	3		1	16	1			
15	DBB FUEL CELL ENGINES	14	9	2	11	11		1	1		

TABLE 16-8. Custom Profile: Top American Universities

Universities	Publication Counts	Authors	Keywords
Univ of Texas [88]	88	Goodenough, JB [33]; Appleby, AJ [18]; Keqin Huang [13]	tubular fuel cell [42]; fuel cells [38]; electrochemical electrodes [15]; electrolytes [9]; electrolytic material [9]
Georgia Tech [81]	81	Min Liu [39]; Winnick, J. [22]; Wepfer, WJ [11]	tubular fuel cell [34]; fuel cells [28]; electrochemical electrodes [23]; solid oxide fuel cells [13]; oxides [8]
Case Western [71]	71	Savinell, RF [47]; Wainwright, JS [20]; Wasmus, S [14]	tubular fuel cell [37]; fuel cells [33]; electrochemistry [13]; electrochemical electrodes [12]; electrolytes [7]

Fourth Basic Analysis Task: Publication Mining—Matrices

We combine two lists to generate "2-D" analyses. Matrices enable you to see interactions (e.g., Table 16-7) or to break out additional details for each member of a chosen list. In VP, handy macros ease the combining of data to produce matrices and tables. In this example we focus largely on "research profiling"—combining information about institution with information about keyword or classification. At the end of the section we highlight a variety of additional matrices that can be tabulated readily and interpreted for fuel cells. These include co-occurrence data as discussed in Chapter 9 (Section 9.3).

Table 16-8 shows the results of running a "custom profile" macro for the top 3 American universities. Here we just asked for the top 3 authors and top 5 keywords for each university, from our fuel cell publications file. This helps spotlight colleagues with whom we want to establish contact. Were our interests in solid oxide fuel cells, based on this information, we might check out websites of Georgia Tech first.*

How might you use such breakouts? Imagine you are helping identify invitees for a special fuel cells workshop. You restrict analyses to publications of the past three years, seeking those currently active in the field or a specialty therein. You then profile, say, Asian and European universities because you are seeking international participants. In addition, you might successively

*See more at http://www.fcbt.gatech.edu/.

search on specific topics such as solid oxide fuel cells, then on proton exchange membrane fuel cells—seeking a balance of interests by type of fuel cells.

We can repeat these sorts of simple analyses with the patent data. Figure 16-1 showed the five companies publishing most on fuel cells. Table 16-9 profiles those companies' patent activity; using classifications, we profile the varying areas of fuel cell activity. We note the following points:

- Siemens and Westinghouse each show extensive patenting activity; one might prefer to analyze these merged companies separately for some purposes. Note also that both show extensive recent patenting, unlike some of the other companies.
- The leading publishers are not the same as the leading patenters—Tokyo Electric shows minimal patenting, whereas International Fuel Cells shows 251 patents in the data set and Honda 186 (not shown here).
- Distinctions in emphases do not appear generally from the Derwent patent classifications, but one could locate activity in particular classes, such as X21—Electric Vehicles.
- Certain inventors are quite prominent—one might well explore further for CTI interests.

Here are additional matrices to stimulate your thinking of candidates to craft for your own tech mining needs:

- Publication Keywords by Year—to identify emerging commercialization opportunities
- Patent Classification by Year—performed for all leading patent assignees. Figure 16-3 shows one type of visualization of such a matrix.
- (Topic or Researcher) by Year—to see trends and what or who is "hot"
- (Topic or Researcher) by Time Slice (e.g., 3-year blocks)—similar, but consolidates time to better see major transitions
- Topic (Keyword or Cluster or Title/Abstract Phrases) by Class or Industry Code—to see which areas are emphasizing which topics
- Author by Times Cited—to see whose work appears to have high influence.

Figure 16-3 shows the evolution of selected companies' patenting activities since 1990. It highlights the current importance of fuel cells. As shown in this figure, Siemens has the longest involvement with this technology, but watch out for Honda, which is dramatically escalating its work.

In addition, co-occurrence (i.e., information derived from terms tending to occur in the same records) can help answer other questions. Suppose you want to decide which conference would be best to attend to learn of current research on solid oxide fuel cells. A co-occurrence matrix of the phrase "solid

TABLE 16-9. Profile of Company Patent Activity

Patent Assignees	Patent Publication Years	Types of priority applications	Derwent Classifications	Inventors
Top 5 Publishing			*Top Terms*	
Siemens-Westinghouse [567]	2002 [148]; 2001 [101]; 2000 [90]	DE [374]; US [169]; EP [22]	X16-Electrochemical Storage [521]; L03-Electro-(in)organic—chemical features of conductors, semiconductor and other materials, batteries, . . . [342] A85-Electrical applications [58]	Mattejat A [26]; Grosse J [26]; Ruka R J [26]
United Technologies Corp [253]	1986 [47]; 1979 [43]; 1978 [40]	US [251]; FR [1]; PCT [1]	X16-Electrochemical Storage [208]; L03-Electro-(in)organic—chemical features of conductors, semiconductor and other materials, batteries, . . . [200] A85-Electrical applications [64]	McElroy J F [9]; Sederquist R A [9]; Trocciola J C [8]
Hitachi Ltd [81]	1987 [22]; 1986 [19]; 1989 [18]	JP [78]; PCT [2]; UK [1]	X16-Electrochemical Storage [72]; L03-Electro-(in)organic—chemical features of conductors, semiconductor and other materials, batteries, . . . [55] A85-Electrical applications [14]	Tamura K [12]; Horiba T [12]; Kahara T [9]
Mitsubishi Electric Corp [63]	1997 [13]; 1987 [13]; 1990 [13]	JP [49]; US[14]	X16-Electrochemical Storage [59]; L03-Electro-(in)organic—chemical features of conductors, semiconductor and other materials, batteries, . . . [41] X21-Electric Vehicles [6]	Fukumoto H [6]; Maeda H [6]; Matsumura M [6]
Tokyo Electric Power Co [3]	1996 [2]; 1993 [2]; 1994 [2]	JP [3]	X12-Power Distribution Components and Converters [1]; X13-Switchgear, Protection, Electric Drives [1]; X16-Electrochemical Storage [1]	Sato H [2]; Sugimura H [1]; Takashima N [1]

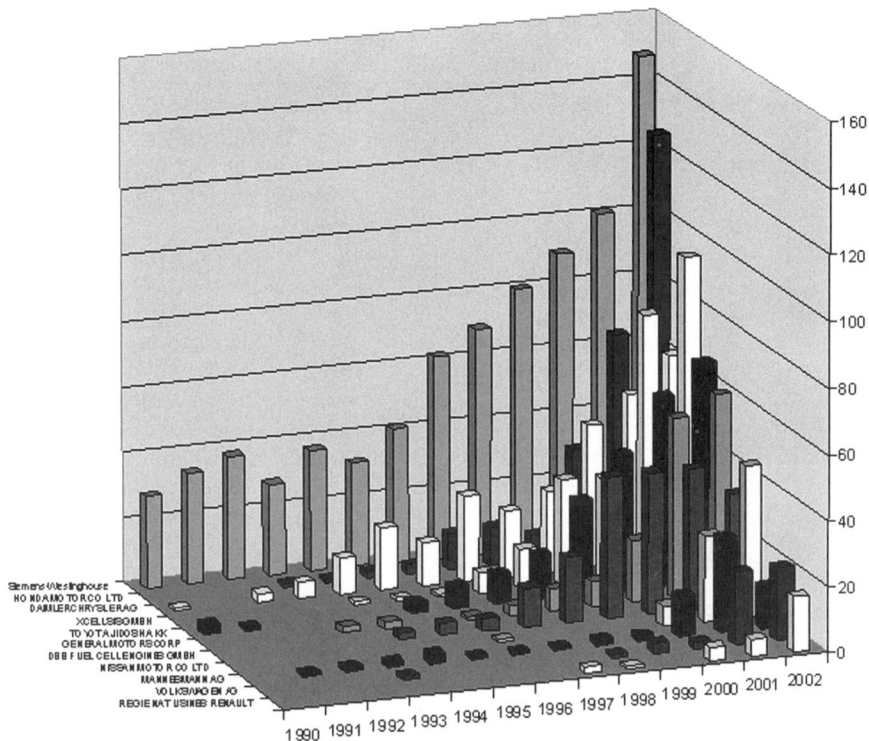

Figure 16-3. Selected companies' patenting activities over time

oxide" versus conferences of the past few years suggests the following as very active.

- Solid Oxide Fuel Cells Conference (e.g., Honolulu, 1999, had 45 papers for which "solid" variations were keywords)
- ICMAT (International Conference for Materials for Advanced Technologies, Singapore, 2001, had 15 such papers)
- Grove Fuel Cell Symposium (London, 2001, had 12 such papers)

16.7. SIXTH STEP: ADVANCED ANALYSES

The goal of our treatment of this step is to introduce a variety of advanced analyses on the fuel cell data, limited by a desire to provide a succinct accounting. Table 16-10 summarizes the analyses that follow in this section. We examine the raw data used, the modeling approach, and the purposes for which we apply the analyses.

TABLE 16-10. Advanced Analyses Used in Fuel Cells Case Study

Data	Models	Purposes
Patent: inventors	Networks	Investigate IP networks for a key fuel cell R&D company.
Patent: inventors	Spatial	Examine relationships among companies based on shared inventors.
Publication: keywords	Spatial	Examine relationships among documents based on shared keywords.
Publication: keywords	Mixed: spatial—network	Create maps to quickly identify key areas of research, and identify points of correspondence with experts.
Publication: keywords	Mixed: spatial—clustering	Discover interesting cross-relationships among technologies using inductive classifications of publications.

Knowledge Networks

A powerful patent analysis approach probes the IP of a target company. As brought up in Chapter 12, we might want to assess the IP capabilities of a company we are considering acquiring. Imagine that we're interested in the third from last company in Table 16-6, Emitec GES Emissionstechnologie MBH. Their 17 patents therein noted expand to 41 when we search our full fuel cell patents data file (not limiting to automotive and year 2000 or later). We now focus on their 22 inventors. We can check years to see who has not patented recently.

We can map "inventors × inventors" to see how activity is concentrated within Emitec, spot core researchers, and denote teaming. Figure 16-4 shows such a map. Checking back, we see that Emitec's seven automotive-related patents all list Brueck, Grosse, and Reizig as inventors; Poppinger appears on six of these seven. Examining the co-occurrence matrix of inventors X inventors (not shown) confirms that these four plus Konieczny team extensively. Beresford only has a single patent, Reizig has 27, Grosse 25, Konieczny 19, and Poppinger 17—every single one of their patents is joint with Brueck, hence the very dense map. So, if our firm seeks Emitec's IP, we want to ascertain the status of this team and its core person, Brueck (38 patents). (Note also a data cleaning tidbit—we suspect that "Bruck, R" is the same as "Brueck, R." Again, depending on your purposes, catching such errors may or may not be critical.)

We can investigate activities of these core inventors through other sources—studying their full patents, websites, and experts. We also check Emitec's patent assignees—lo and behold—27 of them also list Siemens! (See also Fig. 16-8.) We certainly want to speak with knowledgeable persons to learn more of this relationship.

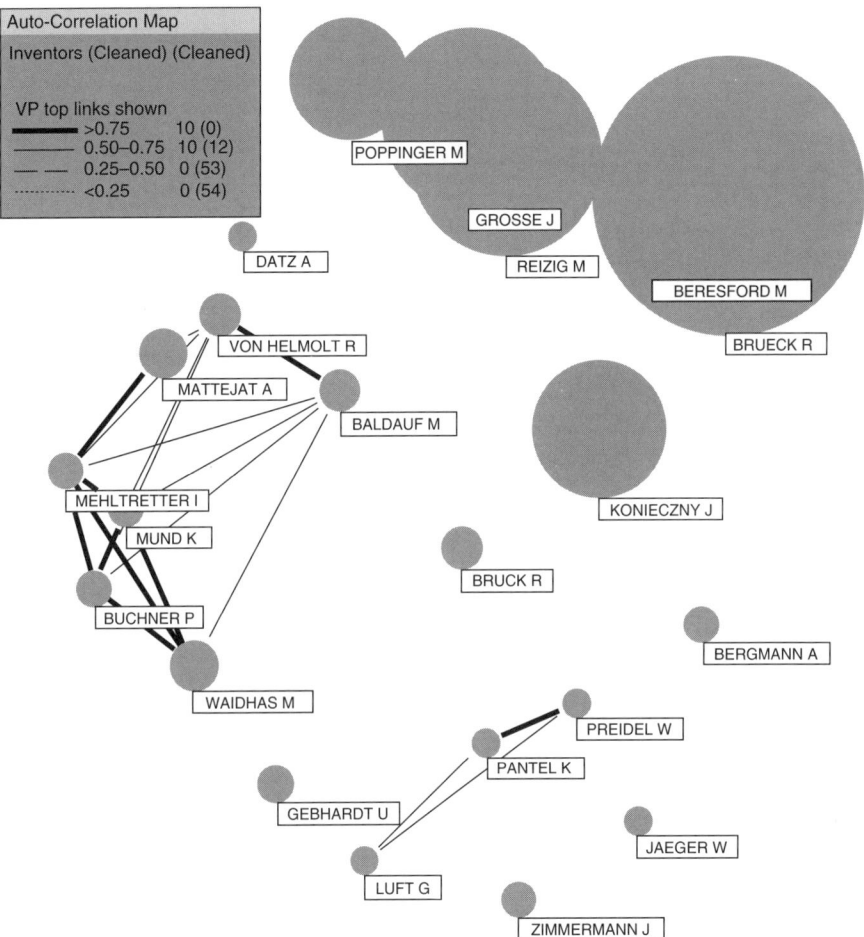

Figure 16-4. Knowledge network: sociometric map showing which inventors work with whom (within Emitec)

Content, Technology, and Topic Analyses

Let's pursue the automotive-related fuel cell development in Western Europe (cf. the 1187-record subfile introduced in the previous section). Imagine that our firm is initiating a novel catalyst development project, so we want to see what emphases current patent publications reflect. We now incrementally import abstract phrases [parsed by VP's natural language processing (NLP) algorithm] to the VP file. This yields over 16,000 noun phrases—a rich avenue along which to explore patent emphases. We can search for who is working on a specific topic of interest, leading inventors, or leading companies.

Suppose our firm's materials scientists ask to see which materials are linked to catalysis in this set of patents. We use "Find" to select 66 abstract phrases

TABLE 16-11. Catalyst Materials Mentioned

Abstract (NLP) (Phrases)	No. of Instances
Copper	7
Cerium oxide	1
Distributed metal polycrystals	1
Electro-catalyst	2
Finely divided alumina	1
Hydrosilylation catalyst	1
Iridium	2
Iridium catalyst	3
Linear fluoropolyether compound	1
Palladium-containing solution	1
Palladium-zinc-cerium-zirconium-based compounds	1
Supported platinum-ruthenium catalyst	1
Unsupported platinum-ruthenium catalyst	1
Zinc-containing solution	2

mentioning catalysis and form a "Group" in the Abstract Phrases List. If we had a "Materials" thesaurus, we could apply it to Abstract Phrases to create a "Materials" group. Another way to accomplish this is to make a co-occurrence matrix of "Abstract Phrases × Abstract Phrases" (Grouped), focusing attention on the "Catalys*" Group to identify materials of interest. For easy presentation here, we chose a third way—creating a "New Data Set" of the 46 patent abstracts that mention catalysis (or catalyst, etc.). From this, we can browse the 1346 Abstract Phrases to grab abstracts mentioning interesting materials for review by our materials scientists. Table 16-11 shows some sample abstract phrases.

This example could be extended to R&D publication analysis. Imagine that the two automotive-related patent publications mentioning "cerium" stimulate questions about research addressing cerium in fuel cells. Impressively, searching for "cerium" in our SCI–INSPEC fuel cells' Keywords List locates 313 records. Investigate these to determine the most active research groups and their key emphases. Ongoing interaction between tech miners, subject experts, and key users can lead to rich, iterative discovery processes.

Mapping of Terms Across Records

VP (and other tech mining software) provides statistically based tools to uncover relationships within a data set. Relationships are based on patterns of terms co-occurring across records. Figure 16-5 presents a high level map of clusters of fuel cell keywords.

Keeping in mind that the keywords are an amalgam of different types of descriptors of the article content, this offers a first cut at the topical content of the research domain. This is a principal components analysis (PCA) map

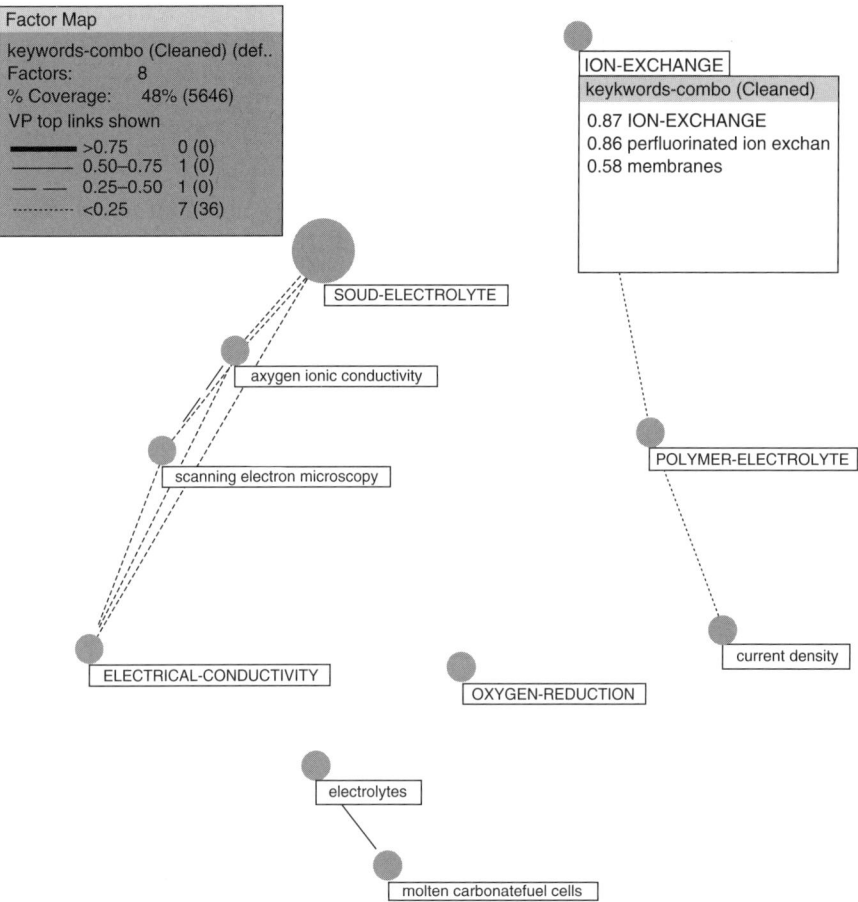

Figure 16-5. Mapping high-level fuel cell topics

that VP generated from the 59 most frequently occurring keywords in the full (11,764 record) SCI–INSPEC dataset, excluding the most generic terms (e.g., "fuel cells," "electrochemistry"). Selection of terms here was judgmental, seeking a very broad overview. One can generate different depictions depending on tech mining aims, number of records, number of factors extracted, and how related the records are. Zhu and Porter (2002) overview factors to consider and show some of the varied maps possible.

This mapping (Fig. 16-5) reflects the following.

- Nodes indicate principal components (a basic form of factor analysis; so we call these factors)—each of these ten reflects a set of keywords that tend to occur with each other in the fuel cell records.
- Node size (to reflect the relative number of records containing any of the high-loading keywords represented by that statistical component)

- Multidimensional scaling (to locate the PCA factors reflecting their degree of relationship to each other)
- Path-erasing algorithm (to indicate the relative strength of relationship among the PCA factors)—darker, more solid lines indicate stronger association.

What does this map show? In the upper left region, we see four nodes interlinked moderately. These pertain to solid oxide fuel cells (SOFCs) and to measurement of electrical properties. The three nodes linked in the upper right relate to proton exchange membrane (PEM) fuel cells. In VP, one can pull down menus to identify terms associated with a given node. Figure 16-6 shows such a pull-down for the ion-exchange factor. This gives the high-loading keywords that make up this factor. The keyword, "ion-exchange" is the most highly loading (most central to the factor), and the factor is given this name. In the VP Reader demonstration posted on the web, you can explore PCA map nodes using "Details Windows."

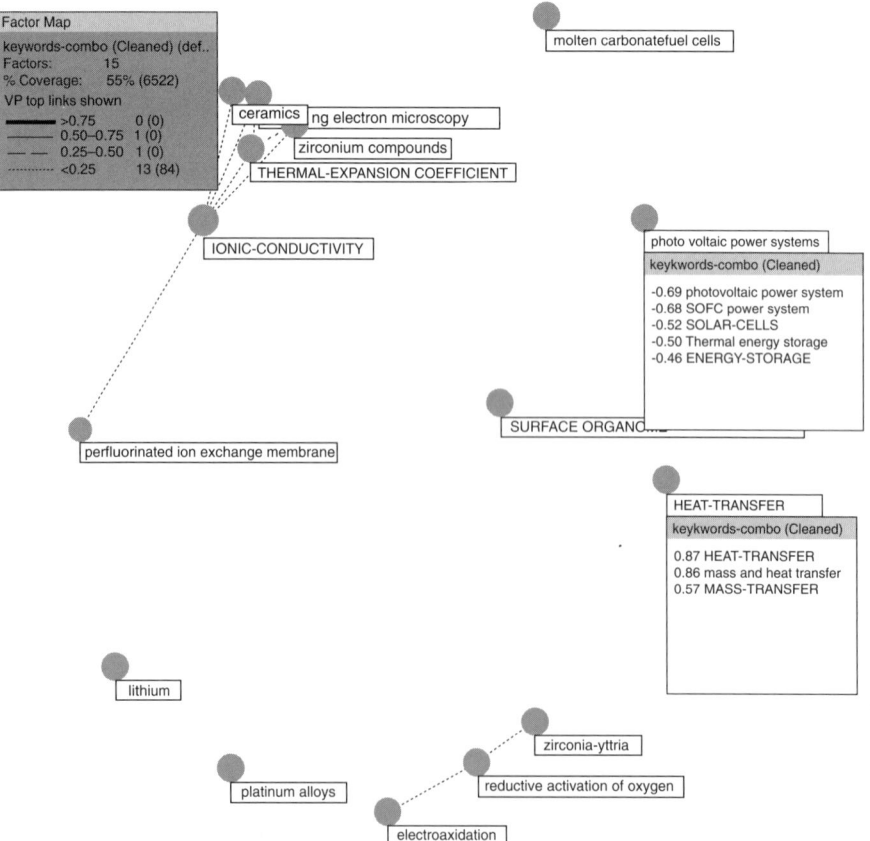

Figure 16-6. Mapping mid-level fuel cell keyword clusters

Figure 16-6 shows another PCA map of the same data set, but based on a selection of more specific fuel cell keywords—just another perspective.

Note that one might choose to rename a factor based on judgment as to a better description of its makeup—the "Heat Transfer" factor to the right pulls together concepts of both "Mass and Heat Transfer." The factor in the upper right with its high-loading keywords shown suggests a relationship between photovoltaic power systems and SOFCs. Maps can help newcomers quickly grasp "what's happening" in a research domain. In contrast, experts can investigate frontiers, get ideas on new tool applications, and identify researchers active in particular subareas.

Let's use the patent data to look at a different kind of map. CTI often probes rather deeply. One way to explore for possible relationships is to see whether any inventor is associated with multiple assignees. Figure 16-7

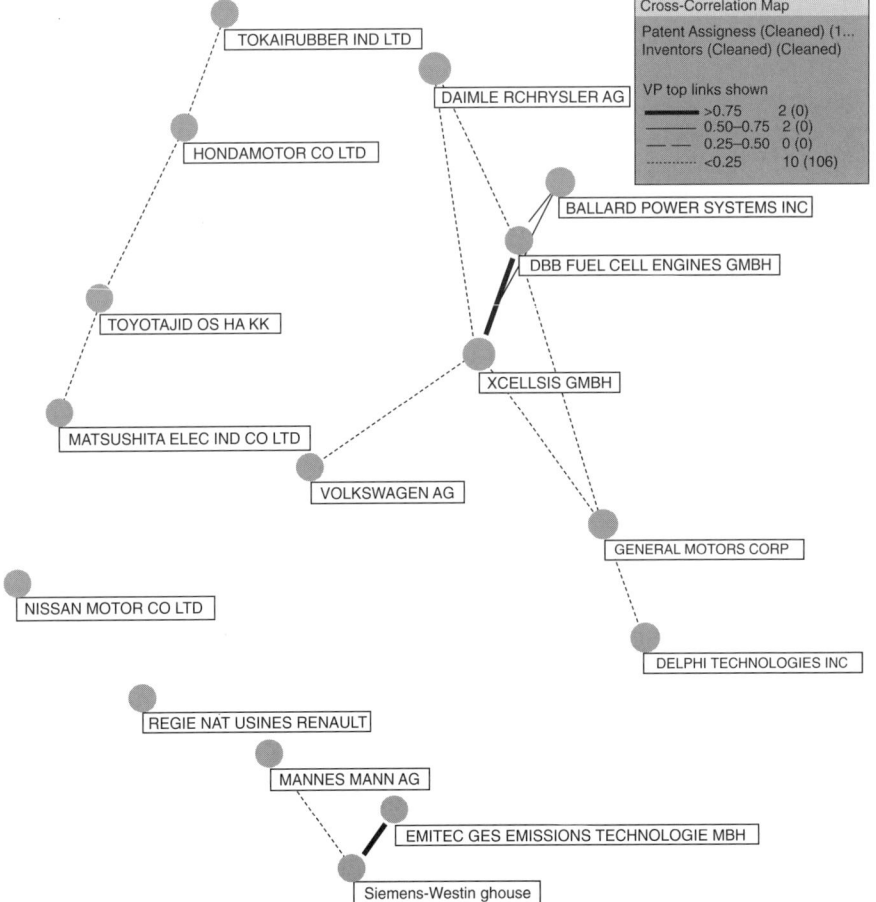

Figure 16-7. Leading automotive-oriented fuel cell patent assignees in leading Western European countries

maps the leading companies patenting in Western Europe based on shared inventors.

Xcellsis, DBB Fuel Cell, and Ballard Power Systems may be working together (or a core inventor team could have shifted companies). Siemens and Emitec also share inventors, suggesting possible collaboration, to be investigated further by examination of their patent records and expert assessment. Chapter 12 pursues a couple of these relationships further via internet searches. "Serious" tech mining would require expert review with in-depth follow-ups. Analogous notions with publications are to explore coauthorship and cocitation patterns.

Mapping of Records Across Terms

As per the discussion relating to Table 16-9, we can also map records based on their shared use of certain terms (the inverse of mapping the terms based on records). This is another variant of co-occurrence mapping. Figure 16-8-a to d shows four "time slices" of the fuel cells publications. This 3-D map sequence depicts concentrations of records for 10,252 of the 11,764 records. Because this map is based on terms in the abstract, not all records were used because not all records have abstracts. In this illustration, we only label selected large "mounds" (concentrations). Such 3-D representations achieved some notoriety in showing how coverage of the 1995 OJ Simpson murder trial dominated U.S. newspaper coverage—"Mount OJ" being by far the tallest peak.

We generated this figure set working with *VxInsight* software (see Chapter Resources). We exported records from *VantagePoint* to *VxInsight* to generate the map. We then selected record groups for each of the main peaks and, returning into *VantagePoint*, listed their leading abstract phrases. Inspecting those, we labeled each peak with the most frequent phrase (or two) that is relatively unique (e.g., ignoring pervasive terms like "fuel cells").

Figure 16-8-a–d presents four time slices chosen to reflect roughly equivalent publication activity. On these, we have plotted a dot for each Georgia Tech paper. We see an intriguing pattern in which this university's research activity seems to almost disappear in the third period (1999–2000), then bounce back sharply (2001–2002). Were one probing this organization, this would warrant further investigation. Inspecting the authorship, we note that Min Liu published 18 papers dated 1995–98 and 20 for 2001–2002, but only 1 in 1999–2000. (We created a "new data set" of the 81 Georgia Tech publications to ascertain this easily.) We also note some active authors ceasing to publish after 1998 with a Georgia Tech affiliation (e.g., Bessette)—Did they leave the university? But we see a cadre of five others with 3–7 publications in 2001–2002, but none before. Might these be new faculty brought in to advance fuel cell research? If so, it seems their interests dovetail closely with existing Georgia Tech research thrusts in that the 2001–2002 dots fall largely in the same terrain as the 1995–98 dots.

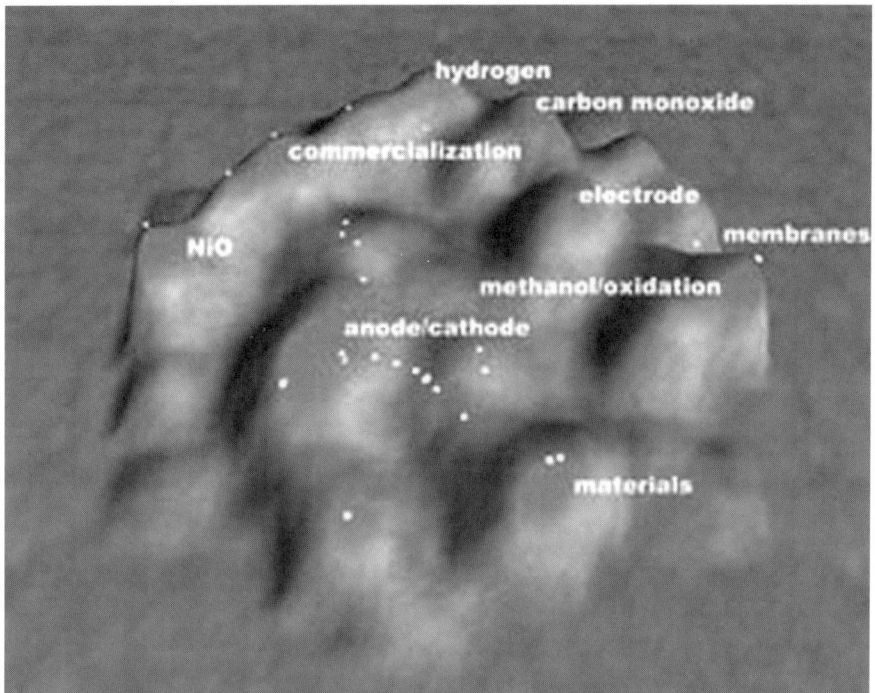

a

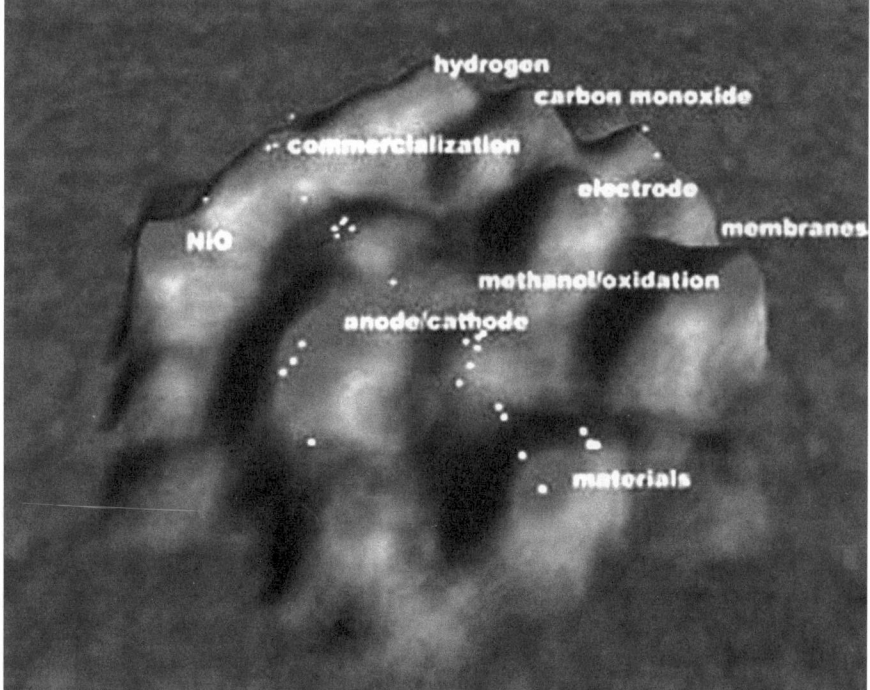

b

Figure 16-8. Fuel cells patent landscapes. a. Time slice 1988–1994. b. Time slice 1995–1998. c. Time slice 1999–2000. d. Time slice 2000–2001

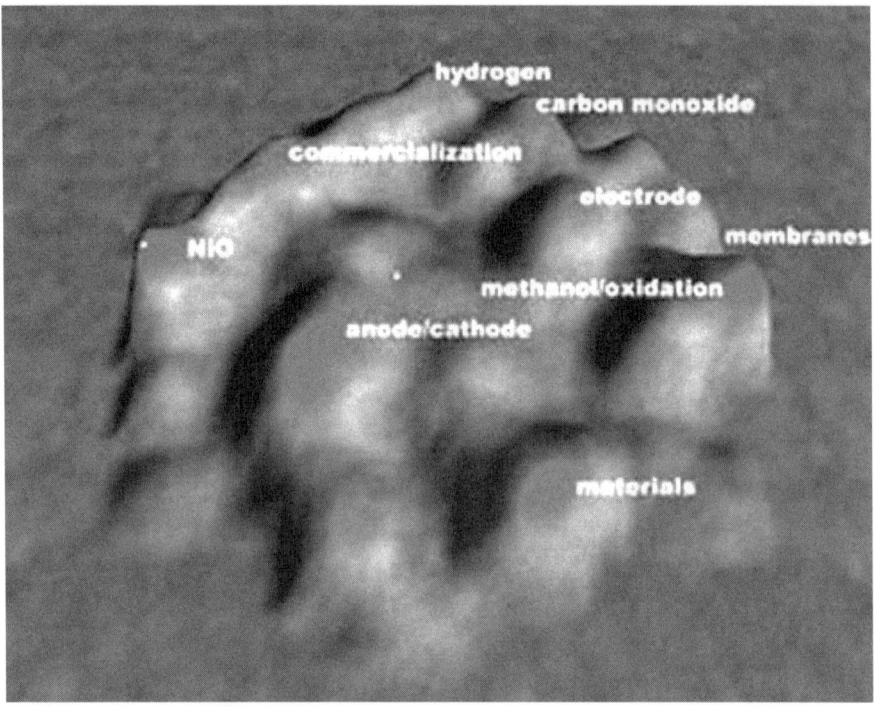

c

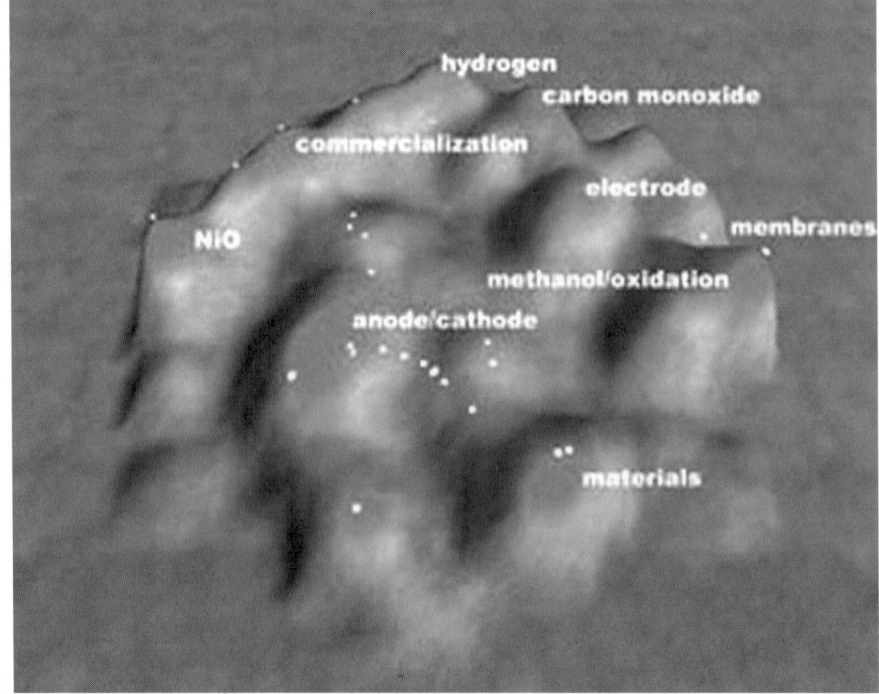

d

Figure 16-8. *Continued*

For this depiction, we have fixed the background for the entire period to focus on shifts in Georgia Tech research emphases. *VxInsight* also enables one to view shifts in overall activity for each time period, and show Georgia Tech changes as overlays on that. Tech mining software enables a great range of ways to cut and depict the data.

As Chapter 13 emphasizes, we strongly recommend that you begin with your technology management questions, to devise suitable indicators, and then work out the empirical manipulations to generate the needed information and its visualization.

Bucketing Records with PCD—Principal Components Decomposition

Another useful text mining function bundles documents (records) based on observed commonalities. This inductive classification contrasts with the use of predefined indexes by which to locate records deductively. "Bucketing" of records can improve access to information when one is dealing with thousands of documents, as we are here.

The PCD algorithm (introduced in Chapter 10) applies metrics that provide a standard solution to an optimization problem (Watts, 2001). The PCD process includes as many of the analyzed records (i.e., abstracts) in the derived "factor groups" as possible. Concurrently, the PCD algorithm strives to maximize both the number of factor groups and group-defining descriptors (i.e., the high-loading "keywords" for each PCA factor group). The algorithm also seeks to minimize the duplication of constituent abstracts between the derived factor groups. This min-max solution approach, conceptually, equates to minimizing the entropy and maximizing the cohesiveness of the PCD-derived factor groups (Borner et al., 2003; Steinbach et al., 2000). Given identical abstract files, the automated PCD analysis will repeatedly derive the same set of factor groups.

Figure 16-9 shows part of this hierarchy for the 2002 publication records. Shown (for space and comprehensibility reasons) are 10 of the 19 main factors. Each of these 10 is named after its most frequently occurring term. The first factor, "ionic conductivity," appears to relate to electrical properties. Four keywords define this cluster (ionic conductivity, oxygen ionic conductivity, electrical resistivity, and cobalt compounds). It contains 190 of the 1902 total records for 2002. Other top-level factors relate to the following typical areas:

- Materials (e.g., lanthanum compounds)
- Research tools (e.g., X-ray diffraction)
- Major types of fuel cells (e.g., PEMFCs, proton-exchange membrane fuel cells)
- Broader issues (e.g., hydrogen economy)

Under each top factor (1–10) are shown the subfactors, giving only the name of the top keyword. In brackets are the number of records in which that

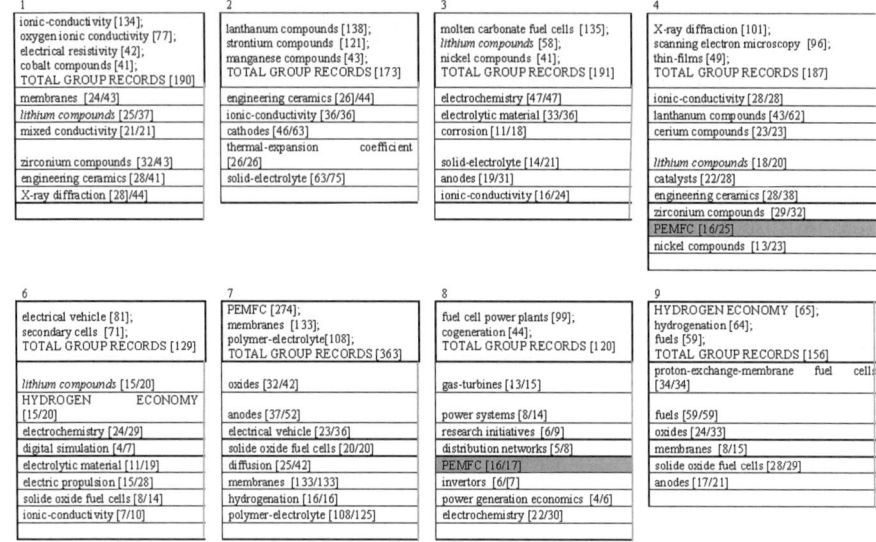

Figure 16-9. "Bucketing" the 2002 fuel cell publications

keyword appears/total number of records in that subfactor. The last subfactor is the "left-overs"—the records not otherwise grouped.

How might you use this? Imagine you are exploring fuel cell R&D published in 2002. You note:

- Hydrogen economy (shown in CAPITAL LETTERS), the defining term of Factor 9, is also a subfactor of Factor 6, with an electrical vehicle focus. Reading the 20 records in the subfactor could provide insights into the association between hydrogen economy and electric vehicles.
- PEMFCs (shaded) show up as the defining term of Factor 8, and also in conjunction with X-ray diffraction (Factor 4) and fuel cell power plants (Factor 8). If you have interests in proton-exchange membrane fuel cells, you might peruse these two clusters to pursue these connections.
- Lithium compounds (in italics) appear as a prominent term in Factor 3 and as sub-factors under ionic-conductivity (Factor 1), X-ray diffraction (Factor 4), and electrical vehicle (Factor 6). These particular bundles of abstract records may hold interest if you are a lithium researcher.
- The "Other" Factor likely reflects a mix of noise and interesting novelties that have not coalesced. Knowledgeable fuel cell researchers may find intriguing leads worth pursuing. "Nafion" was unfamiliar to us, but we quickly discovered 563 of our records mentioning this and a Google search showed that this Dupont membrane dated to the 1970s!
- The Factor "Other"—Subfactor "Other" points to 139 records unclassified through both iterations—great abstracts to explore for new ideas.

In sum, this offers another way to categorize the records with pointers toward subsets of possible interest. PCD can suggest possibly interesting cross-links among technologies, methods, and applications.

Temporal Change Analyses

Temporal markers identify change for various types of information. This can be done in many ways. One can just plot an activity measure versus time (e.g., Figs. 16-3 and 16-8). "Time slicing" into interpretable periods (e.g., before and after a policy shift) can sharpen differences. For other purposes, smoothing trends over time statistically may elucidate what is changing (e.g., using exponential smoothing to reduce the noisy splatter over monthly patenting data).

Tools like VP expedite *list comparisons*. One could compare the keywords from automotive fuel cell articles appearing in 2002 with those seen in prior years to detect new terms. This usually yields a noisy mix from which an expert can glean a few surprises worth following up. Additional examples of this technique are given in Chapter 11.

We have explored ways to compare PCA factor maps over time. Particular maps vary greatly because of the nature of multidimensional scaling, so direct comparison over time slices is not straightforward. Bob Watts is researching metrics (i.e., percentage of keywords included in PCA or PCD factors, entropy, and the richness of linkages among keywords) that could be used as indicators of technological maturation (Watts and Porter, 2003).

Human-interpreted and -drawn maps of the key factors dominant in successive time periods can be very effective. We used this approach in studies for the Army Environmental Policy Institute to track changes in technologies that reduce military noise (e.g., the emergence of active noise suppression methods). To help visualize these changes, we prepared a series of 5-year maps based on PCA. Two future 5-year period maps projected the apparent trends. Thus tech mining helped forecast changes in noise technologies. We did this as well to identify emerging noise issues. The factor maps were presented to experts, whose feedback led to refinements for the final report.

Record topographies, such as Figure 16-8a–d, are wonderful to visualize change over time. For example, we could mark one organization's patents on the landscape by using different colors for different time periods. Interactive presentation could light the earliest period's records; then successive periods would be progressively spotlighted. This can convey the diffusion of that organization's interests over time.

16.8. SEVENTH STEP: REPRESENTATION

Sections 16.6 and 16.7, just previous, deliberately used a wide variety of tabular and graphical forms. The tech miner needs to balance the use of numbers, pictures, and words to best deliver the message to the target users. This should reflect conscious choice, knowing user preferences. We amplify on these issues in the other chapters, so we do not belabor them here.

One option to be considered is interactive representation. On the one hand, providing Figure 16-7 in MS PowerPoint makes it handy to share in presentations. On the other hand, provision via *VP Reader* allows the user to dig down to explore those Honda patent abstracts right then. We invite readers to try this "hands on" via the sample data file on the Wiley website.*

16.9. EIGHTH STEP: INTERPRETATION

Tech mining outputs do not generally speak for themselves. Too often we have fallen into the trap of thinking that they do. Imagine being given Figure 16-5 without clear explanation of what it shows and the meaning to be inferred. Again, other chapters explore interpretation with respect to target users. Be wary of this chapter as an example in that it lacks that compelling reason why one is conducting tech mining. Hence, most of the illustrations do not go all the way to make the case (e.g., should we purchase Company C to obtain that critical IP we need to dominate Application A?).

16.10. NINTH STEP: UTILIZATION

Here as well, we include this heading as a reminder of its importance to tech mining, but we address it elsewhere in the book.

16.11. WHAT CAN WE LEARN?

This chapter set out to illustrate many types of tech mining analyses. It did so without explicit tech mining users to satisfy. Hence, it does not address a main theme of this book—to perform analyses that answer pressing technology management questions, rather than to show off our analytical repertoire. So with that caution, this chapter works through our nine-step tech mining approach. It presents basic and advanced analyses, including trend analyses. Throughout, we deliberately select a variety of representation modes (tables and charts) to get you thinking of options for your own presentations.

We suggest you review Chapter 16 in conjunction with use of the *VP Reader* and sample data available on the web.* Realize that those examples

---

*See at ftp://ftp.wiley.com/public/sci_tech_med/technology_management

will not exactly match the results presented herein because they use approximately 10 percent samples of the publication and patent records. Nonetheless, that should give you a good sense of both large picture summarization and the ability to follow interesting leads back to specific source abstracts.

CHAPTER 16 TAKE-HOME MESSAGES

- The chapter tracks our recommended nine-step approach to tech mining.
- Search quality is required; close consultation with key users can refine the search and facilitate acceptance of the eventual tech mining results.
- Appropriate data cleaning is critical to analysis ease and quality.
- Tech mining finds value in various lists as these convey extent and concentration of R&D activity—"who's doing what."
- Mine lists with "find and group" tools help to focus on key elements (e.g., particular organizations or topics).
- Short lists ("Top 10s") communicate well; don't get fancy unless it's directly useful.
- Augmenting lists with a second dimension (forming a matrix) adds intelligence on interactions, specialization, and capabilities.
- "Profiling" particular organizations or individuals helps answer many user questions.
- Publication and patent data complement each other.
- Tech mining can elucidate "what's new?" and track technological maturation.
- Knowledge networks can be depicted by focusing on individuals and relationships within and among organizations.
- Content analysis gets at particular technologies or topics.
- Mapping provides an array of ways to show relations among *terms* (e.g., technologies, players).
- Other forms of mapping provide ways to show relations among *documents*; these provide compelling landscapes of technological development, particularly good at showing change over time.
- "Bucketing" records can help make cross-topic associations and sift through massive quantities of text records to get to the gold.
- Tech mining should be seen in terms of iterative analyses, where one result prompts further investigations, rather than as once-and-done generation of passively observed outputs.

CHAPTER 16 RESOURCES

We recommend study of this chapter together with use of the *VantagePoint Reader** and sample fuel cell data[†] to try out those methods that seem potentially useful to you.

`

*Available at http://www.thevantagepoint.com.
[†]Available at ftp://ftp.wiley.com/public/sci_tech_med/technology_management

Appendix **A**

Selected Publication and Patent Databases

We note selected science and technology (S&T) information resources.

Databases

Gateways to Access Multiple S&T Publication and Patent Databases:

- Dialog (http://www.dialog.com)
- STN (http://www.fiz-karlsruhe.de)

Patents:

- MicroPatent (http://www.micropat.com)
- Delphion (http://www.delphion.com)
- PatentCafe.com (http://erp.patentcafe.com)
- Questel-Orbit, including citations (http://www.questel.orbit.com)
- WIPS (http://www.wipsglobal.com)
- Derwent World Patent Index and Patent Citation Index (http://www.thomsonderwent.com/)
- IFI CLAIMS (http://www.ificlaims.com)

Tech Mining: Exploiting New Technologies for Competitive Advantage, Edited by Alan L. Porter and Scott W. Cunningham.
ISBN 0-471-47567-X   Copyright © 2005 by John Wiley & Sons, Inc.

S&T Publications

- Engineering Village, including EI Compendex (Engineering Index) and INSPEC (http://www.engineeringvillage2.org)
- Web of Knowledge, including Science Citation Index and Social Science Citation Index (http://www.isinet.com/)
- Chem Abstracts (CAS)—with their special software, SciFinder, that can do considerable profiling of search sets (covering chemical literature and patents): (http://www.cas.org)
- MEDLINE (http://www.ncbi.nlm.nih.gov/PubMed/)
- NTIS, covering over 2 million U.S. government and other government-sponsored research project reports (http://grc.ntis.gov/ntisdb.htm)
- Pascal, covering 5000 multidisciplinary journals (http://www.cas.org/ONLINE/DBSS/pascalss.html)
- RaDiUS, covering U.S. Government projects (http://www.rand.org/services/radius)
- ResearchIndex, free compilation of S&T papers from the Internet (http://citeseer.nj.nec.com)

Below we provide thumbnail sketches of four leading S&T databases to give the flavor of what these contain.

Key Patent Offices

- EPO—European Patent Office (http://www.european-patent-office.org/; esp@cenet; epoline; see also INPADOC at http://www.european-patent-office.org/inpadoc/general.htm)
  Note that their PFS database covers patent publications issued by some 65 patent offices and their PRS database tracks which patents and patent applications are still in force.
- JPO—Japan Patent Office (http://www.jpo.go.jp/)
- USPTO—United States Patent and Trademark Office (http://www.uspto.gov/)
- WIPO—World Intellectual Property Organization (http://www.wipo.org/)

Thumbnail Sketches of Selected S&T Databases

Science Citation Index Science Citation Index (ISI, Thomson Scientific): http://www.isinet.com/products/citation/sci/

The Science Citation Index Expanded™ (SCI) is perhaps the premiere database for coverage of fundamental research. The Institute of Scientific Information (ISI), a subsidiary of Thomson Scientific, produces SCI and companion Social Science Citation Index and Arts & Humanities Citation Index.

All are found at the Web of Knowledge site, by subscription. Researchers at ISI, particularly Henry Small, have been early proponents of using science and technology databases for competitive intelligence and national evaluation or research. The database is a gold standard by which national governments (including United States, Australia, and Britain) evaluate national R&D performance. SCI provides coverage of 5700 journals and exceeds one million records per year (as of 2002). Although these journals nominally cover all of science and technology, SCI maintains a particularly strong physical science emphasis.

Characteristics	Evaluation
Text	SCI is exclusively a text repository.
Density	SCI offers dense coverage of most areas of science.
Quality assurance	SCI identifies peer reviewed journals and distinguishes research articles from commentary.
Intellectual property	SCI contains records and abstracts of individual scientific articles, and is a widely accepted measure of intellectual property
Structured	SCI is cleanly and uniformly structured, with export capabilities to bibliometric management software.
Other points	SCI contains citation information.
	SCI carefully and uniformly notes primary and secondary addresses for authors.
	SCI contains a well-respected classification of the sciences.
	Full text of articles may be available through direct link, depending on service.
	Author abstracts available since 1991; other information from 1945 may be available depending on service.

U.S. Patent Office Database U.S. Patent and Trademark Office: http://www.uspto.gov/main/patents.htm

The U.S. Patent Office (USPTO) provides the world's largest patent database. Many international companies often first seek a U.S. patent, then patent through other world agencies as needed. As a result, the USPTO is a strong indicator of world patent activity. U.S. patents are available directly through the USPTO for a nominal fee (e.g., for CDs). These are also available through premium services that charge for structuring the information and easing download. Key points about this database are noted below.

Characteristics	Evaluation
Text	USPTO contains both diagrams and text; the two sources are clearly distinguished.
Density	USPTO is a rich source for patent activity.
Quality assurance	USPTO clearly distinguishes between patent applications and awards.
Intellectual property	USPTO is backed by U.S. patent laws, a legally binding measure of intellectual property.
Structured	USPTO is well structured; documents are available in XML, a powerful and generic structured format.
Other points	USPTO provides full-text services, through document delivery.
	USPTO contains multiple date fields for dates.
	USPTO contains a well-regarded classification of industrial activity known as "Standard Industrial Codes" (SIC).
	USPTO is searchable on records since 1976.

INSPEC INSPEC (Institution of Electrical Engineers, UK): http://www. iee.org

INSPEC® is provided by IEE, a not-for-profit professional institution headquartered in the United Kingdom. INSPEC comprehensively covers physics, telecommunications, electrical and control engineering, computer science, and information technology. It also provides significant coverage of many other fields of engineering. The applied and industrial focus of INSPEC makes it an attractive source for technology studies. INSPEC covers over 3400 journals and some 2000 conference proceedings. As of 2002 INSPEC reached its 7 million record mark and is growing at a rate of 350,000 records a year in 2002. It explicitly acknowledges the use of its database for current awareness of technology, new product information, competitive intelligence, and technology forecasting.

Characteristics	Evaluation
Text	INSPEC is a text database.
Density	INSPEC provides very dense coverage of computing, telecommunications, and electrical engineering.
Quality assurance	INSPEC allows searchers to distinguish research articles from other sorts of content.
Intellectual property	INSPEC gives a valuable perspective on technology development in industry, including new product development.
Structured	INSPEC is well structured.
Other points	INSPEC offers full text of documents through document delivery.
	INSPEC contains multiple date fields for dates.
	INSPEC provides its own classification for comparison.

MEDLINE   MEDLINE: http://www.ncbi.nlm.nih.gov/PubMed/

Medical articles are qualitatively different than those of other disciplines. Much of medical research involves case studies, which are highly applied research articles discussing concrete intervention on specific medical conditions and cases. Medical science is currently the largest single discipline of science and technology in terms of research publication levels. It is valuable, therefore, when researching medical topics, to examine dedicated medical databases. The leading medical database is MEDLINE, provided as a service by the U.S. National Library of Medicine. This database provides access to over 11 million records. It is free.

Characteristics	Evaluation
Text	MEDLINE is a text database.
Density	MEDLINE provides very dense coverage of medicine and biosciences.
Quality assurance	MEDLINE provides quality assurance by focusing on peer-reviewed journals.
Intellectual property	MEDLINE records represent significant intellectual property in terms of research articles as well as clinical trials.
Structured	MEDLINE is well structured.
Other points	Medline provides its own subject classification system known as MeSH. Medline baselines date from the mid 60s.

Appendix **B**

Text Mining Software

Chapter 14 discusses several strategies for gaining tech mining capabilities. These range from: doing it yourself using manual queries against the S&T database of your choice; utilizing the advanced analysis capabilities of some technology databases; developing your own tech mining "scripts" with programming languages; or using tech mining software. Chapter 9 introduces the tools and software needed to support tech mining. This appendix identifies selected software options.

Many options can support tech mining. These range from dedicated software programs to "toolkits" where you can assemble applications tailored to your needs. Other options include the use of statistical packages to accomplish tech mining analyses. You may also find useful capabilities embedded in database applications (where you may already store data). Others vary along a spectrum from generic text mining applications to those specifically catering to analyzing S&T databases. Still others vary in their emphasis on visualization versus analysis of data. Examples of several kinds of software follow.

Text Mining Software

- ClearResearch from ClearForest (http://www.clearforest.com)

General Text Mining Capability Within Statistical Packages

- LexiQuest-Mine within SPSS (http://www.spss.com/lexiquest)

Tech Mining: Exploiting New Technologies for Competitive Advantage, Edited by Alan L. Porter and Scott W. Cunningham.
ISBN 0-471-47567-X   Copyright © 2005 by John Wiley & Sons, Inc.

- SAS Text Miner, Enterprise Miner, and IntelliVisor (http://www.sas.com)
- WordStat within Simstat (http://www.simstat.com/wordstat.htm)

Other Tools Integrated with Databases*

- MicroPatent's inclusion of Aurigin software (Aureka) (http://www.micropat.com/0/pdf/f_patentweb_091803.pdf)
- Thomson Delphion capabilities [including Citation Link, Snapshot for online lists (bar charts), PatentLab-II for off-line analyses, and Text Clustering] (http://www.delphion.com/)

Software Targeted on Science and Technology Information Resource Mining

- VantagePoint (http://www.theVantagePoint.com)[†]
- Dataview at the University of Marseilles (http://crrm.u-3mrs.fr/)
- Focust (http://wisdomain.com/)

Software Particularly Aimed at Information Visualization

- Anacubis (http://www.anacubis.com/)
- VX Insight (http://www.viswave.com/)

Special "Toolkits"

- IBM's Intelligent Miner for Text (http://www-3.ibm.com/software/data/iminer/)

---

*In 2004, Thomson announced plans to acquire MicroPatent, so further integration appears likely.
[†]You may come upon two variations of this software. *Derwent Analytics* (http://thomsonderwent.com/products/dapt/derwentanalytics/) is specially tailored for use with Delphion, Derwent, and Web of Knowledge data. *TechOASIS* is a U.S. Government version. The software was developed with considerable U.S. support from the Defense Advanced Research Projects Agency (DARPA) and the Army (Tank-automotive and Armaments Command), largely under the guidance of Robert J. Watts. For information, see http://www.searchtech.com.

Appendix **C**

What You Can Do Without Tech Mining Software

Without tech mining software, data "tuning and tallying" is mainly about detecting general activity patterns and building a better query. Your main tool is the search engine itself. By modifying your search strategy, you can vary how many hits you get and record this information.

More complicated analyses are largely constrained by what you already know about the data. For instance, if you know the top research producer in the field, you can construct a query to discover precisely how many articles this individual or institution has published. If you want to check the prevalence of a particular term, you can modify your search strategy to find publications or patents using this term. You could do this for the database as a whole, or within another search set. For instance, you could tally how many of the Borlaug-citing articles mentioned genetic modification. Or, as illustrated in Table 7-3, you could query the on-line database to discover the number of publications occurring in each year. Keep in mind that search engines are designed for researchers, not tech mining analysts. As a result, the interfaces are often poorly adapted to counting, much less actual text mining.

Adept analysts can use other software packages or even their programming skills to generate many tech mining analyses. You don't need tech mining software. In particular, tabulating activity and searching through records can certainly be achieved by using Microsoft Excel and Access. More sophisticated statistical analyses can be handled by any of the major statistics packages. Programming using computer languages such as perl, Java, or VisualBasic can give

Tech Mining: Exploiting New Technologies for Competitive Advantage, Edited by Alan L. Porter and Scott W. Cunningham.
ISBN 0-471-47567-X   Copyright © 2005 by John Wiley & Sons, Inc.

immediate, if limited, returns to those organizations with these skills. Further growth is limited only by time and commitment. The do-it-yourself approach does not seem the best use of analyst skills to us, but we are intimately involved in the development and application of tech mining software. And, it is true that "growing your own" tech mining solutions may permit valuable customization and integration within the existing information technology infrastructure of your firm.

Appendix D

Statistics and Distributions for Analyzing Text Entities

The importance of distributional characteristics to tech mining is addressed in Chapter 9. Many S&T information compilations evidence highly skewed distributions so that assuming normality (or even conditional normality) is inappropriate. Here we provide more details on some of the key distributions that are so important in depicting S&T information.

The Zipf distribution is often used in analyzing content. The distribution relates the number of words, phrases, or other units of content to their relative frequencies in a collection of articles.

Equation D.1. The Zipf distribution

$$P(\omega) = c\omega^{-1}$$

In the equation ω (omega) is a ranked ordering of words and phrases from the most frequent word (rank number one) to the least frequent word (rank N where there are N distinct words in the collection.) The probability of the word occurring when a selection is made from words on the list is given by $P(\omega)$. Finally, the value cc is a constant that varies from collection to collection.

This distribution informs us, for instance, that there are a great many infrequent words in any article collection. The tenth most frequent word in a collection will, for instance, be ten times less frequent than the most common word. The Zipf distribution is widely applicable to scientific as well as nonscientific language.

Tech Mining: Exploiting New Technologies for Competitive Advantage, Edited by Alan L. Porter and Scott W. Cunningham.
ISBN 0-471-47567-X   Copyright © 2005 by John Wiley & Sons, Inc.

Bradford's law tells us that a relatively few sources contain a majority of relevant publications for any given study. It is used, in particular, in analyzing journal coverage. The formula looks familiar:

Equation D.2. Bradford's law

$$P(\varphi) = c\varphi^{-1}$$

In Bradford's law the symbol φ (phi) represents a ranked list of journals from the most useful journals for any given subject matter (ranked 1) to the least useful journals (ranked N). The probability of finding an article of interest from a journal, given a particular article collection, is $P(\varphi)$. The constant c might vary from distribution to distribution. Here again we see a law of scattering where a few journals provide most of the content, but a broad tail of minor journals considerably complement the "core set."

Lotka's law reminds us that relatively few authors will write a majority of scientific publications in most fields.

Equation D.3. Lotka's law

$$P(\alpha) = c\alpha^{-2}$$

Lotka's law relates a ranked list of authors given by α (alpha) where rank 1 is given to the most prolific author, and rank N (out of a list of N authors) is given to the least prolific author. Lotka's law asserts that the probability of finding an author with a given productivity $P(\alpha)$ diminishes with the square of the productivity. The constant may vary from article collection to collection. Unlike the previous laws and distributions, Lotka's law is inverse-squared, emphasizing the extreme rarity of highly prolific authors.

Other authors have examined the formation of scientific networks and also characterized these relationships mathematically, using probability distributions. The law reminds us that there are relatively few "hubs" of knowledge among scientific and technical communities.

Equation D.4. Networked evolution of coauthorship

$$P(\chi) = c\chi^{-3}$$

In this equation P is the frequency of authors with chi (χ) coauthors. The constant c may vary from system to system, collection to collection. In this equation, highly collaborative authors are even more infrequent than highly productive authors. They diminish by the cubic power of the total number of coauthors.

Theories of network formation are of interest to many fields, including tech mining. The original theories postulated that networks formed randomly, and then examined the implication of the assumptions for the structure of the network and for the behavior of "nodes" in the network. More recently, researchers have examined networks in which nodes and linkages evolve dynamically over time—in particular, networks in which the connected have more advantage over time.

These networks have several interesting properties. (1) They are "small-world" networks in which relatively few links and hubs encourage effective intercommunication. (2) These networks bear structural resemblance to networks that occur in real life. (3) They often display power law relationships. The networks are said to be "scale free" because they look very similar structurally across a wide variety of levels of detail.

The strong similarity among these four distributions suggests to some authors that there may be a single explanation for diverse sorts of behavior in publication and collaboration. In all four cases, the laws and distributions show a "power law" form. Underlying similarities have led to attempts to create a single "law" of publishing behavior known to some as the "Bradford–Zipf distribution."

In the Zipf distribution, frequent terms have an advantage in being read, recognized, and adopted by others. Under Bradford's law, journals that begin to specialize in a given domain of knowledge have efficiencies in publishing additional articles of similar character. In Lotka's law of authorship, less prolific authors have a systematic disadvantage because of their lack of name recognition among reviewing and editorial committees. Contrariwise, prolific authors gain specialized knowledge when continuing to publish in a specific science or technology domain. Finally, authors in the center of collaborative networks of scientists gain preferential treatment when new scientists seek out potential collaborators.

The original mathematical treatment of these laws or distributions by their original discoverers (including Zipf, Bradford, and Lotka) differs from the formats given here. This section gives each distribution in the same framework. This eases comparison and, hopefully, increases insight into the commonalities among the four.

Note that in each case the law (or distribution) is shown with an integer power. Actual estimation of these distributions with sample data results in power laws that might differ fractionally from those shown here. Furthermore, actual data often vary systematically from the laws and distributions shown here, particularly at the beginning (most frequent) and end (least frequent) ends of the distribution. By and large these variations from predictions are poorly understood. Some say the power law should continue to hold but limitations in data collection prevent adequate measurement. Other authors suggest that the dynamics apply only in the aggregate—other behavioral regimes may be seen over time and in certain domains.

This section gives some theory to discuss why words, authorship, coauthorship, and journals vary so widely in a collection. Furthermore, it shows that a few words, journals, or authors will always comprise the core of the collection. These laws and distributions are useful therefore in establishing cut-off points: How many authors, journals, or collaborative links need be considered to get a representative measure of the science or technology system being studied?

References

Agichtein, E., Lawrence, S., and Gravano, L. (2004). "Learning to find answers to questions on the web," *ACM Transactions on Internet Technology*, Vol. 4, Issue 2.

Agresti, A. (2002). *Categorical Data Analysis*, New York: John Wiley & Sons.

Akers, L., and Khorsandian, F. (2003). "High Level Overview of Patent Search Types and Approaches," Chicago: *PIUG 2003* (Patent Information Users Group).

Albright, R.E., and Kappel, T.A. (2003). "Roadmapping in the Corporation," *Research Technology Management*, Vol. 46, No. 2, 31–40.

Altshuller, G. (1990). And Then the Inventor Appeared: TRIZ, the Theory of Inventive Problem-Solving, Worcester, MA: Technical Information Center.

Anderson, P., and Tushman, M.L. (1990). "Technological Discontinuities and Dominant Designs: A Cyclical Model of Technological Change," *Administrative Science Quarterly*, Vol. 35, 604–633.

Ascher, W. (1978). *Forecasting: An Appraisal for Policy Makers and Planners*, Baltimore, MD: Johns Hopkins University Press.

Baldi, P., Fasconi, P., and Smyth, P. (2003). *Modeling the Internet and the Web: Probabilistic Methods and Algorithms*, New York: John Wiley & Sons.

Barabási, A.L., Jeong, H., Néda, Z., Ravasz, E., Schubert, A., and Vicsek, T. (2002). "Evolution of the Social Network of Scientific Collaborations," *Physica A*, Vol. 311, 590–614.

Börner, K., Chen, C., and Boyack, K.W. (2003). "Visualizing knowledge domains," *Annual Review of Information Science and Technology*, Vol. 37, 179–255.

Boyack, K. (2003). "An Indicator-Based Characterization of the Proceedings of the National Academy of Sciences," paper presented at the NAS Sackler Colloquium on *Mapping Knowledge Domains*.

Buckley, R. (1998). "Strategic Environmental Assessment," in Porter, A.L., and Fittipaldi, J.J. (Eds), *Environmental Methods Review: Retooling Impact Assessment for the New Century*, p. 77–86, Atlanta, GA: Army Environmental Policy Institute.

Burt, R.S. (1983). *Applied Network Analysis: a Methodological Introduction*, Beverly Hills, CA: Sage Publications.

Campbell, D., and Stanley, J. (1963). *Experimental and Quasi-Experimental Designs for Research*, Chicago, IL: Rand-McNally.

Canter, L.W. (1998). "Methods for Effective Environmental Impact Assessment," in Porter, A.L., and Fittipaldi, J.J. (Eds), *Environmental Methods Review: Retooling Impact Assessment for the New Century*, p. 58–68, Atlanta, GA: Army Environmental Policy Institute.

Carton, L. (2002). "Strengths and Weaknesses of Spatial Language: Mapping Activities as Debating Instrument in a Spatial Planning Process," Washington, DC: *XXII International Congress*.

Chen, C. (2003). *Mapping Scientific Frontiers: The Quest for Knowledge Visualization*, London: Springer.

Clarke, D.W., Sr. (2000). "Strategically Evolving the Future: Directed Evolution and Technological Systems Development," *Technological Forecasting and Social Change*, Vol. 64, 143–153.

Coates, J.E. (1999). "Technology Forecasting for Business Clients," *Futures Research Quarterly*, Fall, 99–109.

Coates, V., Faroque, M., Klavins, R., Lapid, K., Linstone, H.A., Pistorius, C., and Porter, A.L., (2001). "On The Future of Technological Forecasting," *Technological Forecasting and Social Change*, Vol. 67, No. 1, 1–17.

Cohen, H.S., Keller, S., and Streeter, D. (1979). "The Transfer of Technology from Research to Development," *Research Management*, Vol. 22, No. 3, 11–17.

Cook, T.D., and Campbell, D.T. (1979). *Quasi-Experimentation: Design and Analysis Issues for Field Settings*, Boston, MA: Houghton Mifflin Company.

Cuhls, K., Blind, K., and Grupp, H. (2002). *Innovations for Our Future*, Heidelberg: Physica-Verlag.

Cunningham, S. (1996). *The Content Evaluation of British Scientific Research*, Ph.D. Thesis, University of Sussex: Science Policy Research Unit.

Current Science (Vol. 79, No. 5, 10 Sep., 2000). Special section on *Scientometrics*: http://ces.iisc.ernet.in/Current_Science/

Danneels, E. (2002). "The Dynamics of Product Innovation and Firm Competences," *Strategic Management Journal*, Vol. 23, 1095–1121.

de Bruijn, H., and Porter, A.L. (2004). "The Education of a Technology Policy Analyst— to Process Management," *Technology Analysis and Strategic Management*. Vol. 16, No. 2, 261–274.

Deroian, F. (2002). "Formation of Social Networks and the Diffusion of Innovations," *Research Policy*, Vol. 31, No. 5, 835–846.

Devezas, T.C., and Corredine, J.T. (2001). "The Biological Determinants of Long Wave Behavior in Socioeconomic Growth and Development," *Technological Forecasting and Social Change*, Vol. 68, 1–58.

Dror, I. (1989). "Technology Innovation Indicators," *R&D Management* Vol. 19, 243–249.

Dundon, E. (2002). *The Seeds of Innovation*, New York: AMACOM.

Dunphy, S.M., Herbig, P.R., and Howes, M.E. (1996). "The Innovation Funnel," *Technological Forecasting & Social Change* Vol. 53, 279–292.

Edwards, A.W.F. (1992). *Likelihood*, Cambridge: Cambridge University Press.

Ernst, H. (2003). "Patent Information for Strategic Technology Management," *World Patent Information*, Vol. 25, No. 3, 233–242.

Fischhoff, B. (1998). "Risk Perception and Communication Unplugged: Twenty Years of Process," In Löfstedt, R. and Frewer, L. (Eds.), *Risk & Modern Society*, London: Earthscan Publications, 143–145.

Fourez, G.M. (1994). "Technology Assessment: A Pocket Version," *Bulletin of Science, Technology & Society*, Vol. 14, No. 3, 142–143.

Gausemeier, J., Fin, A., and Schlake, O. (1998). "Scenario Management: An Approach to Develop Future Potentials," *Technological Forecasting and Social Change*, Vol. 59, 111–140.

Gaynor, G.H. (2002). *Innovation by Design*, New York: AMACOM.

Georghiou, L., and Meyer-Krahmer, F. (1992). "Evaluation of Socio-economic Effects of Community R&D Programmes—Lessons for Concepts, Methods, and Issues," *Research Evaluation*, Vol. 2, No. 1, 5–15.

Germeraad, P. (2004). "Striking Gold the Patent Way—A Case Study on the Use of Patent Information," Singapore: Europe Asia Patent Information Conference.

Glenn, J., and Gordon, T. (2002). *Futures Research Methods: Version 2.0*, Washington, DC: AC/UNU Millenium Project. http://www.acunu.org/millennium/FRM-v2.html

Granstrand, O. (1999). *The Economics and Management of Intellectual Property*, Cheltenham, UK: Edward Elgar.

Greenspan, A. (1998). "Testimony of Chairman Alan Greenspan before the Sub-committee of Domestic and International Monetary Policy of the Committee on Banking and Financial Services," Washington, DC: U.S. House of Representatives, February 24, 1998. Available on-line in full text: http://www.federalreserve.gov/boarddocs/hh/1998/february/testimony.html

Guinee, J.B. (2002). *Handbook on Life Cycle Assessment*, Dordrecht: Kluwer.

Hagedoorn, J., and Duysters, G. (2002). "Learning in dynamic inter-firm networks: The efficacy of multiple contacts," *Organizational Studies*, Vol. 23, No. 4, 525–548.

Hecker, D. (1999). "High-technology employment: a broader view," *Monthly Labor Review*, June.

Hall, B.H., Jaffe, A., and Trajtenberg, M. (2003). "Market Value and Patent Citations," Berkeley, CA: University of California, Berkeley. http://emlab.berkeley.edu/users/bhhall/

Hansen, M.T., Nohria, N., and Tierney, T. (1999). "What's Your Strategy for Managing Knowledge?," *Harvard Business Review*, Vol. 77, March–April, 106.

Hicks, D., and Katz, S., (1996). "Systemic Indicators for the Knowledge-Based Economy," Paris: *OECD Workshop on New Indicators for the Knowledge-Based Economy*.

Hicks, D., Breitzman, A., Olivastro, D., and Hamilton, K. (2001). "The Changing Composition of Innovative Activity in the US—a Portrait Based on Patent Analysis," *Research Policy*, Vol. 30, 681–703.

Huber, G.P., and McDaniel, R.R. (1986). "The Decision-Making Paradigm of Organizational Design," *Management Science*, Vol. 32, No. 5, 572–589.

Hullman, A., and Meyer, M. (2003). "Publications and Patents in Nanotechnology: An Overview of Previous Studies and the State of the Art," *Scientometrics*, Vol. 58, No. 3, 507–527.

Kash, D.E., and Rycroft, R.W. (2000). "Patterns of Innovating Complex Technologies: A Framework for Adaptive Network Strategies," *Research Policy*, Vol. 29, 819–831.

Kirby, M.R., Mavris, D.N., and Largent, M.C. (2001). "A Process for Tracking and Assessing Emerging Technology Development Programs for Resource Allocation," *AIAA*, Paper No. 2001–5280 (http://www.aiaa.org/research.hfm).

Klavans, J.L., and Klavans, R.A. (2001). "Do Patent Models Reveal Technological Capabilities," *221st American Chemical Society National Meeting*, San Diego, CA.

Kongthon, A. (2003). "Deriving Tree-Structured Networks from Technical Text Using Association Rule Mining: Applied to Thailand's R&D," *Arthur M. Sackler Colloquium on Mapping Knowledge Domains, National Academy of Sciences*.

Kostoff, R.N. (2004). "Various reports on Text Mining, Including Text Discovery Processes." http://www.onr.navy.mil/sci_tech/special/technowatch/

Kostoff, R.N., and E. Geisler (1999). "Strategic Management and Implementation of Textual Data Mining in Government Organizations," *Technology Analysis & Strategic Management*, Vol. 11, No. 4, 493–525.

Lane, P., and Makri, M. (2000). "Responding to Diminishing Technological Opportunities: A Socio-Cognitive Model of Science and Innovation," working paper, Tempe, AZ: Arizona State University, College of Business.

Lawrence, S. (2001). "Access to Scientific Literature," in Butler, D. (Ed.), *The Nature Yearbook of Science and Technology*, 86–88, London: Macmillan.

Lederberg, J. (1997). "Infectious Disease as an Evolutionary Paradigm," *Emerging Infectious Diseases*, Vol. 3, No. 4, 417–423.

Lemons, K.E., and Porter, A.L. (1992). "A Comparative Study of Impact Assessment Methods in Developed and Developing Countries," *Impact Assessment Bulletin*, Vol. 10, No. 3, 57–65.

Lihtenthaler, E. (2003). "Third Generation Management of Technology Intelligence Processes," *R&D Management*, Vol. 33, No. 4, 361–375.

Linstone, H.A. (1999). *Decision Making for Technology Executives: Using Multiple Perspectives to Improve Performance*, Boston, MA: Artech House.

Linstone, H.A., (2002). "Corporate Planning, Forecasting, and the Long Wave," *Futures*, Vol. 34, 317–336.

Mahajan, V., and Muller, E. (1996). "Timing, Diffusion, and Substitution of Successive Generations of Technological Innovations: The IBM Mainframe Case," *Technological Forecasting & Social Change*, Vol. 51, 109–142.

Mansfield, E. (1981). "How Economists See R&D," *Harvard Business Review*, Nov.–Dec., 98–106.

March, J.G. (1991). "Exploration and Exploitation in Organizational Learning," *Organization Science*, Vol. 2, No. 1, 71–81.

Marchau, V.A.W.J. (2000). *Technology Assessment of Automated Vehicle Guidance*, Delft, The Netherlands: Delft University Press.

Martin, B.R., and Irvine, J. (1989). *Research Foresight: Creating the Future*, New York: St. Martin's Press.

Martino, J.P. (1993). *Technological Forecasting for Decision Making*, third edition. New York: McGraw-Hill.

McLachlan, G., and Peel, D. (2002). *Finite Mixture Models*, John Wiley & Sons, New York.

Metcalfe, J.S. (1988). "The Diffusion of Innovations: An Interpretive Survey," in Dorsi, G., Freeman C., Nelson, R., Silverberg, G., and Soete, L. (Eds.), *Technical Change and Economic Theory*, pp. 560–589, London: Pinter.

Michaels, J.V. (1996). *Technical Risk Management*, Upper Saddle River, NJ: Prentice Hall.

Millson, M.R., Raj, S.P., and Wilemon, D. (1992). "A Survey of Major Approaches for Accelerating New Product Development," *Journal of Product Innovation Management*, Vol. 9, 53–69.

Mintzberg, H., Raisinghani, D., and Theoret, A. (1976). "Structure of Unstructured Decision Processes," *Administrative Science Quarterly*, Vol. 21, No. 2, pp. 246–275.

Modis, T. (1993). "Technology Substitutions in the Computer Industry," *Technological Forecasting & Social Change*, Vol. 43, 157–167.

Mogee, M. (1993). "Educating Innovation Mangers: Strategic Issues for Business and Higher Education," *IEEE Transactions on Engineering Management*, Vol. 40, No. 4, 410–417.

Mogee, M.E. (2003). "Integrating Patent Analysis and Technology Intelligence Techniques: A Case Study," Chicago, IL: Patent Information Users Group.

Mohrle, M.G. (2000). TRIZ-based Competitor Analyses, *Ninth International Conference on Management of Technology*, Miami, FL: International Association for Management of Technology (IAMOT).

Naisbitt, J. and Aburdene, P. (1991). *Megatrends 2000*, New York: Morrow Avon.

Narin, F., Alber, M.B., and Smith, V.M. (1992). "Technology Indicators in Strategic Planning," *Science and Public Policy*, Dec., 369–381.

Narin, F., Hamilton, K.S., and Olivastro, D. (1997). "The Increasing Linkage Between U.S. Technology and Public Science," *Research Policy*, Vol. 26, 317–330.

Narin, F., Thomas, P., and Breitzman, A., (2001). "Using Patent Indicators to Predict Stock Portfolio Performance," in Berman, B. (Ed.), *From Ideas to Assets: Investing Wisely in Intellectual Property*, pp. 293–308, New York: Wiley.

National Science Board (2002). *Science and Engineering Indicators—2002*, Arlington, VA: U.S. National Science Foundation.

Newman, N.C., Porter, A.L., and Yang, J. (2001). Information Professionals: Changing Tools, Changing Roles, *Information Outlook*, Vol. 5, No. 3, 24–30.

OECD (2003). *OECD Science, Technology and Industry Scoreboard*, Paris: Organisation for Economic Cooperation and Development.

Office of Technology Assessment (1986). *Research Funding as an Investment: Can We Measure the Returns? A Technical Memorandum*, Washington, DC: U.S. Congress, Office of Technology Assessment, OTA-Tech Mining-SET-36.

Ortt, J.R., and van der Duin, P.A. (2004). *Generations of R&D Management: A Context Dependent Evolution of Best Practices*, Delft, The Netherlands: Technical University of Delft.

Pilkington, A. (2003). Technology Commercialization: Patent Portfolio Alignment and the Fuel Cell, *PICMET—Portland International Conference on Engineering and Technology Management*, Portland, OR.

Popescul, A., Unger, L.H., Pennock, D.M., and Lawrence, S. (2001). "Probabilistic Models for Unified Collaborative and Content-Based Recommendation in Sparse-Data Environments," *Proceedings of the Seventeenth Conference on Uncertainty in Artificial Intelligence*, San Francisco, CA: Morgan Kaufmann.

Porter, A.L. (2002). "Text Mining for Technology Foresight," in Glenn, J., and Gordon, T., *Futures Research Methods*, New York: American Council for the United Nations.

Porter, A.L. (2003). "Iraqi Engineering: Where Has All the Research Gone?," *Science and Public Policy*, Vol. 30, No. 2, 97–105.

Porter, A.L., and Cunningham, S.W. (1995). "Whither Nanotechnology? A Bibliometric Study," *Foresight Update*, Vol. 21,12–15.

Porter, A.L., and Fittipaldi, J.J. (eds.) (1998). *Environmental Methods Review: Retooling Impact Assessment for the New Century*, Atlanta, GA: Army Environmental Policy Institute.

Porter, A.L., Kongthon, A., Lu, J.-C. (2002). "Research Profiling: Improving the Literature Review," *Scientometrics*, Vol. 53, 351–370.

Porter, A.L., and Newman. N.C. (2001). "Why Don't Managers Want our Technological Intelligence? And What Can We Do about it?" Seattle, WA: *Society of Competitive Intelligence Professionals—SCIP*.

Porter, A.L., Rossini, F.A., Carpenter, S.R., and Roper, A.T. (1980). *A Guidebook for Technology Assessment and Impact Analysis*, New York: North-Holland.

Porter, A.L., and Schoeneck, D. (2000). "Mining Electronic R&D Information in Support of Resource Management," Bellingham, WA: *8th International Symposium on Society and Resource Management*.

Porter, A.L., Roper, A.T., Mason, T.W., Rossini, F.A., and Banks, J. (1991). *Forecasting and Management of Technology*, New York: John Wiley & Sons.

Porter, M.E. (1985). *Competitive Advantage: Creating and Sustaining Superior Performance*, New York: Free Press.

Power, D.J. (2002). *Decision Support Systems, Concepts and Resources for Managers*. Westport, CT: Greenwood Publishing.

Price, D.J. (1963). *Little Science, Big Science*, New York: Columbia University Press.

Rasmussen, J. (1997). "Risk Management in a Dynamic Society: A Modelling Problem," *Safety Science*, Vol. 27, No. 2–3, 183–213.

Rivette, K.G., and Kline, D. (2000). *Rembrandts in the Attic*, Boston, MA: Harvard Business School Press.

Rossini, F.A., Porter, A.L., Jacobs, C.C., and Abraham, D.S. (1988). "Trends in Computer Use in Industrial R & D," *Research/Technology Management*, Vol. 31, 36–41, (Sep/Oct).

Ross, S. (2002). *Probability Models*: Eighth Edition, San Diego, CA: Academic Press.

Rouse, W.B. (1994). *Best Laid Plans*, Englewood Cliffs, NJ: Prentice Hall.

Saaty, T.L. (2001). *The Analytic Hierarchy Process: Multicriteria Decision Making: Planning, Priority Setting, Resource Allocation* (revised edition), Pittsburgh, PA: RWS Publications.

Savransky, S.D. (2000). *Engineering of Creativity: Introduction to Triz Methodology of Inventive Problem-Solving*, Boca Raton, FL: CRC Press.

Smith, C.G. (1992). "Understanding Technology Substitution: Generic Types, Substitution Dynamics and Influence Strategies," *Journal of Engineering/Technology Management*, Vol. 9, 279–302.

Souder, W.E., Nashar, A.S., and Padmanabhan, V. (1990). "A Guide to the Best Technology Transfer Practices," *Technology Transfer* (Winter–Spring), 5–16.

Steinbach, M., Karypis, G., and Kumar, V. (2000). "A Comparison of Document Clustering Techniques," Minneapolis, MN: University of Minnesota Technical Report #00-034. http://www.cs.umn.edu/tech_reports/.

Stuart, T., and Sorenson, O. (2003). "The Geography of Opportunity: Spatial Heterogeneity in Founding Rates and the Performance of Biotechnology Firms," *Research Policy*, Vol. 32, No. 2, 229–253.

Swanson, D.R., and Smalheiser, N.R. (1997). "An Interactive System for Finding Complementary Literatures: a Stimulus to Scientific Discovery," *Artificial Intelligence*, Vol. 91, No. 2, 183–203.

Tassey, G. (1999). *R&D Trends in the U.S. Economy: Strategies and Policy Implications*, Washington, DC: U.S. Department of Commerce, Technology Administration, National Institute of Standards and Technology.

Technology Futures Analysis Methods Working Group (2004). "Technology Futures Analysis: Toward Integration of the Field & New Methods," *Technological Forecasting & Social Change*. Vol. 71, No. 3, 287–303.

Teichert, T., and Mittermayer, M.-A. (2002) "Text Mining for Technology Monitoring," *IEEE IEMC 2002*, 596–601.

Terninko, J., Zusman, A., and Zlotin, B. (1998). *Systematic Innovation: An Introduction to TRIZ*, Boca Raton, FL: St. Lucie Press.

Tidd, J., Bessant, J., and Pavitt, K. (1997). *Managing Innovation: Integrating Technological, Market, and Organizational Change*, Chichester, UK: Wiley.

Trippe, A. (2003a). "Patent Analysis: The Once and Future Discipline," Chicago, IL: Patent Information Users Group.

Trippe, A. (2003b in press). "Patinformatics: Tasks to tools," *World Patent Information*, Vol. 25, Issue 3, 211–221.

van Bueren, E.M., Klijn, E.H., and Koppenjan, J.F.M. (2003). "Dealing with Wicked Problems in networks: Analyzing an Environmental Debate from a Network Perspective," *Journal of Public Administration, Research and Theory*, Vol. 13, No. 2, 193–213.

van de Gronden, E.D., van Haeren, J.J.F.M., and Roos, E. (1994). "Use and Effectiveness of Environmental Impact Assessments in Decision Making," Zoetermeer, The Netherlands: Netherlands Ministry of Housing, Spatial Planning, and the Environment.

Van Raan, A.F.J. (Ed.) (1988). *Handbook of Quantitative Studies of Science & Technology*, Dordrecht: North Holland. See also website: http://www.cwts.nl/

Wall, L.T., Christiansen, T., and Orwant, J. (2000). *Programming Perl*, Sebastopol, CA: O'Reilly.

Waterland Neeltje Jans (2000). *The Delta Project*, Burgh-Haamstede, The Netherlands: Waterland Neeltje Jans.

Watts, R. J. (2001). "Knowledge Discovery Using the Tech OASIS: Meeting the Information Infrastructure Needs," Portland, OR: Proceedings of the *Portland International Conference on Management of Engineering and Technology*.

Watts, R.J., and Porter, A.L. (1997). "Innovation Forecasting," *Technological Forecasting and Social Change*, Vol. 56, 25–47.

Watts, R.W., and Porter, A.L. (2003). " R&D Cluster Quality Measures and Technology Maturity," *Technological Forecasting and Social Change*. Vol. 70, No. 8, 735–758.

Webster, R. (1998). "Methods for Environmental Impact Assessment: Selecting a Model and Approach," in Porter, A.L., and Fittipaldi, J.J. (Eds), *Environmental Methods Review: Retooling Impact Assessment for the New Century*, 119–126, Atlanta, GA: Army Environmental Policy Institute.

Xu, G. (2003). "Information for Corporate IP Management," Chicago, IL: *Patent Information Users Group*.

Yardley, H.O. (1957). *The Education of a Poker Player*, New York: Simon and Schuster.

Zhu, D., and Porter, A.L. (2002). "Automated Extraction and Visualization of Information for Technological Intelligence and Forecasting," *Technological Forecasting and Social Change*, Vol. 69, 495–506.

Zhu, D., Porter, A.L., Cunningham, S.W., Carlisle, J., and Nayak, A. (1999). "A Process for Mining Science & Technology Documents Databases, Illustrated for the Case of 'Knowledge Discovery and Data Mining'", *Ciencia da Informacao*, Vol. 28, No. 1, 1–8.

Index

Tech Mining: Exploiting New Technologies for Competitive Advantage, Edited by Alan L. Porter and Scott W. Cunningham.
ISBN 0-471-47567-X   Copyright © 2005 by John Wiley & Sons, Inc.